Lecture Notes in Computer Science　　9777

Commenced Publication in 1973
Founding and Former Series Editors:
Gerhard Goos, Juris Hartmanis, and Jan van Leeuwen

More information about this series at http://www.springer.com/series/7407

Cezar Câmpeanu · Florin Manea
Jeffrey Shallit (Eds.)

Descriptional Complexity of Formal Systems

18th IFIP WG 1.2 International Conference, DCFS 2016
Bucharest, Romania, July 5–8, 2016
Proceedings

Editors
Cezar Câmpeanu
University of Prince Edward Island
Charlottetown
Canada

Florin Manea
Universität Kiel
Kiel
Germany

Jeffrey Shallit
School of Computer Science
University of Waterloo
Waterloo, ON
Canada

ISSN 0302-9743 ISSN 1611-3349 (electronic)
Lecture Notes in Computer Science
ISBN 978-3-319-41113-2 ISBN 978-3-319-41114-9 (eBook)
DOI 10.1007/978-3-319-41114-9

Library of Congress Control Number: 2016942012

LNCS Sublibrary: SL1 – Theoretical Computer Science and General Issues

Printed on acid-free paper

This Springer imprint is published by Springer Nature
The registered company is Springer International Publishing AG Switzerland

Preface

This volume contains the papers presented at DCFS 2016, the 18th International Conference on Descriptional Complexity of Formal Systems, held during July 6–8, 2016, in Bucharest, at the University of Bucharest. DCFS became a working conference in 2016, continuing the former Workshop on Descriptional Complexity of Formal Systems, which was a merger in 2002 of two other workshops: FDSR (Formal Descriptions and Software Reliability) and DCAGRS (Descriptional Complexity of Automata, Grammars and Related Structures).

DCAGRS was previously held in Magdeburg (1999), London (2000), and Vienna (2001). FDSR was previously held in Paderborn (1998), Boca Raton (1999), and San Jose (2000).

Since 2002, DCFS has been successively held in London, Ontario, Canada (2002), Budapest, Hungary (2003), London, Ontario, Canada (2004), Como, Italy (2005), Las Cruces, USA (2006), Novy Smokovek (High Tatras), Slovakia (2007), Charlottetown, Canada (2008), Magdeburg, Germany (2009), Saskatoon, Canada (2010), Giessen, Germany (2011), Porto, Portugal (2012), London, Ontario, Canada (2013), Turku, Finland (2014), and Waterloo, Ontario, Canada (2015).

This conference was an official event of the International Federation for Information Processing and IFIP Working Group 1.2 (Descriptional Complexity) and was jointly organized by the IFIP WG 1.2 and the Faculty of Mathematics and Computer Science of the University of the Bucharest.

The working conference was sponsored by the Department of Computer Science of the University of Bucharest and other sponsors.

Descriptional complexity is a field in computer science that deals with the size of all kinds of objects that occur in computational models, such as Turing machines, finte automata, grammars, splicing systems and others. The topics of this conference are related to all aspects of descriptional complexity and include, but are not limited to:

- Various modes of operations and complexity measures for automata, grammars, languages, and related systems
- Succinctness of description of objects, state-explosion-like phenomena
- Trade-offs between descriptional complexity and mode of operation
- Circuit complexity of Boolean functions and related measures
- Succinctness of description of (finite) objects
- Descriptional complexity in resource-bounded or structure-bounded environments
- Complexity aspects related to the combinatorics of words
- Structural complexity of formal systems as related to descriptional complexity
- Descriptional complexity of formal systems for applications (e.g., software reliability, software and hardware testing, modelling of natural languages)
- Descriptional complexity aspects of nature-motivated (bio-inspired) architectures and unconventional models of computing
- Frontiers between decidability and undecidability

- Universality and reversibility
- Blum static (a.k.a. Kolmogorov/Chaitin) complexity, algorithmic information

The working conference of DCFS 2016 included four invited lectures, 13 contributed papers, discussion sessions, and a visit of the surroundings of Bucharest city, concluded by the conference dinner.

The proceedings of DCFS 2016, published in this volume of the *Lecture Notes in Computer Science* series, were available at the workshop and contain the invited lectures and the contributed papers.

There were 21 submissions to DCFS 2016 by a total of 47 authors from 15 different countries – Canada, Germany, India, Italy, Portugal, Slovakia, South Africa, Brazil, Russia, Austria, Czech Republic, Romania, France, Poland, and the UK.

On the basis of at least three reviews for each contribution, an international committee selected 13 papers – which accounts for an acceptance rate of approximately 60 % – for inclusion in the workshop program and this proceedings volume. The submission and refereeing process was supported by the EasyChair conference management system.

We warmly thank those who contributed to the success of DCFS 2016:

- The invited speakers James Currie (University of Winnipeg, Winnipeg/Manitoba, Canada), Gabriel Istrate (Timioara, Romania), Galina Jirásková (Mathematical Institute Slovak Academy of Sciences, Kosice, Slovak Republic), and Mikhail V. Volkov (Ural Federal University, Ekaterinburg, Russia).
- The authors of contributed and discussion papers.
- The reviewers and the Program Committee for their excellent work in making this selection.
- The members of the Organizing Committee for their commitment in the preparation of the scientific sessions and social events
- The staff of Springer and, in particular, Computer Science Editorial, for the extremely helpful and efficient collaboration in making this volume available before the conference. As volume editors, we value their experience, advice, and instructions, which were very helpful for the preparation of this volume.
- All the speakers and participants for attending the DCFS workshop.

Special thanks go to the "Asociaţia Alumni Universităţii din Bucureşti" for their financial and logistic support. We gratefully acknowledge the generous direct financial support of the Faculty of Mathematics and Computer Science of the University of Bucharest and the valuable in-kind support from Springer. Without this support, for which we are thankful, it would have been very difficult to conduct DCFS 2016.

We hope, as in the previous years, that DCFS 2016 has initiated new scientific discussions and stimulated research and scientific cooperation in the area of descriptional complexity, and trust that this volume will contribute to raising the interest in this field.

We look forward to seeing this year's participants and many others at DCFS in 2017!

May 2016

Cezar Câmpeanu
Florin Manea
Jeffrey Shallit

Organization

Steering Committee

Cezar Câmpeanu   University of Prince Edward Island, Charlottetown/Prince Edward Island, Canada

Cezar Câmpeanu	University of Prince Edward Island, Charlottetown/Prince Edward Island, Canada
Erzsébet Csuhaj-Varjú	Budapest, Hungary
Jürgen Dassow	Otto von Guericke University of Magdeburg, Germany
Helmut Jürgensen	Western University, London/Ontario, Canada
Martin Kutrib	Giessen University, Germany
Giovanni Pighizzini	University of Milan, Italy, Chair
Rogério Reis	University of Porto, Portugal

Program Committee

Valerie Berthe	CNRS LIAFA, Paris, France
Cristian S. Calude	University of Auckland, New Zealand
Cezar Câmpeanu	University of Prince Edward Island, Canada
Zoltan Esik	University of Szeged, Hungary
Yo-Sub Han	Yonsei University, South Korea
Christos Kapoutsis	Carnegie Mellon University in Qatar
Jarkko Kari	University of Turku, Finland
Lila Kari	The University of Waterloo, Ontario, Canada
Stavros Konstantinidis	Saint Mary's University, Canada
Martin Kutrib	Institut für Informatik, Universität Giessen, Germany
Florin Manea	Kiel University, Germany
Ian McQuillan	University of Saskatchewan, Canada
Carlo Mereghetti	Università degli Studi di Milano, Italy
Nelma Moreira	CMUP, Faculdade de Ciências da Universidade do Porto, Portugal
Dirk Nowotka	Christian-Albrechts-Universität zu Kiel, Germany
Alexander Okhotin	University of Turku, Finland
Dana Pardubska	Comenius University, Bratislava, Slovakia
Andrei Păun	University of Bucharest, Romania
Giovanni Pighizzini	Università degli Studi di Milano, Italy
Jean-Eric Pin	LIAFA, CNRS and University of Paris 7, France
Marinella Sciortino	University of Palermo, Italy
Jeffrey O. Shallit	University of Waterloo, Canada
Bianca Truthe	Justus-Liebig-Universität Gießen, Germany

Additional Reviewers

Carton, Olivier
choffrut, christian
Digulescu, Mircea
Eom, Hae-Sung
Guillon, Bruno
Hirvensalo, Mika

Holzer, Markus
Ivan, Szabolcs
Jalonen, Joonatan
Kopra, Johan
Li, Lvzhou
López, Damián

Malcher, Andreas
Meduna, Alexander
Mercas, Robert
Rao, Michael
Reis, Rogério
Schmid, Markus L.

Abstracts of Invited Talks

Completely Reachable Automata

Eugenija A. Bondar and Mikhail V. Volkov

Institute of Mathematics and Computer Science,
Ural Federal University, Lenina 51, 620000 Yekaterinburg, Russia
bondareug@gmail.com, mikhail.volkov@usu.ru

Abstract. We present a few results and several open problems concerning complete deterministic finite automata in which every non-empty subset of the state set occurs as the image of the whole state set under the action of a suitable input word.

Supported by the Russian Foundation for Basic Research, grant no. 16-01-00795, the Ministry of Education and Science of the Russian Federation, project no. 1.1999.2014/K, and the Competitiveness Program of Ural Federal University. The paper was written during the second author's stay at Hunter College of the City University of New York as Ada Peluso Visiting Professor of Mathematics and Statistics with a generous support from the Ada Peluso Endowment

Words Avoiding Patterns, Enumeration Problems and the Chomsky Hierarchy

James D. Currie

Department of Mathematics and Statistics
University of Winnipeg
515 Portage Avenue
Winnipeg, Manitoba R3B 2E9, Canada
j.currie@uwinnipeg.ca

Abstract. The study of words avoiding patterns is a mature branch of combinatorics on words. Patterns are themselves words, but their alphabets may be partitioned into variables, constants, function symbols such as reversal, or other tokens. As in the classical case of overlap-free words, one typically begins with the problem of whether pattern p is avoidable by an infinite string over alphabet $\sum$, and then moves on to sharper questions, such as language-theoretic properties of the set L of finite words over $\sum$ avoiding p, and the problem of enumerating words of L of length n.

Strong techniques for the enumeration of regular or context-free languages are well-known, following Schützenberger's foundational work. However, because of the pumping lemma, the language of binary overlap-free words is not context-free; nevertheless, there is a sharp description of the language of binary overlap-free words due to Cassaigne, via regular languages coding a sequence of operator applications. This leads to sharp characterization of the growth of the number of binary overlap-free words of length n, which turns out to be polynomial. The growth of the language L of finite words over $\sum$ avoiding p has been studied in various cases, and has generally been exponential, but in a few instances polynomial.

With this background, it was natural for Shallit et al. to ask whether the language of binary words avoiding xxx^R grows polynomially, or exponentially. The surprising answer turns out to be 'neither'. It follows that the language in question is not context-free; interestingly, no more direct proof of this is known. The language of binary words avoiding xx^Rx also turns out to have growth intermediate between polynomial and exponential, but the analysis is simpler. Given these surprising results involving patterns over $\{x, x^R\}$, it is natural to study binary avoidability of patterns over $\{x,x^R,y, y^R\}$, and the related growth questions. Studying growth questions for 2-avoidable patterns over $\{x, x^R,y, y^R\}$ leads to consideration of an under-utilized tool originally due to Shelton, the method of fixing block inequalities.

This talk will give an overview of the above matters, ending with recent results and open problems.

Heapability, Interactive Particle Systems, Partial Orders: Results and Open Problems

Gabriel Istrate[1,2] and Cosmin Bonchiş[1,2]

[1] Department of Computer Science, West University of Timişoara,
Timişoara, Romania
gabrielistrate@acm.org
[2] e-Austria Research Institute, Bd. V. Pârvan 4, cam. 045 B,
300223 Timişoara, Romania

Abstract. We outline results and open problems concerning partitioning of integer sequences and partial orders into heapable subsequences (previously defined and established by Byers et al.).

Self-Verifying Finite Automata
and Descriptional Complexity

Galina Jirásková

Mathematical Institute, Slovak Academy of Sciences,
Grešákova 6, 040 01 Košice, Slovakia
jiraskov@saske.sk

Abstract. We survey recent results on the descriptional complexity of self-verifying finite automata. In particular, we discuss the cost of simulation of self-verifying finite automata by deterministic finite automata, and the complexity of basic regular operations on languages represented by self-verifying finite automata.

Research supported by VEGA grant 2/0084/15

Contents

Completely Reachable Automata

Eugenija A. Bondar and Mikhail V. Volkov[✉]

Institute of Mathematics and Computer Science,
Ural Federal University, Lenina 51, 620000 Yekaterinburg, Russia
bondareug@gmail.com, mikhail.volkov@usu.ru

Abstract. We present a few results and several open problems concerning complete deterministic finite automata in which every non-empty subset of the state set occurs as the image of the whole state set under the action of a suitable input word.

Keywords: Deterministic finite automaton · Complete reachability · Transition monoid · Syntactic complexity · PSPACE-completeness

1 Background and Overview

We consider the most classical species of finite automata, namely, complete deterministic automata. Recall that a *complete deterministic finite automaton* (DFA) is a triple $\mathscr{A} = \langle Q, \Sigma, \delta \rangle$, where Q and Σ are finite sets called the *state set* and the *input alphabet* respectively, and $\delta \colon Q \times \Sigma \to Q$ is a totally defined map called the *transition function*. Let Σ^* stand for the collection of all finite words over the alphabet Σ, including the empty word. The function δ extends to a function $Q \times \Sigma^* \to Q$ (still denoted by δ) in the following natural way: for every $q \in Q$ and $w \in \Sigma^*$, we set $\delta(q, w) := q$ if w is empty and $\delta(q, w) := \delta(\delta(q, v), a)$ if $w = va$ for some word $v \in \Sigma^*$ and some letter $a \in \Sigma$. Thus, via δ, every word $w \in \Sigma^*$ induces a transformation of the set Q.

Let $\mathcal{P}(Q)$ stand for the set of all non-empty subsets of the set Q. The function δ can be further extended to a function $\mathcal{P}(Q) \times \Sigma^* \to \mathcal{P}(Q)$ (again denoted by δ) by letting $\delta(P, w) := \{\delta(q, w) \mid q \in P\}$ for every non-empty subset $P \subseteq Q$. Thus, the triple $\mathcal{P}(\mathscr{A}) := \langle \mathcal{P}(Q), \Sigma, \delta \rangle$ is a DFA again; this DFA is referred to as the *powerset automaton* of $\mathscr{A}$.

Whenever we deal with a fixed DFA, we simplify our notation by suppressing the sign of the transition function; this means that we may introduce the DFA

Supported by the Russian Foundation for Basic Research, grant no. 16-01-00795, the Ministry of Education and Science of the Russian Federation, project no. 1.1999.2014/K, and the Competitiveness Program of Ural Federal University. The paper was written during the second author's stay at Hunter College of the City University of New York as Ada Peluso Visiting Professor of Mathematics and Statistics with a generous support from the Ada Peluso Endowment.

© IFIP International Federation for Information Processing 2016
Published by Springer International Publishing Switzerland 2016. All Rights Reserved
C. Câmpeanu et al. (Eds.): DCFS 2016, LNCS 9777, pp. 1–17, 2016.
DOI: 10.1007/978-3-319-41114-9_1

as the pair $\langle Q, \Sigma \rangle$ rather than the triple $\langle Q, \Sigma, \delta \rangle$ and may write $q.w$ for $\delta(q, w)$ and $P.w$ for $\delta(P, w)$.

Given a DFA $\mathscr{A} = \langle Q, \Sigma \rangle$, we say that a non-empty subset $P \subseteq Q$ is *reachable* in $\mathscr{A}$ if $P = Q.w$ for some word $w \in \Sigma^*$. A DFA is called *completely reachable* if every non-empty subset of its state set is reachable.

Let us start with an example that served as a first spark which ignited our interest in completely reachable automata. A DFA $\mathscr{A} = \langle Q, \Sigma \rangle$ is called *synchronizing* if it has a reachable singleton, that is, $Q.w$ is a singleton for some word $w \in \Sigma^*$. Any such word w is said to be a *reset word* for the DFA. The minimum length of reset words for $\mathscr{A}$ is called the *reset threshold* of $\mathscr{A}$. In 1964 Černý [8] constructed for each $n > 1$ a synchronizing automaton $\mathscr{C}_n$ with n states, 2 input letters, and reset threshold $(n-1)^2$. Recall the definition of $\mathscr{C}_n$. If we denote the states of $\mathscr{C}_n$ by $1, 2, \ldots, n$ and the input letters by a and b, the actions of the letters are as follows:

$$i.a := \begin{cases} i & \text{if } i < n, \\ 1 & \text{if } i = n; \end{cases} \quad i.b := \begin{cases} i+1 & \text{if } i < n, \\ 1 & \text{if } i = n. \end{cases}$$

The automaton $\mathscr{C}_n$ is shown in Fig. 1.

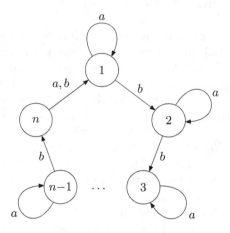

Fig. 1. The automaton $\mathscr{C}_n$

The automata in the Černý series are well-known in the connection with the famous Černý conjecture about the maximum reset threshold for synchronizing automata with n states, see [18]. The automata $\mathscr{C}_n$ provide the lower bound $(n-1)^2$ for this maximum, and the conjecture claims that these automata represent the worst possible case since it has been conjectured that every synchronizing automaton with n states can be reset by a word of length $(n-1)^2$. The automata $\mathscr{C}_n$ also have other interesting properties, including the one registered here:

Example 1. Each automaton $\mathscr{C}_n$, $n > 1$, is completely reachable.

The result of Example 1 was first observed by Maslennikova [15, Proposition 2], see also [16], in the course of her study of the so-called reset complexity of regular ideal languages. Later, Don [9, Theorem 1] found a sufficient condition for complete reachability that applies to the automata $\mathscr{C}_n$. In Sect. 2 we present another sufficient condition that both simplifies and generalizes Don's one. We provide an example showing that our condition is not necessary but we conjecture that it may be necessary for a stronger version of complete reachability.

In Sect. 3 we discuss the problem of recognizing completely reachable automata. We show PSPACE-completeness of the following decision problem: given a DFA $\mathscr{A} = \langle Q, \Sigma \rangle$ and a subset $P \subseteq Q$, decide whether or not P is reachable in $\mathscr{A}$. We also outline a polynomial algorithm that recognizes completely reachable automata with 2 input letters modulo the conjecture from Sect. 2.

Given a DFA $\mathscr{A} = \langle Q, \Sigma \rangle$, its *transition monoid* $M(\mathscr{A})$ is the monoid of all transformations of the set Q induced by the words in Σ^*. By the *syntactic complexity* of $\mathscr{A}$ we mean the size of $M(\mathscr{A})$. Clearly, the syntactic complexity of a completely reachable automaton $\mathscr{A}$ with n states cannot be less than $2^n - 1$ since, for each non-empty subset P of the state set, the transition monoid of $\mathscr{A}$ must contain a transformation whose image is P. In Sect. 4 we address the question of the existence and classification of *minimal* completely reachable automata, i.e., completely reachable automata with minimum possible syntactic complexity. This question has been recently investigated in the realm of transformation monoids by the first author [3,4]; here we translate her results into the language of automata theory and augment them by determining the input alphabet size of minimal completely reachable automata.

The present paper is in fact a work-in-progress report, and therefore, each of Sects. 2, 3, and 4 includes some open questions. Several additional open questions form Sect. 5; they mostly deal with synchronization properties of completely reachable automata.

We assume the reader's acquaintance with some basic concepts of graph theory, monoid theory, and computational complexity.

2 A Sufficient Condition

If Q is a finite set, we denote by $T(Q)$ the *full transformation monoid* on Q, i.e., the monoid consisting of all transformations $\varphi \colon Q \to Q$. For $\varphi \in T(Q)$, its *defect* is defined as the size of the set $Q \setminus Q\varphi$. Observe that the defect of a product of transformations is greater than or equal to the defect of any of the factors and is equal to the defect of a factor whenever the other factors are permutations of Q. In particular, if a product of transformations has defect 1, then one of the factors must have defect 1.

Let $\mathscr{A} = \langle Q, \Sigma \rangle$ be a DFA. The defect of a word $w \in \Sigma^*$ with respect to $\mathscr{A}$ is the defect of transformation induced by w. Consider a word w of defect 1. For such a word, the set $Q \setminus Q.w$ consists of a unique state, which is called the *excluded state* for w and is denoted by $\mathrm{excl}(w)$. Further, the set $Q.w$ contains a unique state p such that $p = q_1.w = q_2.w$ for some $q_1 \neq q_2$; this state p is called the *duplicate state* for w and is denoted by $\mathrm{dupl}(w)$. Let $D_1(\mathscr{A})$ stand for the set of all words of defect 1 with respect to $\mathscr{A}$, and let $\Gamma_1(\mathscr{A})$ denote the directed graph having Q as the vertex set and the set

$$E_1 := \{(\mathrm{excl}(w), \mathrm{dupl}(w)) \mid w \in D_1(\mathscr{A})\}$$

as the edge set. Since we consider only directed graphs in this paper, we call them just graphs in the sequel. Recall that a graph is *strongly connected* if for every pair of its vertices, there exists a directed path from the first vertex to the second.

Theorem 1. *If a DFA $\mathscr{A} = \langle Q, \Sigma \rangle$ is such that the graph $\Gamma_1(\mathscr{A})$ is strongly connected, then $\mathscr{A}$ is completely reachable.*

Proof. Take an arbitrary non-empty subset $P \subseteq Q$. We prove that P is reachable in $\mathscr{A}$ by induction on $k := |Q \setminus P|$. If $k = 0$, then $P = Q$ and nothing is to prove as Q is reachable via the empty word. Now let $k > 0$ so that P is a proper subset of Q. Since the graph $\Gamma_1(\mathscr{A})$ is strongly connected, there exists an edge $(q, p) \in E_1$ that connects $Q \setminus P$ and P in the sense that $q \in Q \setminus P$ while $p \in P$. By the definition of E_1, there exists a word w of defect 1 with respect to $\mathscr{A}$ for which q is the excluded state and p is the duplicate state. By the definition of the duplicate state, $p = q_1.w = q_2.w$ for some $q_1 \neq q_2$, and since the excluded state q for w does not belong to P, for each state $r \in P \setminus \{p\}$, there exists a unique state $r' \in Q$ such that $r = r'.w$. Now letting $R := \{q_1, q_2\} \cup \{r' \mid r \in P \setminus \{p\}\}$, we conclude that $P = R.w$ and $|R| = |P| + 1$. Then $|Q \setminus R| = k - 1$, and the induction assumption applies to the subset R whence $R = Q.v$ for some word $v \in \Sigma^*$. Then $P = Q.vw$ so that P is reachable as required.

Don [9] has formulated a sufficient condition for complete reachability in the terms of what he called a state map. Consider a DFA $\mathscr{A} = \langle Q, \Sigma \rangle$ with n states in which every subset of size $n - 1$ is reachable. Let W be a set of n words of defect 1 with respect to $\mathscr{A}$ such that for every subset $P \subset Q$ with $|P| = n - 1$ there is a unique word $w \in W$ with $P = Q.w$. (Such a set is termed a 1-*contracting collection* in [9]). The *state map* $\sigma_W \colon Q \to Q$ induced by W is defined by

$$q\sigma_W := \mathrm{dupl}(w) \quad \text{for } w \in W \text{ such that } q = \mathrm{excl}(w).$$

The following is one of the main results in [9]:

Theorem 2. *A DFA $\mathscr{A}$ is completely reachable if it admits a 1-contracting collection such that the induced state map is a cyclic permutation of the state set of $\mathscr{A}$.*

Even though Theorem 2 is stated in different terms, it is easily seen to constitute a special case of Theorem 1. Indeed, if W is a 1-contracting collection and σ_W is the corresponding state map, then each pair $(q, q\sigma_W)$ can be treated as an edge in E_1. Therefore, if σ_W is a cyclic permutation of Q, then the set of edges $\{(q, q\sigma_W) \mid q \in Q\}$ forms a directed Hamiltonian cycle in the graph $\Gamma_1(\mathscr{A})$ whence the latter is strongly connected.

We believe that Theorem 1 may have strongly wider application range than Theorem 2 even though at the moment we do not have any example confirming this conjecture. If the conditions of the two theorems were equivalent, every strongly connected graph of the form $\Gamma_1(\mathscr{A})$ would possess a directed Hamiltonian cycle, and this does not seem to be likely.

Now we demonstrate that the condition of Theorem 1 is not necessary.

Example 2. Consider the DFA $\mathscr{E}_3$ with the state set $\{1, 2, 3\}$ and the input letters $a_{[1]}, a_{[2]}, a_{[3]}, a_{[1,2]}$ that act as follows:

$$i.a_{[1]} := \begin{cases} 2 & \text{if } i = 1, 2, \\ 3 & \text{if } i = 3; \end{cases} \qquad i.a_{[2]} := \begin{cases} 1 & \text{if } i = 1, 2, \\ 3 & \text{if } i = 3; \end{cases}$$

$$i.a_{[3]} := \begin{cases} 1 & \text{if } i = 1, 2, \\ 2 & \text{if } i = 3; \end{cases} \qquad i.a_{[1,2]} := 3 \text{ for all } i = 1, 2, 3.$$

The automaton $\mathscr{E}_3$ is shown in Fig. 2 on the left. The graph $\Gamma_1(\mathscr{E}_3)$ is shown in Fig. 2 on the right; it is not strongly connected. However, it can be checked by a straightforward computation that the automaton $\mathscr{E}_3$ is completely reachable.

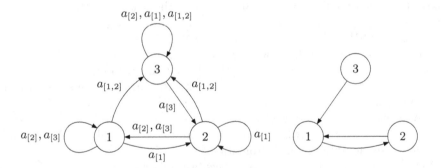

Fig. 2. The automaton $\mathscr{E}_3$ and the graph $\Gamma_1(\mathscr{E}_3)$

The reason of why the converse of Theorem 1 fails becomes obvious if one analyzes the above proof. In fact, we have proved more than we have formulated, namely, our proof shows that if a DFA $\mathscr{A}$ is such that the graph $\Gamma_1(\mathscr{A})$ is strongly connected, then *every proper non-empty subset of the state set of $\mathscr{A}$ is reachable via a product of words of defect* 1. Of course, this stronger property has no reason to hold in an arbitrary completely reachable automaton. For instance, in the automaton $\mathscr{E}_3$ of Example 2 the singleton $\{3\}$ is not an image of any product of words of defect 1. On the other hand, for the stronger property italicized above, the condition of Theorem 1 may be not only sufficient but also necessary. We formulate this guess as a conjecture.

Conjecture 1. If for every proper non-empty subset P of the state set of a DFA $\mathscr{A}$ there is a product w of words of defect 1 with respect to $\mathscr{A}$ such that $P = Q.w$, the graph $\Gamma_1(\mathscr{A})$ is strongly connected.

One can formulate further sufficient conditions for complete reachability in terms of strong connectivity of certain *hypergraphs* related to words of defect 2.

3   Complexity of Deciding Reachability

Given a DFA, one can easily decide whether or not it is completely reachable considering its powerset automaton: a DFA $\mathscr{A} = \langle Q, \Sigma \rangle$ is completely reachable if and only if Q is connected with every its non-empty subset by a directed path in the powerset automaton $\mathcal{P}(\mathscr{A})$, and the latter property can be recognized by breadth-first search on $\mathcal{P}(\mathscr{A})$ starting at Q. This algorithm is however exponential with respect to the size of $\mathscr{A}$, and it is natural to ask whether or not complete reachability can be decided in polynomial time. First, consider the following decision problem:

REACHABLE SUBSET: *Given a DFA* $\mathscr{A} = \langle Q, \Sigma, \delta \rangle$ *and a non-empty subset* $P \subseteq Q$, *is it true that* P *is reachable in* $\mathscr{A}$?

Theorem 3. *The problem* REACHABLE SUBSET *is* PSPACE-*complete.*

Proof. The fact that REACHABLE SUBSET is in the class PSPACE is easy and known, see, e.g., [5, Lemma 6,item 1].

To prove PSPACE-hardness of REACHABLE SUBSET, we reduce to it in logarithmic space the well-known PSPACE-complete problem FAI (FINITE AUTOMATA INTERSECTION, see [14]). Recall that an instance of FAI consists of k DFAs $\mathscr{A}_j = \langle Q_j, \Sigma, \delta_j \rangle$, $j = 1, \ldots, k$, with disjoint state sets and a common input alphabet. In each DFA $\mathscr{A}_j$ an *initial state* $s_j \in Q_j$ and a *final state* $t_j \in Q_j$ are specified; a word $w \in \Sigma^*$ is said to be *accepted* by $\mathscr{A}_j$ if $\delta_j(s_j, w) = t_j$. The question of FAI asks whether or not there exists a word $w \in \Sigma^*$ which is simultaneously accepted by all automata $\mathscr{A}_1, \ldots, \mathscr{A}_k$.

Now, given an instance of FAI as above, we construct the following instance $(\mathscr{A}, P)$ of REACHABLE SUBSET. The state set of the DFA $\mathscr{A}$ is $Q := \bigcup_{j=1}^{k} Q_j$; the input alphabet of $\mathscr{A}$ is Σ with one extra letter ρ added. The transition function $\delta \colon Q \times (\Sigma \cup \{\rho\}) \to Q$ is defined by the rule

$$\delta(q, a) := \begin{cases} \delta_j(q, a) & \text{if } a \in \Sigma \text{ and } q \in Q_j, \\ s_j & \text{if } a = \rho \text{ and } q \in Q_j. \end{cases} \tag{1}$$

Expressing this rule less formally, it says that, given a state $q \in Q$, one first should find the index $j \in \{1, \ldots, k\}$ such that q belongs to Q_j; then every letter $a \in \Sigma$ acts on q in the same way as it does in the automaton $\mathscr{A}_j$ while the added letter ρ sends q to the initial state s_j of $\mathscr{A}_j$ (so ρ artificially 'initializes' each $\mathscr{A}_j$). Observe that each set Q_j is closed under the action of each letter in $\Sigma \cup \{\rho\}$. Finally, we set $P := \{t_1, \ldots, t_k\}$, that is, P consists of the final states of $\mathscr{A}_1, \ldots, \mathscr{A}_k$.

We claim that the subset P is reachable in $\mathscr{A}$ if and only if there exists a word $w \in \Sigma^*$ which is simultaneously accepted by all automata $\mathscr{A}_1, \ldots, \mathscr{A}_k$. Indeed, if such a word w exists, then $\delta(Q, \rho w) = P$ since we have $\delta(Q, \rho) = \{s_1, \ldots, s_k\}$ by (1) and $\delta(s_j, w) = \delta_j(s_j, w) = t_j$ for each $j = 1, \ldots, k$ by the choice of w. Conversely, suppose that P is reachable in $\mathscr{A}$, that is, $\delta(Q, u) = P$ for some word $u \in (\Sigma \cup \{\rho\})^*$. Then we must have $\delta_j(Q_j, u) = \{t_j\}$ for each $j = 1, \ldots, k$. If the word u has no occurrence of the letter ρ, then $u \in \Sigma^*$ and $\delta_j(s_j, u) = \{t_j\}$ for each $j = 1, \ldots, k$ so that u is simultaneously accepted by all automata $\mathscr{A}_1, \ldots, \mathscr{A}_k$. Otherwise we fix the rightmost occurrence of ρ in u and denote by w the suffix of u following this occurrence so that $w \in \Sigma^*$ and $u = v\rho w$ for some $v \in (\Sigma \cup \{\rho\})^*$. Then $\delta_j(Q_j, v\rho) = \{s_j\}$ and $\delta_j(s, w) = \delta(Q_j, v\rho w) = \{t_j\}$ for each $j = 1, \ldots, k$. We conclude that w is simultaneously accepted by all automata $\mathscr{A}_1, \ldots, \mathscr{A}_k$. This completes the proof of our claim and establishes the reduction which obviously can be implemented in logarithmic space.

The reduction used in the above proof is an adaptation of a slightly more involved log-space reduction used by Brandl and Simon [5, Section 3] to show PSPACE-hardness of a natural problem about transformation monoids presented by a bunch of generating transformations.

In connection with Theorem 3, an interesting result by Goralčík and Koubek [13, Theorem 1] is worth being mentioned. If stated in the language adopted in the present paper, their result says that, given a DFA $\mathscr{A} = \langle Q, \Sigma \rangle$ with $|Q| = n$, $|\Sigma| = m$ and a subset $P \subseteq Q$ with $|P| = k$, one can decide in $O((k+1)n^{k+1}m)$ time whether or not there exists a word $w \in \Sigma^*$ such that $P = Q.w = P.w$. (The difference from our definition of reachability is that here one looks for a word not only having the subset P as its image but also acting on P as a permutation.) Thus, if the size of the target set P is treated as a parameter, the algorithm from [13] becomes polynomial. One can ask if a similar result holds for the parameterized version of REACHABLE SUBSET formulated as follows:

REACHABLE SUBSET$_k$: *Given a DFA $\mathscr{A} = \langle Q, \Sigma, \delta \rangle$ and a non-empty subset $P \subseteq Q$ of size k, is it true that P is reachable in $\mathscr{A}$?*

For $k = 1$, the cited result by Goralčík and Koubek applies since, for P being a singleton, any word $w \in \Sigma^*$ such that $P = Q.w$ automatically satisfies the additional condition $P.w = P$. For $k > 1$, the question about the complexity of REACHABLE SUBSET$_k$ is open. The reduction from the proof of Theorem 3 cannot help here because the size k of the subset P in this reduction is equal to the number of DFAs in the instance of FAI from which we depart, and for each fixed k, there is a polynomial algorithm that decides on all instances of FAI with k automata.

Now we return to the question of whether or not complete reachability can be decided in polynomial time. It should be noted that Theorem 3 does not imply any hardness conclusion here: while checking reachability of individual subsets is PSPACE-complete, checking reachability of all non-empty subsets may still be polynomial even though the latter problem consists of exponentially many individual problems! One can illustrate this phenomenon of 'simplification due to collectivization' with the following example. If is known [14] that the following *membership problem* for transition monoids of DFAs is PSPACE-complete: given a DFA $\mathscr{A} = \langle Q, \Sigma \rangle$ and a transformation $\varphi \colon Q \to Q$, does φ belongs to the transition monoid $M(\mathscr{A})$, i.e., is there a word $w \in \Sigma^*$ such that $q\varphi = q.w$ for all $q \in Q$? On the other hand, one can decide in polynomial time whether or not *every* transformation of the state set belongs to the transition monoid of a given DFA. Indeed, given a DFA $\mathscr{A} = \langle Q, \Sigma \rangle$, we partition the alphabet Σ as $\Sigma = \Pi \cup \Delta$, where Π consists of all letters that act on Q as permutations and Δ contains all letters with non-zero defect. First we inspect Δ: if no letter in Δ has defect 1, then it is clear that the monoid $M(\mathscr{A})$ contains no transformation of defect 1 (see the observation registered at the beginning of Sect. 2). Further, we invoke twice the polynomial algorithm by Furst et al. [11] for the membership problem in permutation groups: we fix a cyclic permutation and a transposition of Q and check if they belong to the permutation group on Q generated by the permutations induced by the letters in Π. If the answers to all these queries are affirmative, then $M(\mathscr{A})$ contains a cyclic permutation, a transposition, and a transformation of defect 1, and it is well-known that any such trio of transformations generates the full transformation monoid $T(Q)$, see, e.g., [12, Theorem 3.1.3].

Thus, the complexity of deciding complete reachability for a given DFA remains unknown so far. We expect this problem to be computationally hard for automata over unrestricted alphabets while for automata with a fixed number of letters a polynomial algorithm may exist. For instance, if Conjecture 1 holds true, there exists a polynomial algorithm that recognizes completely reachable automata among DFAs with 2 input letters. Indeed, let $\mathscr{A} = \langle Q, \{a, b\} \rangle$ be a DFA with n states, $n > 1$. Every subset of the form $Q.w$, where w is a nonempty word over $\{a, b\}$, is contained in either $Q.a$ or $Q.b$. At least one of the letters must have defect 1 since no subset of size $n - 1$ is reachable otherwise, and if the other letter has defect greater than 1, only one subset of size $n - 1$ is reachable. Hence, if $\mathscr{A}$ is a completely reachable automaton, one of its letters has defect 1 while the other has defect at most 1. Therefore for each proper reachable subset $P \subset Q$, there is a product w of words of defect 1 with respect to $\mathscr{A}$ such that $P = Q.w$. In view of Theorem 1, if Conjecture 1 holds true, then complete reachability of $\mathscr{A}$ is equivalent to strong connectivity of the graph $\Gamma_1(\mathscr{A})$. It remains to show that for automata with 2 input letters, the latter condition can be verified in polynomial time.

Once the graph $\Gamma_1(\mathscr{A})$ is constructed, checking its strong connectivity in polynomial time makes no difficulty. However, it is far from being obvious that $\Gamma_1(\mathscr{A})$, even though it definitely has polynomial size, can always be constructed in polynomial time. Indeed, by the definition, the edges of $\Gamma_1(\mathscr{A})$ arise from transformations of defect 1 in the transition monoid of $\mathscr{A}$, and for an automaton with n states, the number of transformations of defect 1 in $M(\mathscr{A})$ may reach $n!\binom{n}{2}$. Our algorithm depends on some peculiarities of automata with 2 input letters. It incrementally appends edges to a spanning subgraph of $\Gamma_1(\mathscr{A})$ in a way such that one can reach a conclusion about strong connectivity of $\Gamma_1(\mathscr{A})$ by examining only polynomially many transformations of defect 1. In the following brief description of the algorithm, we use the notation introduced in Sect. 2 in the course of defining the graph $\Gamma_1(\mathscr{A})$.

Thus, again, let $\mathscr{A} = \langle Q, \{a, b\}\rangle$ be a DFA with n states, $n > 1$. For certainty, let a stand for the letter of defect 1. If b also has defect 1, then at most two subsets of size $n - 1$ are reachable (namely, $Q.a$ and $Q.b$), and $\mathscr{A}$ can only be completely reachable provided that $n = 2$. The automaton $\mathscr{A}$ is then nothing but the classical flip-flop, see Fig. 3. Beyond this trivial case, b must be a permutation of Q whence b^n acts on Q as the identity transformation. Then the set $\{\mathrm{excl}(w) \mid w \in D_1(\mathscr{A})\}$ of the states at which edges of $\Gamma_1(\mathscr{A})$ may originate is easily seen to coincide with the set $\{\mathrm{excl}(a), \mathrm{excl}(ab), \ldots, \mathrm{excl}(ab^{n-1})\}$. For $\Gamma_1(\mathscr{A})$ to be strongly connected, it is necessary that every vertex is an origin of an edge whence the latter set must be equal to Q. Taking into account that $\mathrm{excl}(ab^k) = \mathrm{excl}(a).b^k$ for each $k = 1, \ldots, n - 1$, we conclude that b must be a cyclic permutation of Q. It is easy to show that $\mathrm{excl}(w).b = \mathrm{excl}(wb)$ and $\mathrm{dupl}(w).b = \mathrm{dupl}(wb)$ for every word w of defect 1, and therefore, b acts as a permutation on the edge set E_1 of $\Gamma_1(\mathscr{A})$.

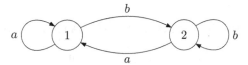

Fig. 3. Filp-flop

The set E_1 contains the edges

$$(\mathrm{excl}(a), \mathrm{dupl}(a)), \ldots, (\mathrm{excl}(ab^{n-1}), \mathrm{dupl}(ab^{n-1})). \tag{2}$$

Since $\mathrm{dupl}(ab^k) = \mathrm{dupl}(a).b^k$ for each $k = 1, \ldots, n - 1$, the edges in (2) are the 'translates' of the edge $(\mathrm{excl}(a), \mathrm{dupl}(a))$. Any two edges in (2) start at different vertices and end at different vertices, whence for some d such that $d < n$ and d divides n, the edges in (2) form d directed cycles, each of size $\frac{n}{d}$. If $d = 1$, we can already conclude that the graph $\Gamma_1(\mathscr{A})$ is strongly connected. If $d > 1$, denote the cycles by $C_1, \ldots, C_d$ and consider the words $a^2, aba, \ldots, ab^{n-1}a$. It can be easily shown that exactly two of them have defect 1; let us denote these two words by w_1 and w_2. Since w_1 and w_2 end with a, we have $Q.w_1 = Q.w_2 = Q.a$ whence $\mathrm{excl}(w_1) = \mathrm{excl}(w_2) = \mathrm{excl}(a)$. Thus, the edges $(\mathrm{excl}(w_1), \mathrm{dupl}(w_1))$ and $(\mathrm{excl}(w_2), \mathrm{dupl}(w_2))$ start at the vertex $\mathrm{excl}(a)$ which can be assumed to belong to the cycle C_1. If also the ends of these edges lie in C_1, one can show that no further edge in E_1 can connect C_1 with another cycle whence C_1 forms a strongly connected component of $\Gamma_1(\mathscr{A})$. We then conclude that $\Gamma_1(\mathscr{A})$ is not strongly connected.

Now suppose that the edge $(\mathrm{excl}(w_i), \mathrm{dupl}(w_i))$ where $i = 1$ or $i = 2$ connects the vertex $\mathrm{excl}(a)$ with a vertex from the cycle C_j where $j > 1$. Then we append the edge and all its translates $(\mathrm{excl}(w_i b^k), \mathrm{dupl}(w_i b^k))$, $k = 1, \ldots, n - 1$, to $C_1, \ldots, C_d$; in the case where both $(\mathrm{excl}(w_1), \mathrm{dupl}(w_1))$ and $(\mathrm{excl}(w_2), \mathrm{dupl}(w_2))$ leave C_1, we append both these edges and all their translates. After that, we get larger strongly connected subgraphs $D_1, \ldots, D_\ell$ isomorphic to each other, where $\ell < d$ and ℓ divides d. If $\ell = 1$, then the graph $\Gamma_1(\mathscr{A})$ is strongly connected. If $\ell > 1$, we iterate by considering the words $w_i a, w_i b a, \ldots, w_i b^{n-1} a$. Eventually, either we reach a strongly connected spanning subgraph of $\Gamma_1(\mathscr{A})$, and then the graph $\Gamma_1(\mathscr{A})$ is strongly connected as well, or on some step the process gets stacked, which means that $\Gamma_1(\mathscr{A})$ has a proper strongly connected component, and therefore, is not strongly connected.

The described process branches, and in the worst case the number of words of defect 1 to be analyzed doubles at each step. On the other hand, since the steps are indexed by a chain of divisors of n, the number of steps does not exceed $\log_2 n + 1$. Thus, executing the algorithm, we have to analyze at most

$$1 + 2 + 4 + \cdots + 2^{\lceil \log_2 n \rceil + 1} = O(n)$$

words of maximum length $O(n \log_2 n)$, and therefore, the algorithm can be implemented in polynomial time.

4 Minimal Completely Reachable Automata

Syntactic complexity of a regular language is a well established concept that has attracted much attention lately, see, e.g., [6,7]. It can be defined as the size of the transition monoid of the minimal DFA recognizing the language. It appears to be worthwhile to extend this concept to automata by defining the *syntactic complexity* of an arbitrary DFA $\mathscr{A}$ as the size of its transition monoid $M(\mathscr{A})$. In fact, if one thinks of a DFA as a computational device rather than acceptor, its transition monoid can be thought of as the device's 'software library' since the monoid contains exactly all programs (transformations) that the automaton can execute. From this viewpoint, measuring the complexity of an automaton by the size of its 'software library' is fairly natural.

As already mentioned in Sect. 1, the syntactic complexity of a completely reachable automaton with n states cannot be less than $2^n - 1$. It turns out that this lower bound is tight if one considers automata over unrestricted alphabet. We present now a construction for completely reachable automata with n states and syntactic complexity $2^n - 1$; for short, we call them *minimal completely reachable automata*.

Our construction produces minimal completely reachable automata from full binary trees satisfying certain subordination conditions. Recall that a binary tree is said to be *full* if each its vertex v either is a leaf or has exactly two children that we refer to as the *left child* or the *son* of v and the *right child* or the *daughter* of v. (Thus, all vertices except the root have a gender.) It is well known (and easy to verify) that a full binary tree with n leaves has $2n - 1$ vertices. As full binary trees are the only trees occurring in this paper, we call them just trees in the sequel.

If Γ is a tree and v is a vertex in Γ, we denote by Γ_v the subtree of Γ rooted at v. The *span* of v, denoted span(v), is the number of leaves in the subtree Γ_v. Figure 4 shows a tree with vertices labelled by their spans.

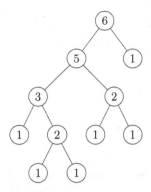

Fig. 4. An example of a tree with spans of its vertices shown

By a *homomorphism* between two trees Γ_1 and Γ_2 we mean a map from the vertex set of Γ_1 into the vertex set of Γ_2 that sends the root of Γ_1 to the root of Γ_2 and preserves the parent–child relation. Given two trees Γ_1 and Γ_2, we say that Γ_1 *subordinates* Γ_2 if there exists a 1-1 homomorphism $\xi \colon \Gamma_1 \to \Gamma_2$ such that span(v) $\leq$ span($v\xi$) for every vertex v of Γ_1. If u and v are two vertices of the same tree Γ, we say that u *subordinates* v if the subtree Γ_u subordinates the subtree Γ_v. A tree is said to be *respectful* if it satisfies two conditions:

(S1) if a male vertex has a nephew, the nephew subordinates his uncle;
(S2) if a female vertex has a niece, the niece subordinates her aunt.

For an illustration, the tree shown in Fig. 4 satisfies (S1) but fails to satisfy (S2): the daughter of the root has a niece but this niece does not subordinates her aunt. On the other hand, the tree shown in Fig. 5 is respectful. (In order to ease the inspection of this claim, we have shown the uncle–nephew and the aunt–niece relations in this tree with dotted and dashed arrows respectively.)

It is easy to show that there exist respectful trees with any number of leaves. In the following table (borrowed from [4]) we present the numbers of respectful trees with up to 10 leaves.

Number of leaves	1	2	3	4	5	6	7	8	9	10
Number of respectful trees	1	1	2	3	6	10	18	32	58	101

We are not aware of any closed formula for the number of respectful trees with a given number of leaves.

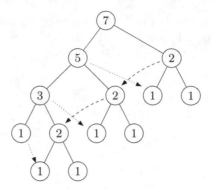

Fig. 5. An example of a respectful tree

In our construction, we use certain markings of trees by intervals of the set $\mathbb{N}$ of positive integers considered as a chain under the usual order:

$$1 < 2 < \cdots < n < \dots .$$

If $i, j \in \mathbb{N}$ and $i \leq j$, the *interval* $[i, j]$ is the set $\{k \in X_n \mid i \leq k \leq j\}$. We write $[i]$ instead of $[i, i]$. By the *span* of an interval we mean the number of its elements. Now, a *faithful interval marking* of a tree Γ is a map μ from the vertex set of Γ into the set of all intervals in $\mathbb{N}$ such that for each vertex v,

- the span of the interval $v\mu$ is equal to $\mathrm{span}(v)$;
- if $v\mu = [i, j]$ and s and d are respectively the son and the daughter of v, then $s\mu = [i, k]$ and $d\mu = [k + 1, j]$ for some k such that $i \leq k < j$.

It easy to see that every tree Γ admits a faithful interval marking which is unique up to an additive translation: given any two markings μ, μ' of Γ, there is an integer m such that $v\mu = v\mu' + m$ for every vertex v. Observe that if μ is a faithful interval marking of a tree Γ and v is a vertex of Γ, then the restriction of μ to the subtree Γ_v is a faithful interval marking of the latter. Figure 6 demonstrates a faithful interval marking of the tree from Fig. 5.

We have prepared everything and can now present our construction.

Construction T2A (trees to automata). For each respectful tree Γ with n leaves and each its faithful interval marking μ, we construct an automaton denoted by $\mathscr{A}_\mu(\Gamma)$. The states of $\mathscr{A}_\mu(\Gamma)$ are the elements of the interval $r\mu$, where r stands for the root of Γ, and the input alphabet of $\mathscr{A}_\mu(\Gamma)$ consists of $2n - 2$ letters a_v, one for each non-root vertex v of Γ. To define the action of the letters, we proceed by induction on n. For $n = 1$, that is, for the trivial tree Γ with one vertex r and no edges, $\mathscr{A}_\mu(\Gamma)$ is the trivial automaton with one state and no transitions, so that nothing has to be defined.

Now suppose that $n > 1$. Take any non-root vertex v of Γ; we have to define the action of the letter a_v on the elements of the interval $r\mu$. If s and d are respectively the son and the daughter of r, the interval $r\mu$ is the disjoint union

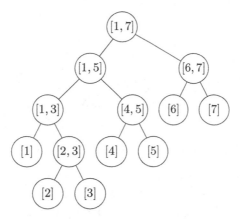

Fig. 6. A faithful interval marking of the tree from Fig. 5

of $s\mu$ and $d\mu$. If $v \neq s$ and $v \neq d$, then v is a non-root vertex in one of the subtrees Γ_s or Γ_d. These two cases are symmetric, so that we may assume that v belongs to Γ_s. By the induction assumption applied to Γ_s and its marking induced by μ, the action of a_v is already defined on the states from the interval $s\mu$; we extend this action to the whole interval $r\mu$ by setting $y.a_v := y$ for each $y \in d\mu$.

It remains to define the action of the letters a_s and a_d. Again, by symmetry, it suffices to handle one of these cases, so that we define that action of a_s. If s has no nephew in Γ, then d is a leaf and $d\mu = [m]$ for some $m \in \mathbb{N}$. Then we let $x.a_s := m$ for each $x \in r\mu$. Otherwise let t be the nephew of s. The subordination condition (S1) implies that there exists a 1-1 homomorphism $\xi \colon \Gamma_t \to \Gamma_s$. It is easy to see that the intervals $(\ell\xi)\mu$, where ℓ runs over the set of all leaves of the tree Γ_t, form a partition of the interval $s\mu$. Now we define the action of a_s on $s\mu$ as follows: if a number $x \in s\mu$ belongs to $(\ell\xi)\mu$ for some leaf ℓ of Γ_t and $\ell\mu = [y]$ for some $y \in \mathbb{N}$, we let $x.a_s := y$.

By the induction assumption applied to the subtree Γ_d and its marking induced by μ, the action of the letter a_t is already defined on the states from the interval $d\mu$; now we define the action of a_s on $d\mu$ by setting $y.a_s := y.a_t$ for all $y \in d\mu$. This completes our construction.

The reader may find it instructive to work out Construction T2A on a concrete example. For the tree from Figs. 5 and 6 used for illustrations above, computing all 12 input letters of the corresponding automaton would be rather cumbersome but one can check, for instance, that the letters a_s and a_d act on the set $[1, 7]$ as follows:

$$a_s = \begin{pmatrix} 1\,2\,3\,4\,5\,6\,7 \\ 6\,6\,6\,6\,6\,6\,7 \end{pmatrix} \quad a_d = \begin{pmatrix} 1\,2\,3\,4\,5\,6\,7 \\ 1\,1\,1\,2\,3\,4\,5 \end{pmatrix}.$$

Those who prefer a complete example can look at the DFA $\mathscr{E}_3$ from Example 2: the automaton was in fact derived by Construction T2A from the respectful tree

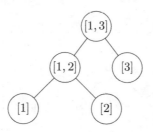

Fig. 7. The tree behind the automaton $\mathscr{E}_3$

with 3 leaves shown in Fig. 7. In particular, this explains our choice of notation for the input letters of $\mathscr{E}_3$ that perhaps had slightly puzzled the reader when she or he encountered this automaton in Sect. 2.   By the way, the flip-flop in Fig. 3 also can be obtained by Construction T2A (from the unique tree with 2 leaves).

Observe that all automata constructed from different markings of the same respectful tree are isomorphic since passing to another marking only results in a change of the state names. Taking this into account, we omit the reference to μ in the notation and denote the automaton derived from any marking of a given respectful tree Γ simply by $\mathscr{A}(\Gamma)$.

We say that two DFAs $\mathscr{A} = \langle Q, \Sigma, \delta \rangle$ and $\mathscr{B} = \langle Q, \Delta, \zeta \rangle$ are *syntactically equivalent* if their transition monoids coincide. Now we are ready for the main result of this section.

Theorem 4. *1. For each respectful tree Γ, the automaton $\mathscr{A}(\Gamma)$ is a minimal completely reachable automaton.*
2. Every minimal completely reachable automaton is syntactically equivalent to an automaton of the form $\mathscr{A}(\Gamma)$ for a suitable respectful tree Γ.
3. Every minimal completely reachable automaton with n states has at least $2n-2$ input letters.

Claims 1 and 2 in Theorem 4 are essentially equivalent to the main results of the papers [3,4] by the first author who has used a slightly different construction expressed in the language of transformation monoids: given a marking of a respectful tree Γ she constructs the transition monoid of $\mathscr{A}(\Gamma)$ rather than the automaton itself. Claim 3 is new but we have not included its proof here due to the space limitations because the only proof we have at the moment requires reproducing several concepts and results from [3,4] and restating them in the language adopted in the present paper. It is very tempting to invent a direct proof of this claim that would bypass rather bulky considerations from [3,4].

Theorem 4 leaves widely open the question about lower bounds for syntactic complexity of completely reachable automata with restricted alphabet. In particular, the case of completely reachable automata with 2 input letters both is of interest and seems to be tractable. The latter conclusion follows from our analysis of completely reachable automata with 2 input letters at the end of Sect. 3 which demonstrates that such DFAs have rather a specific structure.

We say that a DFA $\mathscr{A} = \langle Q, \Sigma, \delta \rangle$ *induces* a DFA $\mathscr{B} = \langle Q, \Delta, \zeta \rangle$ on the same state set if the transition monoid of $\mathscr{A}$ contains that of $\mathscr{B}$. Equivalently, this means that for every letter $b \in \Delta$, there exists a word $w \in \Sigma^*$ such that $\zeta(q, b) = \delta(q, w)$ for every $q \in Q$. This relation between automata plays an essential role in the theory of synchronizing automata, see, e.g., [2]. With respect to completely reachable automata, the following question is of interest: is it true that every completely reachable automaton induces a minimal completely reachable automaton? In other words, is it true that an automaton of the form $\mathscr{A}(\Gamma)$ 'hides' within every completely reachable automaton?

5   More Open Questions

Since completely reachable automata are synchronizing, it is natural to ask what is the maximum reset threshold for completely reachable automata with n states. In view of Example 1, the lower bound $(n-1)^2$ for this maximum is provided by the Černý automata $\mathscr{C}_n$. For completely reachable automata with 2 input letters this bound is tight because, except for the flip-flop, such automata have a letter that acts as a cyclic permutation of the state set, and therefore, Dubuc's result [10] applies to them. Some partial results about synchronization of completely reachable automata can be found in [9], but the general problem of finding the maximum reset threshold for completely reachable automata with n states and unrestricted alphabet remains open.

The problem discussed in the previous paragraph basically asks what is the minimum length of a word that reaches a singleton. For completely reachable automata, a similar question makes sense for an arbitrary non-empty subset. Thus, we suggest to investigate the minimum length of a word that reaches a subset with m element in a completely reachable automaton with n states as a function of n and m. Don [9, Conjecture 2] has formulated a very strong conjecture that implies the upper bound $n(n - m)$ on this length. Observe that if this upper bound indeed holds, then completely reachable automata satisfy the Černý conjecture. To see this, take a completely reachable automaton $\mathscr{A} = \langle Q, \Sigma \rangle$ with n states; it should possess a letter $a \in \Sigma$ such that $q.a = q'.a$ for two different states $q, q' \in Q$. If a word $w \in \Sigma^*$ of length at most $n(n - 2)$ is such that $Q.w = \{q, q'\}$, the word wa is a reset word for $\mathscr{A}$ and has length at most $n(n - 2) + 1 = (n - 1)^2$.

Another intriguing problem about completely reachable automata suggested by the theory of synchronizing automata is a variant of the Road Coloring Problem. We recall notions involved there. A *road coloring* of a finite graph Γ consists in assigning non-empty sets of labels (colors) from some alphabet Σ to edges of Γ such that the label sets assigned to the outgoing edges of each vertex form a partition of Σ. Colored this way, Γ becomes a DFA over Σ; every such DFA is called a *coloring* of Γ. Figure 8 shows a graph and two of its colorings by $\Sigma = \{a, b\}$, one of which is the Černý automaton $\mathscr{C}_4$. The Road Coloring Problem, recently solved by Trahtman [17], had asked which strongly connected graphs admit *synchronizing colorings*, i.e., colorings that are synchronizing automata. It turns

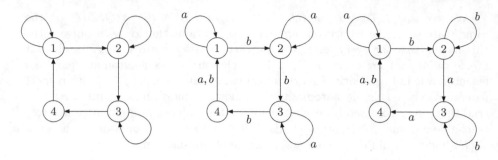

Fig. 8. A graph and two of its colorings

out that, as it was conjectured in [1], the necessary and sufficient condition for a strongly connected graph to possess a synchronizing coloring is that the greatest common divisor of lengths of all directed cycles in the graph should be equal to 1. The latter property is called *aperiodicity* or *primitivity*.

An analogous question makes sense for completely reachable automata. Namely, call a coloring of a graph *completely reachable* if it yields a completely reachable automaton. Our problem then consists in characterising graphs that admit completely reachable colorings. Such graphs must be strongly connected and primitive since every completely reachable automaton is strongly connected and synchronizing. However, it is easy to produce an example of a strongly connected primitive graph that has no completely reachable coloring; such a graph is shown in Fig. 9 on the left. Moreover, there are interesting phenomena that have no parallel in the theory of synchronizing automata; for instance, there exist graphs that have no completely reachable coloring with 2 letters but admit such a coloring with 3 letters; an example of such a graph is presented in the center of Fig. 9 while the corresponding coloring is shown on the right.

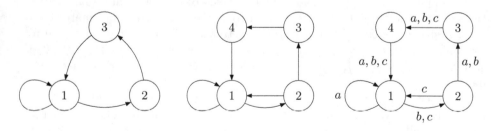

Fig. 9. The left graph has no completely reachable coloring; the central graph has no completely reachable coloring with 2 letters but has a completely reachable coloring with 3 letters shown in the right

References

1. Adler, R.L., Goodwyn, L.W., Weiss, B.: Equivalence of topological Markov shifts. Isr. J. Math. **27**, 49–63 (1977)
2. Ananichev, D.S., Gusev, V.V., Volkov, M.V.: Primitive digraphs with large exponents and slowly synchronizing automata. J. Math. Sci. **192**(3), 263–278 (2013)
3. Bondar, E.: $\mathcal{L}$-cross-sections of the finite symmetric semigroup. Algebra and Discrete Math. **18**(1), 27–41 (2014)
4. Bondar, E.: Classification of $\mathcal{L}$-cross-sections of $\mathcal{T}_n$. Algebra Discrete Math. **21**(1), 1–17 (2016)
5. Brandl, C., Simon, H.U.: Complexity analysis: transformation monoids of finite automata. In: Potapov, I. (ed.) DLT 2015. LNCS, vol. 9168, pp. 143–154. Springer, Heidelberg (2015)
6. Brzozowski, J.A., Li, B.: Syntactic complexity of $\mathcal{R}$ and $\mathcal{J}$-trivial regular languages. Int. J. Found. Comput. Sci. **25**(7), 807–822 (2014)
7. Brzozowski, J., Szykuła, M.: Upper bound on syntactic complexity of suffix-free languages. In: Shallit, J., Okhotin, A. (eds.) DCFS 2015. LNCS, vol. 9118, pp. 33–45. Springer, Heidelberg (2015)
8. Černý, J.: Poznámka k homogénnym eksperimentom s konečnými automatami. Matematicko-fyzikalny Časopis Slovensk. Akad. Vied **14**(3), 208–216 (1964). (in Slovak)
9. Don, H.: The Černý conjecture and 1-contracting automata. CoRR, abs/1507.06070 (2015). http://arxiv.org/abs/1507.06070
10. Dubuc, L.: Sur les automates circulaires et la conjecture de Černý. RAIRO Inform. Théor. Appl. **32**(1–3), 21–34 (1998). (in French)
11. Furst, M.L., Hopcroft, J.E., Luks, E.M.: Polynomial-time algorithms for permutation groups. In: 21st Annual Symposium on Foundations of Computer Science, pp. 36–41. IEEE Computer Society, Washington (1980)
12. Ganyushkin, O., Mazorchuk, V.: Classical Finite Transformation Semigroups: An Introduction. Springer, Heidelberg (2009)
13. Goralčík, P., Koubek, V.: Rank problems for composite transformations. Int. J. Algebra Comput. **5**(3), 309–316 (1995)
14. Kozen, D.: Lower bounds for natural proof systems. In: 18th Annual Symposium on Foundations of Computer Science, pp. 254–266. IEEE Computer Society, Washington (1977)
15. Maslennikova, M.I.: Reset complexity of ideal languages. In: Bieliková, M., Friedrich, G., Gottlob, G., Katzenbeisser, S., Špánek, R., Turán, G. (eds.) 38th International Conference on Current Trends in Theory and Practice of Computer Science, SOFSEM 2012, vol. II, pp. 33–44. Institute of Computer Science Academy of Sciences of the Czech Republic, Prague (2012)
16. Maslennikova, M.I.: Reset complexity of ideal languages. CoRR, abs/1404.2816 (2014). http://arxiv.org/abs/1404.2816
17. Trahtman, A.N.: The road coloring problem. Isr. J. Math. **172**, 51–60 (2009)
18. Volkov, M.V.: Synchronizing Automata and the Černý Conjecture. In: Martín-Vide, C., Otto, F., Fernau, H. (eds.) LATA 2008. LNCS, vol. 5196, pp. 11–27. Springer, Heidelberg (2008)

Heapability, Interactive Particle Systems, Partial Orders: Results and Open Problems

Gabriel Istrate[1,2]($\boxtimes$) and Cosmin Bonchiş[1,2]

[1] Department of Computer Science, West University of Timişoara,
Timişoara, Romania
gabrielistrate@acm.org
[2] e-Austria Research Institute, Bd. V. Pârvan 4, cam. 045 B,
300223 Timişoara, Romania

Abstract. We outline results and open problems concerning partitioning of integer sequences and partial orders into heapable subsequences (previously defined and established by Byers et al.).

Keywords: Heapable sequences · Posets

1   Introduction

Suppose $a_1, a_2, \ldots, a_n$ is a sequence of integers. Can one insert the elements of the sequence, successively, as the leaves of a binary tree that satisfies the min heap property? This is possible, for instance, for sequence 1 3 2 7 6 5 4 but not for sequence 5 4 3 2 1. Byers et al. [1] (who introduced the notion), called such a sequence *heapable*. They provided a polynomial time algorithm to recognize heapability (though, interestingly, *complete heapability*, i.e. heapability on a complete binary tree is NP-complete).

One can view the notion of heapability as a (parametric) relaxation of the notion of *monotonicity*. Indeed, heapability of a sequence requires the fact that the smallest element comes first. The next two elements may, however, arive in any order and the constraints on element ordering become progressively looser. The view of heapability as a generalization of monotonicity, connects the study of heapable sequences to the rich theory built in connection with longest increasing subsequence [2].

In [3] we studied the partition of random permutations into heapable sequences. Similar results were obtained independently in [4]. Perhaps the most exciting finding was the scaling of the number of classes in a partition of a random permutation into heapable subsequences, conjectured to scale as $\phi \cdot \ln(n)$, with ϕ the golden ratio: in Sect. 5 we explain and motivate this conjecture.

This extended abstract continues this line of inquiry. We present some results and outline several open questions related to the problem of extending notions related to heapability from numbers to partial orders. More topics will be mentioned in the conference presentation.

© IFIP International Federation for Information Processing 2016
Published by Springer International Publishing Switzerland 2016. All Rights Reserved
C. Câmpeanu et al. (Eds.): DCFS 2016, LNCS 9777, pp. 18–28, 2016.
DOI: 10.1007/978-3-319-41114-9_2

2   Preliminaries

A *(binary min-)heap* is a binary tree, not necessarily complete for the purposes of this paper, such that $A[parent[x]] \leq A[x]$ for every non-root node x. If instead of binary we require the tree to be k-ary we get the concept of k-ary min-heap.

A partially ordered set $P = (X, \prec)$ is called *k-heapable* if there exists some k-ary tree T whose nodes are in bijection with the elements of X, such that for every non-root node X_i and parent X_j, $X_j \prec X_i$ and $j < i$. In particular a 2-heapable partial order will simply be called *heapable*.

We easily recover the case of permutations, dealt with in [3], as follows: given permutation $\pi \in S_n$, we define partial order $\prec$ on $\{1, 2, \ldots, n\}$ by $i \prec j$ iff $i < j$ and $\pi[i] < \pi[j]$.

The *height of partial order* P, denoted by $h(P)$, is the length of the longest *chain* (totally ordered subset) of P. The *width of P* is defined as the size of the largest antichain of P. By Dilworth's Theorem [5], $w(P)$ is equal to the smallest number of elemenst in a partition of P into chains. Finally, the *dimension of P* is the smallest number r such that the partial order is the intersection of r permutations.

Example 1. Let $X = \{I_1, I_2, \ldots I_k\}$ be a finite set of closed intervals on the real line, with the partial order $I \preceq J$ given by $end(I) \leq start(J)$. By the Gallai theorems for intervals [6], $height(P)$ is equal to the minimal number of points that *pierce* (i.e. intesect) every interval in P. On the other hand $width(P)$ is equal to the maximum cardinality of a set of intervals with nonempty joint intersection.

We give a parametric generalization of $height(P)$ and $width(P)$ as follows:

Definition 1. *Given an integer $k \geq 1$, a subset $Q \subset P$ is a k-chain if nodes of Q are the vertices of a k-ary $\preceq$-ordered subtree of P (not necessarily induced).*

The k-height of P is defined to be the size of the largest k-ary chain of P. The k-width of P is defined as the minimal number of classes in a partition of P into k-chains.

We will employ random models of partial orders of fixed dimension. A complete discussion is beyond the scope of the paper [7]. Instead, we recall the following popular model $P_d(n)$ [8]: given constant $d \geq 1$ we choose random partial order $\prec$ as the intersection of d permutations $\pi_1, \pi_2, \ldots, \pi_d$ chosen uniformly at random with repetitions from S_n. In other words, given $i, j \in \{1, 2, \ldots, n\}$ define

$$i \prec j \iff \pi_1(i) < \pi_1(j), \pi_2(i) < \pi_2(j), \ldots, \pi_d(i) < \pi_d(j).$$

An equivalent mode to generate a partial order P from $P_d(n)$ is the following: choose n points $P_1, P_2, \ldots P_n,$, $P_i = (x_1^i, \ldots, x_d^i)$, uniformly at random from the hypercube $[0, 1]^d$. Define

$$i \prec j \iff \pi_1(i) < \pi_1(j), \pi_2(i) < \pi_2(j), \ldots, \pi_d(i) < \pi_d(j).$$

We will refer to this alternate description as model (II).

3 The Computational Complexity of Generalized Height and Width

Open Problem 1. *What is the computational complexity of the following decision problem:*

- *[GIVEN:] Partial order $P = (X, \prec)$ and integer $r \geq 1$.*
- *[TO DECIDE:] Can X be partioned into at most r k-chains? That is, is inequality $k\text{-}w(P) \leq r$ true?*

Even the case $k = 1$ (a.k.a. the *longest heapable subsequence* of a random permutation) is still open [1]. In contrast, the k-width of a finite partial order can be computed in polynomial time:

Theorem 1. *For every fixed $k \geq 1$ there is a polynomial time algorithm that, given finite partial order $P = (X, \preceq)$ as input, computes the value $k\text{-}w(P)$.*

Proof. Define the following boolean integer programming problem: define a variable $X_{p,q}$ for every pair $p \prec q \in P$. Intuitively $X_{p,q} = 1$ if p is the parent of q in the k-chain decomposition of P, 0 otherwise.

Every integral solution to this system correponds to a decomposition of P into k-ary trees: indeed, every node has at most one parent in the decomposition induced by variables $X_{p,q} = 1$, and at most k children.

Since in each tree the number of edges is one less than the number of vertices, in any decomposition of P into k-chains, the number of such chains is $n - \sum_{p \prec q} X_{p,q}$.

So to compute the k-width of P we have to solve the following integer program:

$$
\begin{cases}
max(\sum_{p \prec q} X_{p,q}) \\[2mm]
\sum_{q:p \prec q} X_{p,q} \leq k, \forall p \in X \\
\sum_{p:p \prec q} X_{p,q} \leq 1, \forall q \in X \\[2mm]
X_{p,q} \in \{0,1\}
\end{cases}
$$

Consider the linear programming relaxation of the system above, obtained by replacing condition $X_{p,q} \in \{0,1\}$ by $X_{p,q} \geq 0$. The matrix of the system is totally unimodular, since it coincides with the vertex-edge incidence matrix of the bipartite graph induced by partial order $\prec$. Such bipartite matrices are well-known to be totally unimodular [9]. So linear programming will find an integral solution to the system in polynomial time. □

Remark 1. The argument above owes much to a discussion with János Balogh from Szeged: we told him a restricted version of the problem, that of scheduling intervals on binary trees. This amounts to the setting of Example 1. At the time we had a direct (somewhat complicated) proof of this special case. He came up with a (different but related) argument, using network flows. Subsequently we came with this third proof for the general setting, obviously related to his.

Both our original argument and his extend to the general case, and will be jointly presented somewhere else. In retrospect, the fact that there are several distinct proofs is not surprising: Theorem 1 is obviously related to Dilworth's Theorem, and the three existing proofs (direct, using network flows, using linear programming) can be seen as extensions of the corresponding arguments for proving this latter result.

4   The Asymptotic Behavior of the Average k-height and k-width

The problem of computing the 1-width of a random partial order of dimension 2 is a variant of the classical problem of computing the longest increasing subsequence of a random permutation. The correct asymptotic behavior is $2\sqrt{n}$, [10–13] and substantially more is known.

The (1-)width and (1-)height of a partial order have also been studied in other dimensions: notable partial results are due to Winkler [8], who showed that the correct order of magnitude for the height of a partial order of dimension k is $\Theta(n^{1/k})$. Further results were obtained by Brightwell [14].

As for the height, the 1-height of a d-dimensional partial order was considered by Winkler [8], and then determined by Bollobás and Winkler [15] to be approximately $c_k \cdot n^{1/k}$ for some constant $c_k > 0$.

In [3] we gave a simple simple lower bound valid for all values of the k-width(P), where P is a random permutation of width 2. We extend this argument to all dimensions as follows:

Theorem 2. *For every fixed* $k, n, d \geq 1$

$$E_{P \in P_d(n)}[k\text{-}w(P)] \geq \frac{ln^{k-1}(n)}{(k-1)!} \cdot (1 + o(1)). \tag{1}$$

Proof. For $P \in P_d(n)$, generated according to model (II) as a sequence of random points $P = (P_1, P_2, \ldots, P_n) \in [0, 1]^d$ we define the set of its *minima* as

$$Min(P) = \{j \in [n] : P_i < P_j \text{ for no } 1 \leq i < j\}.$$

Clearly k-width(P)$\geq |Min(P)|$. Indeed, every minimum of P must determine the starting of a new heap, no matter what k is. Now we use an inequality proved by Winkler [8]:

$$E_{P \in P_d(n)}[|Min(P)|] \geq \frac{ln^{k-1}(n)}{(k-1)!} \cdot (1 + o(1)).$$

$\square$

Open Problem 2. *Is there a constant $c_{k,d} > 0$ such that*

$$\lim_{n \to \infty} \frac{E_{P \in P_d(n)}[k\text{-}w(P)]}{ln^{k-1}(n)} = c_{k,d} ? \tag{2}$$

As for the k-height, a result from Byers et al. can be recast as $h(P) = n - o(n)$ for almost all $\pi \in S_n$. We easily generalize this result to random d-dimensional partial orders as follows:

Theorem 3. *For all $d \geq 2, k \geq 1$ and almost all permutations $P \in P_d(n)$ we have $k\text{-}h(P) = n - o(n)$.*

Proof. A straightforward adaptation of the argument of Byers et al. [1]. Rather than with k-dimensional permutations, we will work with random points in $[0, 1]^d$ (model II).

First one shows that w.h.p. $k\text{-}h(P) = \Omega(n)$, using a similar idea to the one in [1]: we consider division of P into subcubes $[0, 1/2]^d$ and $[1/2, 1]^d$, respectively. Let A_1 be the suborder of P determined by the restriction to the first $n/2$ elements and first subcube. W.h.p. $LHS(A_1) = \Theta(n^{1/d})$. This follows from the result of Bollobás and Winkler [15], together with the result of Bollobás and Brightwell [16], that provides concentration of measure for $LIS(A_1)$.

Now we organize the subsequence A_1 into a k-ary tree W with $\Omega(n^{1/d})$ leaves and continue to add elements of subsequence A_2, corresponding to points in the second half; we assume we add elements greedily, in the first possible subheap rooted at a node of A_1 on the frontier of W, stopping when we can no longer place a node in the tree. With high probability this happens after adding $\Omega(n)$ nodes from A_2: to see this we employ the observation that the stopping of the algorithm implies the existence of a decreasing sequence of A_2 of size $\Omega(n^{1/d})$. We then apply the concentration inequality [16] for $LDS(A_2)$.

For the second, rescaled part of the proof, we search for constants $\alpha, \beta > 0$ such that w.h.p. the subsequence B_1, consisting of points among the first n^α ones that belong to the rectangle $[0, n^{-\beta}]^d$ has w.h.p. k-width $\Omega(n^{1/d+\epsilon})$. For this to happen, we take α, β so that $\alpha - d \cdot \beta > 1/d$. It is always possible to find some positive α, β with this property, e.g. $\alpha = 1 - \frac{1}{2d^2}, \beta = \frac{1}{2d^3}$. Now subsequence B_2 consisting of numbers in the rectangle $[n^{-\beta}, 1]^d$ among the last $n - n^\alpha$ ones has w.h.p. its LDS of size $\Theta(n^{1/d})$. Thus sequence B_2 can w.h.p. be placed in its entirety on the tree W. Ther remaining parallelipipeds have $o(1)$ volume, hence a sublinear number of points. The rest of the details are as in [1].   □

Let us note that a random d-dimensional partial order P can be regarded, by definition, as a subset (thinning) of a $(d - 1)$-dimensional partial order Q: if $P_1, P_2, \ldots, P_d$ are the permutations defining P, simply define Q to be the intersection of $P_1, P_2, \ldots, P_{d-1}$. So the previous result can be interpreted as the statement that no constant amount of thinning is enough to reduce the width of a random permutation to sublinear.

5 The Special Case $d = 2$

In the special case of heapable sequences and random permutations ($d = 2$) we have better insights on the constants $c_{k,d}$ from the above open problem:

Conjecture 1. We have $c_{2,2} = \phi$, with $\phi = \frac{1+\sqrt{5}}{2}$ the golden ratio. More generally

$$c_{k,2} = \frac{1}{\phi_k}, \tag{3}$$

where ϕ_k is the unique root in $(0, 1)$ of equation $X^k + X^{k-1} + \ldots + X = 1$.

Open Problem 3. *Prove this conjecture.*

In the next session we sketch some of the experimental and nonrigorous theoretical evidence for this result. The calculations are nonrigorous, "physics-like", and have yet to be converted to a rigorous argument.

5.1 The Connection with the Multiset Hammersley Process

One of the most rewarding ways to analyze the asymptotic behavior of the LIS of a random permutation is the connection with a model from Nonequilibrium Statistical Physics called *the Hammersley process*.

The easiest way to describe the Hammersley process is via a sequence of random numbers $X_1, X_2, \ldots, X_n \ldots \in (0, 1)$ (note that this combinatorial description is good for our purposes; the general Hammersley process assumes a unit intensity Poisson process on the real line).

We interpret X_i's as *particles*. At each moment the insertion of a new particle removes (*kills*) the smallest (if any) particle X_j, $X_j > X_i$. Intuitively, particles correspond to pile heads in *patience sorting*, a well-known algorithm for computing LIS. The piles are nondecreasing, hence putting a new particle on a pile with head X_j "kills" X_j. Particles that are the largest at the moment when inserted do not kill any particle but simply start a new pile.

A sequence Y of n random particles corresponds naturally to a random n-dimensional permutation. The live particles in the Hammersley process correspond to piles in patience sorting. Therefore $LIS(Y)$ is equal to the number of live particles.

The correspondance between live particles and trees in an optimal decomposition of a random permutation carries on to the framework of heapability as well, with a twist: the multiset generalization of the Hammersley process (defined in [3] and denoted by HAD_k) sees every particle come with a fixed number of k *lives*. A particle does X_i does not kill outright the smallest particle $X_j > X_i$: it simply removes one of its lives.

The infinite-time limit of the multiset Hammersley process with two lives (so-called *hydrodynamic behavior* [17]) seems experimentally to be the so-called *compound Poisson process*. This can be understood combinatorially as follows:

- At stage n the "typical" configuration of the HAM_2 process is characterized by n particles holding 0,1 or two lives.
- The number of particles holding λ lives, for $\lambda \in \{0,1,2\}$ is approximately equal to $d_\lambda \cdot n$, for some constants $0 < d_\lambda < 1$. That is, the global density of particles with λ lives converges asymptotically to d_λ.
- Moreover, particles with λ lives are distributed approximately uniformly at random throughout interval $(0, 1)$, so that the relative densities are valid not only globally, but throughout each bin.

The heuristic explanation given above is confirmed experimentally by Fig. 1. Here we have divided interval (0,1) into 200 bins, and we plot the relative densities (for each bin, represented on the x axis as the corresponding point in [0,1]) of average number of particles in that bin holding 0,1,2 lives, respectively. We simulated each realization of the HAM_2 process for 100.000 steps, and average each value over 100 realizations. The densities seem to be approximately constant among bins. Moreover $d_0 = d_2 \sim 0.38...$, whereas $d_1 \sim 0.23....$ End bin differences appear to be simulation artifacts: larger simulations reduce this difference.

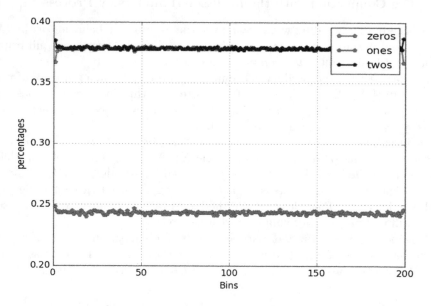

Fig. 1. Relative densities of particles in the HAM_2 process. (Color figure online)

But what are constants d_0, d_1, d_2? Clearly $d_0 + d_1 + d_2 = 1$. The number of particles with two lives grows by one at each step. On the other hand, except in the (probabilistically rare) cases the new particle is the largest live one, it takes a life from a particle counted by d_1 or d_2. Assuming well-mixing the probability that it takes a life of particle with two lives is $\frac{d_2}{d_1+d_2}$. We get, therefore, a "mean-field" equation for d_2:

$$d_2 = 1 - \frac{d_2}{d_1 + d_2}. \tag{4}$$

As for d_1, the flow into d_1 has rate $\frac{d_2}{d_1+d_2}$. However, with probability $\frac{d_1}{d_1+d_2}$ there is a flow from d_1 to d_0, decreasing d_1. The "mean-field" equation for d_1 is:

$$d_1 = \frac{d_2 - d_1}{d_1 + d_2} \tag{5}$$

Solving the system of equations for d_0, d_1, d_2 yields

$$d_0 = d_2 = \frac{3 - \sqrt{5}}{2} \sim 0.381\ldots, d_1 = \sqrt{5} - 2 \sim 0.236\ldots \tag{6}$$

a prediction matching the experimental evidence in Fig. 1.

So how does this hydrodynamical limit predict the claimed scaling behavior, $E[2 - w(P)] \sim \frac{1+\sqrt{5}}{2}$?

In the compound Poisson process the density of live particles is $d_1 + d_2 = \frac{\sqrt{5}-1}{2}$. If the first n particles were sampled *exactly* from this distribution, the expected value of the largest live particle would be $1 - \frac{\sqrt{5}+1}{2} \cdot \frac{1}{n}$. A new particle would start a new heap precisely when it is larger than all live particles (hence it does not kill anyone). The probability of this happening is $\frac{\sqrt{5}+1}{2} \cdot \frac{1}{n}$. Thus, "on the average", in the first $n + 1$ stages the number of created heaps is

$$1 + \frac{\sqrt{5}+1}{2} \cdot H_n = \phi \ln(n) + O(1),$$

with H_n the Harmonic number. Since process HAM_2 is asymptotically a compound Poisson process, we expect the high-order terms to be correct. Similar but more complicated calculations can be performed in the case $d = 2$ with k arbitrary.

6   High-Dimensional Permutations

Linial has initiated [18], under the slogan of "high dimensional combinatorics", a multidimensional analog of permutations. A *p-dimensional permutation of order* n is a $n \times n \times \ldots \times n = [n]^{p+1}$ array of 0/1 values in which each *line* (obtained by setting p indices to values in $[n]$ and leaving free the remaining coordinate) contains *exactly* a one. Ordinary permutations correspond to the one-dimensional case, whereas two-dimensional permutations are essentially latin squares.

Recently, Linal and Simkin [19] have considered notions of monotonicity in high-dimensional permutations, proving a high-dimensonal analog of the Erdős-Székeres theorem. They studied afterwards the scaling of LIS of a random multidimensional permutation, obtaining the scaling $E[LIS(\pi)] = \Theta(n^{p/p+1})$ for a random p-dimensional permutation.

Open Problem 4. *Study the heapability (2-width and 2-height) of random high-dimensional permutations.*

7 Partition into (un)equal Parts: Entropy and Compression

So far we have been interested into the partition of a sequence of numbers into a minimal number of k-chains.

One may want, instead, a partition that insists on parts as equal/unequal as possible. Porfilio [4] showed that the problem of dividing a sequence of integers into a number of equal parts is NP-complete.

One may look for the opposite kind of division, that into mostly unbalanced parts. One way to measure the imbalance is via *entropy* of the distribution induced on the poset by a partition into k-chains. Of course, of all distributions with finite support the uniform distribution has the largest entropy. Minimizing entropy is an objective of recent interest in combinatorial optimization [20–26].

Open Problem 5. *Study the complexity of partitioning a poset P into k-chains leading to a distribution of minimal entropy.*

The open problem is easily seen to be related to the minimum entropy coloring problem for interval graphs. Chromatic entropy is a natural measure with important applications to coding [20, 27, 28].

On the other hand we can state the following natural greedy algorithms:

- for $k = 1, d = 2$: compute a longest increasing subsequence L_1 of P using patience sorting (or dynamic programming).
- for other values of pair (k, d): use instead the Byers et al. algorithm for finding a longest heapable subsequence with $n - o(n)$ elements.
- remove L_1 from P and proceed recursively.

Open Problem 6. *Can one give guarantees on the approximation performance of these algorithms?*

Finally, the decomposition of permutations into components (e.g. runs) forms the basis of the recent theory of data structures and methods for compressing permutations [29, 30] and partial orders. A question that arose during a conversation with Travis Gagie at CPM'2015, and that we would like to state here as an open question is

Open Problem 7. *Is the decomposition of sequences into trees, of the sort employed in computing the 2-width of a partial order, relevant to compression as well?*

Acknowledgments. This research has been supported by CNCS IDEI Grant PN-II-ID-PCE-2011-3-0981 "Structure and computational difficulty in combinatorial optimization: an interdisciplinary approach".

References

1. Byers, J., Heeringa, B., Mitzenmacher, M., Zervas, G.: Heapable sequences and subseqeuences. In: Proceedings of ANALCO, pp. 33–44 (2011)
2. Romik, D.: The Surprising Mathematics of Longest Increasing Subsequences, vol. 4. Cambridge University Press, New York (2015)
3. Istrate, G., Bonchiş, C.: Partition into heapable sequences, heap tableaux and a multiset extension of hammersley's process. In: Cicalese, F., Porat, E., Vaccaro, U. (eds.) CPM 2015. LNCS, vol. 9133, pp. 261–271. Springer, Heidelberg (2015)
4. Porfilio, J.: A combinatorial characterization of heapability. Master's thesis. Williams College, May 2015. http://library.williams.edu/theses/pdf.php?id=790
5. Dilworth, R.P.: A decomposition theorem for partially ordered sets. Ann. Math. **51**, 161–166 (1950)
6. Gyárfás, A., Lehel, J.: Covering and coloring problems for relatives of intervals. Discrete Math. **55**(2), 167–180 (1985)
7. Brightwell, G.: Models of random partial orders. In: Surveys in Combinatorics, pp. 53–83 (1993)
8. Winkler, P.: Random orders. Order **1**(4), 317–331 (1985)
9. Schrijver, A.: Theory of linear and integer programming. Wiley, New York (1998)
10. Hammersley, J.M., et al.: A few seedlings of research. In: Proceedings of the Sixth Berkeley Symposium on Mathematical Statistics and Probability, Volume 1: Theory of Statistics (1972)
11. Logan, B.F., Shepp, L.A.: A variational problem for random Young tableaux. Adv. Math. **26**(2), 206–222 (1977)
12. Vershik, A., Kerov, S.V.: Asymptotics of plancherel measure of symmetrical group and limit form of Young tables. Doklady Akademii Nauk SSSR **233**(6), 1024–1027 (1977)
13. Aldous, D., Diaconis, P.: Hammersley's interacting particle process and longest increasing subsequences. Probab. Theory Relat. Fields **103**(2), 199–213 (1995)
14. Brightwell, G.: Random k-dimensional orders: Width and number of linear extensions. Order **9**(4), 333–342 (1992)
15. Bollobás, B., Winkler, P.: The longest chain among random points in euclidean space. Proc. Am. Math. Soc. **103**(2), 347–353 (1988)
16. Bollobás, B., Brightwell, G.: The height of a random partial order: Concentration of measure. Ann. Appl. Probab. **2**(4), 1009–1018 (1992)
17. Groeneboom, P.: Hydrodynamical methods for analyzing longest increasing subsequences. J. Comput. Appl. Math. **142**(1), 83–105 (2002)
18. Linial, N., Luria, Z.: An upper bound on the number of high-dimensional permutations. Combinatorica **34**(4), 471–486 (2014)
19. Linial, N., Simkin, M.: Monotone subsequences in high-dimensional permutations. arXiv preprint arXiv:1602.02719 (2016)
20. Cardinal, J., Fiorini, S., van Assche, G.: On minimum entropy graph colorings. In: Proceedings of ISIT 2004, The International Symposium on Information Theory, p. 43 (2004)
21. Halperin, E., Karp, R.: The minimum entropy set cover problem. Theor. Comput. Sci. **348**(2–3), 340–350 (2005)
22. Cardinal, J., Fiorini, S., Joraet, G.: Tight results on minimum entropy set cover. Algorithmica **51**(1), 49–60 (2008)
23. Cardinal, J., Fiorini, S., Joret, G.: Minimum entropy orientations. Oper. Res. Lett. **36**(6), 680–683 (2008)

24. Cardinal, J., Fiorini, S., Joret, G.: Minimum entropy combinatorial optimization problems. In: Ambos-Spies, K., Löwe, B., Merkle, W. (eds.) CiE 2009. LNCS, vol. 5635, pp. 79–88. Springer, Heidelberg (2009)
25. Kovačević, M., Stanojević, I., Šenk, V.: On the entropy of couplings. Inf. Comput. **242**, 369–382 (2015)
26. Istrate, G., Bonchiş, C., Dinu, L.P.: The minimum entropy submodular set cover problem. In: Dediu, A.-H., Janoušek, J., Martín-Vide, C., Truthe, B. (eds.) LATA 2016. LNCS, vol. 9618, pp. 295–306. Springer, Heidelberg (2016)
27. Alon, N., Orlitsky, A.: Source coding and graph entropies. IEEE Trans. Inform. Theory **42**, 1329–1339 (1995). Citeseer
28. Doshi, V., Shah, D., Médard, M., Effros, M.: Functional compression through graph coloring. IEEE Trans. Inf. Theory **56**(8), 3901–3917 (2010)
29. Barbay, J., Munro, J.I.: Succinct encoding of permutations : Applications to text indexing. In: Encyclopedia of Algorithms, pp. 915–919. Springer (2008)
30. Barbay, J., Navarro, G.: On compressing permutations and adaptive sorting. Theoret. Comput. Sci. **513**, 109–123 (2013)

Self-Verifying Finite Automata
and Descriptional Complexity

Galina Jirásková[(✉)]

Mathematical Institute, Slovak Academy of Sciences,
Grešákova 6, 040 01 Košice, Slovakia
jiraskov@saske.sk

Abstract. We survey recent results on the descriptional complexity of self-verifying finite automata. In particular, we discuss the cost of simulation of self-verifying finite automata by deterministic finite automata, and the complexity of basic regular operations on languages represented by self-verifying finite automata.

1 Introduction

A self-verifying finite automaton is a nondeterministic automaton whose state set consists of three disjoint groups of states: accepting states, rejecting states, and neutral states. On every input string, at least one computation must end in either an accepting or in a rejecting state. Moreover, there is no input string with both accepting and rejecting computations.

The existence of an accepting computation on an input string proves the membership of the string to the language. This is the same as in a nondeterministic finite automaton (NFA). However, in a self-verifying finite automaton (SVFA), the existence of a rejecting computation definitely proves that the input is not in the language. This is in contrast with NFAs, where the existence of a non-final computation leaves open the possibility that the input may be accepted by a different computation. Thus, even if the transitions are nondeterministic, when a computation of an SVFA ends in an accepting or in a rejecting state, the automaton "can trust" the outcome of that computation, and accept or reject the input. The name "self-verifying" comes from this property. SVFAs were introduced in [4], and were considered mainly in connection with probabilistic Las Vegas computations. However, as pointed in [8], they are also interesting *per se.*

Every SVFA can be converted to an equivalent deterministic finite automaton (DFA) by the standard subset construction [19]. On the other hand, every complete DFA may be viewed as a self-verifying finite automaton with all the final states being accepting, and all the non-final states being rejecting. Hence SVFAs recognize exactly the class of regular languages.

From the descriptional point of view, every n-state NFA can be simulated by a DFA of at most 2^n states [19]. This bound is known to be tight in the

Research supported by VEGA grant 2/0084/15.

Published by Springer International Publishing Switzerland 2016. All Rights Reserved
C. Câmpeanu et al. (Eds.): DCFS 2016, LNCS 9777, pp. 29–44, 2016.
DOI: 10.1007/978-3-319-41114-9_3

binary case [5,14,18]. However, Assent and Seibert [1] proved that in the DFA obtained by applying the subset construction to an SVFA some states must be equivalent. As a consequence, they obtained an upper bound for the conversion of self-verifying automata to deterministic automata in $O(2^n/\sqrt{n})$. Later this result was strengthened in [11], where the tight bound for such a conversion was given by a function $g(n)$ which grows like $3^{n/3}$. The witness languages meeting the bound $g(n)$ were defined over a binary alphabet.

The investigation of self-verifying automata was further deepened by Jirásek et al. [9]. Using the tight bound $g(n)$ from [11], it was shown that a minimal SVFA for a regular language may not be unique. Then the authors introduced an sv-fooling set lower bound technique for the number of states in SVFAs. Using this technique, they obtained tight upper bounds on the complexity of reversal, boolean operations, star, left and right quotients, and asymptotically tight upper bound for concatenation of languages represented by SVFAs.

Here we survey these results. We deal with SVFA-to-DFA conversion in Sect. 2, and discuss the complexity of basic regular operations on SVFAs in Sect. 3. To conclude this introduction, let us recall some basic notions and preliminary results. For further details, the reader may refer to [21].

All DFAs in this paper are assumed to be complete, and NFAs have a unique initial state. Sometimes we also consider NNFAs — nondeterministic finite automata with a nondeterministic choice of the initial state [22] — where we admit multiple initial states.

A *self-verifying finite automaton* (SVFA) is a tuple $A = (Q, \Sigma, \delta, s, F^a, F^r)$, where Q, Σ, δ, and s are the same as in an NFA, F^a is the set of accepting states, F^r is the set of rejecting states, and $F^a \cap F^r = \emptyset$; the remaining states in Q are called neutral. It is required that for each input string w in Σ^*, there exists at least one computation ending in an accepting or in a rejecting state, and there are no strings w such that both $\delta(s, w) \cap F^a$ and $\delta(s, w) \cap F^r$ are nonempty.

The language accepted by the SVFA A, denoted as $L^a(A)$, is the set of all input strings having a computation ending in an accepting state, while the language rejected by A, denoted as $L^r(A)$, is the set of all input strings having a computation ending in a rejecting state. It follows directly from the definition that $L^a(A) = (L^r(A))^c$ for each SVFA A. Hence, when we say that an SVFA A *accepts* a language L, we mean that $L = L^a(A)$ and $L^c = L^r(A)$.

The *state complexity* of a regular language L, sc(L), is defined as the smallest number of states in any DFA for L. The *state complexity of a regular operation* is the maximal state complexity of languages resulting from the operation, considered as a function of the state complexities of the operands. Similarly, the *nondeterministic state complexity* and *self-verifying state complexity* of a regular language L, denoted by nsc(L) and svsc(L), is defined as the smallest number of states in any NFA (with a unique initial state) and SVFA, respectively, for L.

Every NNFA $A = (Q, \Sigma, \delta, I, F)$ can be converted to an equivalent DFA $A' = (2^Q, \Sigma, \cdot, I, F')$, where $R \cdot a = \delta(R, a)$ for each R in 2^Q and each a in Σ, and $F' = \{R \in 2^Q \mid R \cap F \neq \emptyset\}$ [19]. The DFA A' is called the *subset automaton* of the NFA A. Let us recall two observations from [8,11].

Proposition 1 ([8,11]). *Let a language L be accepted by an n-state SVFA. Then the languages L and L^c are accepted by n-state NFAs.* □

Proof. Let L be accepted by an SVFA $A = (Q, \Sigma, \delta, s, F^a, F^r)$. Then L is accepted by NFA $(Q, \Sigma, \delta, s, F^a)$, while L^c is accepted by NFA $(Q, \Sigma, \delta, s, F^r)$. □

Proposition 2 ([8,11]). *Let languages L and L^c be accepted by an m-state and n-state NNFAs, respectively. Then $\mathrm{svsc}(L) \leq m + n + 1$.* □

Proof. Let L be accepted by an m-state NNFA $N = (Q, \Sigma, \delta, I, F)$ and L^c be accepted by an n-state NNFA $N' = (Q', \Sigma, \delta', I', F')$. Then we can get an SVFA A for L with $m + n + 1$ states from NFAs N and N' as follows. We add a new initial state s going to $\delta(I, a) \cup \delta'(I', a)$ on each a in Σ. The state s is accepting if $\varepsilon \in L$, and it is rejecting otherwise. All the states in F are accepting in SVFA A, and all the states in F' are rejecting in A. □

2   SVFA-to-DFA Conversion and Minimal SVFAs

The SVFA-to-DFA conversion was first studied by Assent and Seibert [1]. Then Jiráskova and Pighizzini [11] obtained a tight upper bound for such a conversion.

Proposition 3 ([1, **Theorem 2.1**]). *Every n-state SVFA can be converted to an equivalent DFA of at most $O(2^n/\sqrt{n})$ states.*

Proof (Proof Idea). If S and T are two reachable subset of the subset automaton of an SVFA A such that $S \subseteq T$, then S and T are equivalent. This gives an upper bound $\binom{n}{\lfloor n/2 \rfloor} \in O(2^n/\sqrt{n})$. □

Theorem 4 ([11, **Theorem 9**]). *Every n-state SVFA can be converted to an equivalent DFA of at most $g(n)$ states, where*

$$g(n) = \begin{cases} 1 + 3^{(n-1)/3}, & \text{if } n \bmod 3 = 1 \text{ and } n \geqslant 4, \\ 1 + 4 \cdot 3^{(n-5)/3}, & \text{if } n \bmod 3 = 2 \text{ and } n \geqslant 5, \\ 1 + 2 \cdot 3^{(n-3)/3}, & \text{if } n \bmod 3 = 0 \text{ and } n \geqslant 3, \\ n, & \text{if } n \leqslant 2. \end{cases} \tag{1}$$

Moreover, the bound $g(n)$ is tight, and can be met by a binary n-state SVFA.

Proof (Proof Idea). To an n-state SVFA A, we assign an undirected graph $G(A)$ whose vertex set is Q, and which contains an edge $\{p, q\}$ if and only if two computations starting from p and q cannot give contradictory answers on the same string. Then each reachable subset in the subset automaton of A is represented by a clique in $G(A)$. Moreover, if S and T are two subsets such that $S \cup T$ is a clique in $G(A)$, then S and T are equivalent [11, Lemma 4]. Hence the number of states in the minimal DFA for $L(A)$ is given by the number of maximal cliques in $G(A)$. Next, in $G(A)$ there is exactly one maximal clique containing the initial state of A. This results in at most $1 + f(n-1)$ states, where $f(n)$ denotes the

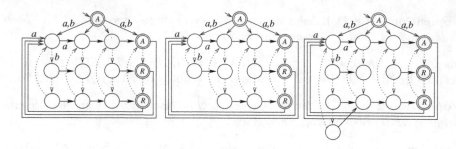

Fig. 1. The witnesses for SVFA-to-DFA conversion; $n = 13, 12$, and 14.

maximum number of possible maximal cliques in a graph with n nodes and, as shown by Moon and Moser [17, Theorem 1], we have $f(n) = 3^{n/3}$ if $n \bmod 3 = 0$, $f(n) = 4 \cdot 3^{\lfloor n/3 \rfloor - 1}$ if $n \bmod 3 = 1$, and $f(n) = 2 \cdot 3^{\lfloor n/3 \rfloor}$ if $n \bmod 3 = 2$. This gives the upper bound. For tightness, let $A = (Q, \{a, b\}, \delta, q_0, F^a, F^r)$, where $n = 1 + 3m$ and $m \geqslant 2$, see Fig. 1 (left) for $m = 13$, be an automaton defined by

$Q = \{q_0\} \cup \{(i, j) \mid 0 \leqslant i \leqslant 2, 1 \leqslant j \leqslant m\}$,

$\delta(q_0, a) = \delta(q_0, b) = \{(0, 1), (0, 2), \ldots, (0, m)\}$,

and for all i, j with $0 \leqslant i \leqslant 2$ and $1 \leqslant j \leqslant m$,

$\delta((i, j), a) = \begin{cases} \{(i, j + 1)\}, & \text{if } j < m, \\ \{(0, 1)\}, & \text{otherwise}, \end{cases}$

$\delta((i, j), b) = \{((i + 1) \bmod 3, j)\}$,

$F^a = \{q_0, (0, m)\}$, and $F^r = \{(1, m), (2, m)\}$.

It is shown in [11, Lemma 8] that A is an SVFA whose minimal DFAs requires $g(n)$ states. To get witnesses for $n = 3k$ or $n = 3k + 2$, we modify the SVFA A as shown in Fig. 1 (middle and right).   □

Thus if we know that the minimal DFA for a language L has more then $g(n)$ states, then by Theorem 4, every SVFA for L must have at least $n + 1$ states. We use this result to show that a minimal SVFA may not be unique.

Example 5. Consider the two 7-state non-isomorphic SVFAs shown in Fig. 2. Apply the subset construction to both of them. In both cases, the subset

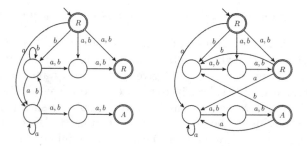

Fig. 2. Two non-isomorphic minimal SVFAs for the language $(a + b)^* a (a + b)^2$.

automata restricted to the reachable states are the same. These subset automata, and therefore also the two SVFAs, accept the language $(a + b)^*a(a + b)^2$, the minimal DFA for which has 8 states. Since we have $g(6) = 7$, every SVFA for this language has at least 7 states. Hence both SVFAs in Fig. 2 are minimal. □

3   Lower Bound Methods and Operations on SVFAs

To prove that a DFA is minimal, we only need to show that all its states are reachable from the initial state, and that no two distinct states are equivalent. To prove minimality of NFAs, a fooling set lower bound method may be used [2,6]. A *fooling set* for a language L is a set of pairs of strings $\{(u_1, v_1), \ldots, (u_n, v_n)\}$ satisfying two conditions:
  (i) for each i, $u_i v_i \in L$, and
  (ii) if $i \neq j$, then $u_i v_j \notin L$ or $u_j v_i \notin L$.
  In the case of SVFAs, we change the two conditions. The first condition can be removed since we have either an accepting or rejecting computation on every string. Before modifying the second condition, consider the following example.

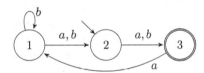

Fig. 3. An NFA for the language L in Example 6.

Example 6. Let L be accepted by the 3-state NFA shown in Fig. 3. Let A be an SVFA for L. Let us show that A has at least 6 states. Consider the following pairs of strings:

$(u_1, v_1) = (a^2, a^2)$	Acc		$(u_4, v_4) = (a^2ba, a^2)$	Rej
$(u_2, v_2) = (a^2b, a)$	Acc		$(u_5, v_5) = (a, a)$	Rej
$(u_3, v_3) = (a^2b^2, \varepsilon)$	Acc		$(u_6, v_6) = (ab, \varepsilon)$	Rej

The strings $u_1 v_1$, $u_2 v_2$, and $u_3 v_3$ are in L, while $u_4 v_4$, $u_5 v_5$, and $u_6 v_6$ are not in L, so we must have accepting and rejecting computations in A on these strings:

$$s \xrightarrow{u_1} p_1 \xrightarrow{v_1} f_1 \in F^a, \qquad s \xrightarrow{u_4} p_4 \xrightarrow{v_4} f_4 \in F^r,$$
$$s \xrightarrow{u_2} p_2 \xrightarrow{v_2} f_2 \in F^a, \qquad s \xrightarrow{u_5} p_5 \xrightarrow{v_5} f_5 \in F^r,$$
$$s \xrightarrow{u_3} p_3 \xrightarrow{v_3} f_3 \in F^a, \qquad s \xrightarrow{u_6} p_6 \xrightarrow{v_6} f_6 \in F^r.$$

Since $u_1 v_2 = a^3$, and a^3 is not in L, we must have $p_1 \neq p_2$ because otherwise $s \xrightarrow{u_1} p_1 = p_2 \xrightarrow{v_2} f_2$ would be an accepting computation on $u_1 u_2$. Similarly, $u_1 v_3$ and $u_2 v_3$ are not in L, so p_1, p_2, and p_3 must be pairwise distinct. On the other hand, the strings $u_4 v_5$, $u_4 v_6$, and $u_5 v_6$ are in L, and therefore the states p_4, p_5, p_6 must be pairwise distinct. Next, let $1 \leq i \leq 3$ and $1 \leq j \leq 3$. If $i \leq j$, then $u_j v_i$ is not in L, and therefore $p_i \neq p_j$ because otherwise $s \xrightarrow{u_j} p_j = p_i \xrightarrow{v_i} f_i$ would be an accepting computation on $u_j v_i$. Finally, if $i > j$, then $u_i v_j$ is in L, and therefore $p_i \neq p_j$. Thus all the state p_i are pairwise distinct, so the SVFA A has at least 6 states.   □

Notice that in the previous example, we were able to interchange the right sides of two pairs (u_i, v_i) and (u_j, v_j) so that at least one of the resulting strings $u_i v_j$ and $u_j v_i$ had a "different finality" than the concatenations $u_j v_j$ and $u_i v_i$, respectively. We formalize this in the following definition.

Definition 7. *A set of pairs of strings* $\mathcal{F} = \{(u_1, v_1), (u_2, v_2), \ldots, (u_n, v_n)\}$ *is called an* sv-fooling set *for a language* L *if for all* i, j *with* $i \neq j$ *at least one of the following two conditions holds:*
   (i) exactly one of the strings $u_i v_j$ *and* $u_j v_j$ *is in* L, *or*
   (ii) exactly one of the strings $u_j v_i$ *and* $u_i v_i$ *is in* L.

Lemma 8 (Lower Bound Method for SVFAs). *Let* $\mathcal{F}$ *be an sv-fooling set for a language* L. *Then every SVFA for the language* L *has at least* $|\mathcal{F}|$ *states.*

Proof. Let A be an SVFA for the language L with the initial state s. Then for each $u_i v_i$, there is an accepting or a rejecting computation of SVFA A on $u_i v_i$. Fix such a computation for each $u_i v_i$. Let p_i be the state in this computation that is reached after reading u_i, and let f_i be the final state reached after reading v_i. Let us show that the states $p_1, p_2, \ldots, p_n$ must be pairwise distinct.

Assume for contradiction that there are i and j with $i \neq j$ such that $p_i = p_j$. Then we have
$$s \xrightarrow{u_i} p_i = p_j \xrightarrow{v_j} f_j \quad \text{and} \quad s \xrightarrow{u_j} p_j \xrightarrow{v_j} f_j; \text{ and}$$
$$s \xrightarrow{u_j} p_j = p_i \xrightarrow{v_i} f_i \quad \text{and} \quad s \xrightarrow{u_i} p_i \xrightarrow{v_i} f_i.$$
It follows that there are computations on $u_i v_j$ and on $u_j v_j$ that end in state f_j. Thus either both this strings are in L, or both of them are in L^c. Moreover, there are computations on $u_j v_i$ and $u_i v_i$ that end in state f_i, so either both these strings are in L, or both of them are in L^c. Hence neither (i) nor (ii) in the definition of an sv-fooling set holds, which is a contradiction.   □

Notice that the lemma above may also be applied to a model of SVFAs with multiple initial states [11, Sect. 5]. Hence if a language L is accepted by an n-state SVFA with multiple initial states, we cannot have an sv-fooling set of size more than n. In such a case, we can use the following observation to prove that an SVFA with a unique initial state needs one more state.

Lemma 9. *Let* $\mathcal{F} = \{(u_1, v_1), (u_2, v_2), \ldots, (u_n, v_n)\}$ *be an sv-fooling set for* L. *For each* i, *let there exist a string* w_i *such that* $\{(u_i, v_i)\} \cup \{(\varepsilon, w_i)\}$ *is an sv-fooling set for* L. *Then every SVFA for* L *has at least* $|\mathcal{F}| + 1$ *states.*

Proof. For each pair in $\mathcal{F}$, fix an accepting or a rejecting computation as in Lemma 8. Then the unique initial state, reached after reading ε, must be different from all the states reached after reading the left part of any pair in $\mathcal{F}$. It follows that the SVFA has at least $|\mathcal{F}| + 1$ pairwise distinct states.   □

Example 10. Let us continue our previous example. Define $w_1 = w_2 = w_3 = a^2$, $w_4 = w_5 = \varepsilon$, and $w_6 = a$. Notice that we have

$u_i \cdot a^2 \in L$ and $\varepsilon \cdot a^2 \notin L$ for $i = 1, 2, 3$,

$u_i \cdot \varepsilon \in L$ and $\varepsilon \cdot \varepsilon \notin L$ for $i = 4, 5$,

$u_6 \cdot a \notin L$ and $\varepsilon \cdot a \in L$.

By Lemma 9, every SVFA for L has at least 7 states.   □

In what follows we use this simple methods to get tight upper bounds on the self-verifying complexity of reversal, boolean operations, star, and left and right quotients. In the case of concatenation, we get an asymptotically tight upper bound.

3.1 Reversal

If a language L is accepted by an n-state DFA A, then the language L^R is accepted by an n-state NNFA A^R obtained from A by swapping the role of the initial and final states of A, and by reversing all the transitions. By applying the subset construction to NNFA A^R, we get a DFA for L^R of at most 2^n states. The bound 2^n is known to be tight [14,16], and the witness languages can be defined over a binary alphabet [12,13].

If a language L is represented by an n-state NFA A, then we can construct an NNFA A^R for L^R in the same way as for DFAs. An equivalent NFA may require one more state. The upper bound $n + 1$ is known to be tight, with worst-case examples defined over a binary alphabet [7,10]. Our next result shows that the self-verifying state complexity of the reversal operation is given by the function $2n + 1$. Notice that the reverse of our worst-case example is a generalization of our NFA language in Example 6.

Theorem 11 ([9]). *Let $n \geq 3$. Let L be a regular language over an alphabet Σ with $\mathrm{svsc}(L) = n$. Then $\mathrm{svsc}(L^R) \leq 2n + 1$, and the bound is tight if $|\Sigma| \geq 2$.*

Proof. Let $A = (Q, \Sigma, \delta, s, F^a, F^r)$ be an SVFA for L. Then L is accepted by the n-state NFA $N = (Q, \Sigma, \delta, s, F^a)$, and L^c is accepted by the n-state NFA $N' = (Q, \Sigma, \delta, s, F^r)$ by Proposition 1. By swapping the role of initial and final states in NFAs N and N', and by reversing all the transitions, we get n-state NNFAs for languages L^R and $(L^c)^R = (L^R)^c$. By Proposition 2, we have $\mathrm{svsc}(L^R) \leq 2n+1$. This proves the upper bound.

For tightness, let L be the language accepted by the DFA A shown in Fig. 4. Construct an NFA A^R for the language L^R as described above. Denote by $[i, j]$ the set of integers $\{k \mid i \leq k \leq j\}$; notice that $[i, j] = \emptyset$ if $i > j$. Consider the following family of $2n$ subsets

$$\mathcal{R} = \big\{[1, i] \mid 1 \leq i \leq n\big\} \cup \big\{[i+1, n] \mid 1 \leq i \leq n\big\}.$$

Notice that each set in $\mathcal{R}$ is reachable in the subset automaton of the NFA A^R from the initial subset $\{n - 1\}$. Thus for each subset S in $\mathcal{R}$, there is a string

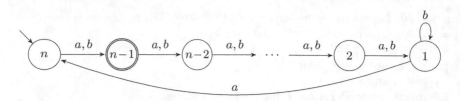

Fig. 4. The binary witness for reversal meeting the bound $2n + 1$.

u_S by which the initial state $\{n - 1\}$ of the subset automaton of A^R goes to S. Consider the following set of $2n$ pairs of strings:

$$\mathcal{F} = \{(u_{[1,i]}, a^{n-i}) \mid 1 \le i \le n\} \cup \{(u_{[j+1,n]}, a^{n-j}) \mid 1 \le j \le n\}.$$

Let us show that the set $\mathcal{F}$ is an sv-fooling set for the language L^R.

First, notice that the string a^{n-i} is accepted by the NFA A^R from a subset S of $[1, n]$ if and only if the state i is in the subset S. To show that $\mathcal{F}$ is an sv-fooling set for L, we have three cases to consider:

(1) Let $1 \le i < k \le n$. Then $u_{[1,i]} \cdot a^{n-k} \notin L^R$ and $u_{[1,k]} \cdot a^{n-k} \in L^R$.

(2) Let $1 \le j < \ell \le n$. Then $u_{[j+1,n]} \cdot a^{n-\ell} \in L^R$ and $u_{[\ell+1,n]} \cdot a^{n-\ell} \notin L^R$.

(3) Let $1 \le i \le n$ and $1 \le j \le n$. Here we have two subcases:

    (3a) If $i \le j$, then $u_{[j+1,n]} \cdot a^{n-i} \notin L^R$ and $u_{[1,i]} \cdot a^{n-i} \in L^R$.

    (3b) If $i > j$, then $u_{[1,i]} \cdot a^{n-j} \in L^R$ and $u_{[j+1,n]} \cdot a^{n-j} \notin L^R$.

Hence we have shown that $\mathcal{F}$ is an sv-fooling set for the language L^R. Now, we use Lemma 9 to show that one more state is necessary for an SVFA to accept L^R. To this aim, let $w_i = a^{n-1}$ for $i = 1, 2, \ldots, n$, $w_{n+j} = \varepsilon$ for $j = 1, 2 \ldots, n - 1$, and $w_{2n} = a$. Then we have

$\varepsilon \cdot a^{n-1} \notin L^R$ while $u_{[1,i]} \cdot a^{n-1} \in L^R$ if $1 \le i \le n$ and $n \ge 3$,

$\varepsilon \cdot \varepsilon \notin L^R$ while $u_{[j+1,n]} \cdot \varepsilon \in L^R$ if $1 \le j \le n - 1$.

$\varepsilon \cdot a \in L^R$ while $u_\emptyset \cdot a \notin L^R$ since $n \ge 3$.

By Lemma 9, every SVFA for L^R has at least $2n + 1$ states.   □

3.2 Boolean Operations

To get a DFA for the complement of a given regular language, we only need to interchange the final and non-final states in a DFA for the given language. Formally, if a regular language L is accepted by a DFA $A = (Q, \Sigma, \delta, s, F)$, then the language L^c is accepted by the DFA $A' = (Q, \Sigma, \delta, s, Q \setminus F)$. Moreover, if A is minimal, then A' is minimal as well. It follows that the state complexities of a regular language and its complement are the same.

    On the other hand, if a language is represented by an NFA, we first apply the subset construction to this NFA, and only after that we can interchange the final and non-final states. This gives an upper bound 2^n. This upper bound is known to be tight [2,20], and witness languages can be defined over a binary alphabet [10]. Our first observation shows that the self-verifying complexity of a language and its complement are the same.

Then we consider the following four Boolean operations: intersection, union, difference, and symmetric difference. In the general case of all regular languages, the state complexity of all four operations is given by the function mn, and the worst-case examples are defined over a binary alphabet [3,15,19,23]. The nondeterministic state complexity of intersection and union is mn and $m+n+1$, respectively, with witness languages defined over a binary alphabet [7].

The difference and symmetric difference on languages represented by NFAs have not been studied yet. Since both these operations require complementation, the nondeterministic state complexities $m \cdot 2^n$ and $m \cdot 2^n + n \cdot 2^m$ of difference and symmetric difference, respectively, could be expected. In the case of self-verifying state complexity, we obtain a tight upper bound mn for all four operations, with worst-case examples defined over a binary alphabet.

Theorem 12. ([9]). *Let K and L be languages over an alphabet Σ with* $\mathrm{svsc}(K) = m$ *and* $\mathrm{svsc}(L) = n$. *Then*
  (i) $\mathrm{svsc}(L^c) = n$,
  (ii) $\mathrm{svsc}(K \cap L), \mathrm{svsc}(K \cup L), \mathrm{svsc}(K \setminus L), \mathrm{svsc}(K \oplus L) < mn$,
  and all the bounds are tight if $|\Sigma| \geq 2$. □

Proof. (i) Let L be accepted by an SVFA A. To get an SVFA A' for the language L^c, we only need to interchange the accepting and rejecting states in the SVFA A. Moreover, if A is minimal, then A' is minimal as well.

(ii) Now we consider intersection. Let K and L be accepted by SVFAs $A = (Q_A, \Sigma, \delta_A, s_A, F_A^a, F_A^r)$ and $B = (Q_B, \Sigma, \delta_B, s_B, F_B^a, F_B^r)$ of m and n states. Construct the product automaton $A \times B = (Q, \Sigma, \delta, s, F^a, F^r)$, where
  $Q = Q_A \times Q_B$; $s = (s_A, s_B)$;
  $F^a = \{(p,q) \mid p \in F_A^a \text{ and } q \in F_B^a\}$ and $F^r = \{(p,q) \mid p \in F_A^r \text{ or } q \in F_B^r\}$;
  $\delta((p,q),a) = \delta_A(p,a) \times \delta_B(q,a)$ for each (p,q) in Q and each a in Σ.
The product automaton $A \times B$ accepts $K \cap L$, and it is self-verifying.

For tightness, consider languages $K = \{w \in \{a,b\}^* \mid \#_a(w) \equiv 0 \bmod m\}$ and $L = \{w \in \{a,b\}^* \mid \#_b(w) \equiv 0 \bmod n\}$ accepted by an m-state and n-state DFAs, so also SVFAs, respectively. Then the set of pairs $\mathcal{F} = \{(a^i b^j, a^{m-i} b^{n-j}) \mid 0 \leq i \leq m-1 \text{ and } 0 \leq j \leq n-1\}$ is an sv-fooling set of size mn for $K \cap L$.

Let us continue with union and difference. Since $K \cup L = (K^c \cap L^c)^c$, and self-verifying state complexity of a language and its complement are the same, we can get an SVFA for the union of K and L as follows. We first construct SVFAs for K^c and L^c. Then we construct an SVFA for $K^c \cap L^c$. Finally, we take an SVFA for the complement of the resulting language. As witness languages, we can take the complements of the witnesses for intersection. Similar considerations can be done also for difference since $K \setminus L = K \cap L^c$.

Finally, we consider symmetric difference. To get the upper bound, we construct a product automaton for symmetric difference in a similar way as for intersection. However, now the sets of accepting and rejecting states are

$F^a = \{(p,q) \mid p \in F_A^a \text{ and } q \in F_B^r\} \cup \{(p,q) \mid p \in F_A^r \text{ and } q \in F_B^a\};$
$F^r = \{(p,q) \mid p \in F_A^a \text{ and } q \in F_B^a\} \cup \{(p,q) \mid p \in F_A^r \text{ and } q \in F_B^r\}.$
This is an mn-state SVFA for the symmetric difference of given languages.

For tightness, let K and L be languages accepted by DFAs A and B shown in
Fig. 5. Construct a product automaton for $K \oplus L$ as described above, and notice
that the set $\mathcal{F} = \{(a^i b^j, a^{m-1-i} b^{n-1-j}) \mid 0 \le i \le m-1 \text{ and } 0 \le j \le n-1\}$
is an sv-fooling set for the language $K \oplus L$. $\qquad\square$

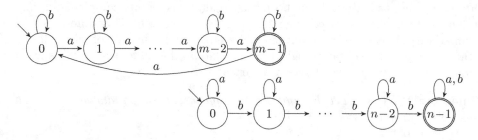

Fig. 5. The binary witnesses for symmetric difference meeting the bound mn.

3.3   Star

The state complexity of the star operation is $3/4 \cdot 2^n$ with binary witness languages [10,15,23]. In the unary case, the tight bound on the state complexity of
star is $(n-1)^2 + 1$ [23,24]. The nondeterministic state complexity of star is $n+1$,
with witnesses defined over a unary alphabet [7]. In this section we show that
the self-verifying state complexity of star is $3/4 \cdot 2^n$. Our worst-case examples
are defined over an alphabet which grows exponentially with n. However, for a
four-letter alphabet, we still get an exponential lower bound $2^{n-1} - 1$.

Theorem 13 ([9]). *Let L be a language over Σ with* $\mathrm{svsc}(L) = n$. *Then*
  (i) $\mathrm{svsc}(L^*) \le 3/4 \cdot 2^n$, *and the bound is tight if* $|\Sigma| \ge 3/4 \cdot 2^n + 1$;
  (ii) the bound $2^{n-1} - 1$ *can be met by a quaternary language.*

Proof. (i) Let $A = (Q, \Sigma, \delta, s, F^a, F^r)$ be an SVFA for L. If only the initial
state s is accepting, then $L^* = L$. Assume that A has k accepting states that are
different from s. Construct an NFA A^* for L^* from A as follows. First, add a new
initial and final state q_0 and for each symbol a in Σ, add a transition from q_0 to
$\delta(s,a)$ if $\delta(s,a) \cap F^a = \emptyset$, and to $\{s\} \cup \delta(s,a)$ otherwise. Next, for each state q in
Q and each symbol a, add a transition from q to s on a whenever $\delta(q,a) \cap F^a \ne \emptyset$.
The initial state of A^* is q_0, and the set of final states is $\{q_0\} \cup F^a$. Now consider
the subset automaton of A^*. Notice that no set containing a state in F^a and not
containing s is reachable in the subset automaton. Next, we can show that the
empty set is unreachable. Hence the subset automaton has at most $2^{n-1} + 2^{n-1-k}$
reachable subsets. The maximum is attained if $k = 1$, and it is equal to $3/4 \cdot 2^n$.

To prove tightness, consider the following family of $3/4 \cdot 2^n - 1$ subsets:
$\mathcal{R} = \{S \mid S \subseteq \{0, 1, \ldots, n - 1\}$ and $0 \in S\} \cup \{S \mid \emptyset \neq S \subseteq \{1, 2, \ldots, n - 2\}\}$.
Let $\Sigma = \{a, b\} \cup \{c_S \mid S \in \mathcal{R}\}$ be an alphabet consisting of $3/4 \cdot 2^n + 1$ symbols.
Let L be accepted by an n-state DFA $A = (\{0, 1, \ldots, n - 1\}, \Sigma, \delta, 0, \{n - 1\})$,
where the transitions are defined as follows: $\delta(i, a) = (i + 1) \bmod n$; $\delta(0, b) = 0$,
$\delta(i, b) = i + 1$ if $1 \leq i \leq n - 2$, and $\delta(n - 1, b) = n - 1$; and for each set S in $\mathcal{R}$,

$$\delta(i, c_S) = \begin{cases} 0, & \text{if } i \in S, \\ n - 1, & \text{if } i \notin S. \end{cases}$$

The transitions on a and b in A are shown in Fig. 6 (top-left), and the transitions
on the symbol $c_{\{1,3\}}$ in the case of $n = 5$ are shown in Fig. 6 (bottom-right).

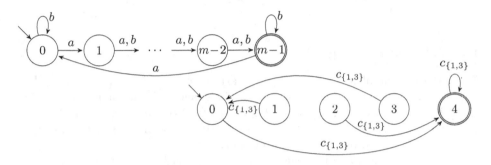

Fig. 6. The witness for star; symbols a a b (top-left) and symbol $c_{\{1,3\}}$ for $n = 5$.

Construct an NFA A^* for the language L^* as described above. Notice that
each subset in $\mathcal{R}$ is reachable in the subset automaton of A^*, that is, for each
subset S in $\mathcal{R}$, there is a string u_S, by which $\{q_0\}$ goes to the subset S. Then
the set $\mathcal{F} = \{(u_S, c_S) \mid S \in \mathcal{R}\}$ is an sv-fooling set of size $3/4 \cdot 2^n - 1$ for L^*.
Finally, by setting $w_S = \varepsilon$ if $n - 1 \notin S$ and $w_S = b$ otherwise, we use Lemma 9
to show that one more state is necessary in every SVFA for the language L^*.

(ii) Consider the language L accepted by the quaternary DFA B shown in
Fig. 7. Notice that the transitions on symbols a and b are the same as in the
DFA A above. It follows that all the subsets of $\{0, 1, \ldots, n - 1\}$, that have
been shown to be reachable in the subset automaton of A^*, are reachable in
the subset automaton of B^* as well. In particular, all the non-empty subsets of
$\{0, 1, \ldots, n - 2\}$ are reachable. Similarly as in the proof above, let u_S be a string
over $\{a, b\}$ by which the initial subset $\{q_0\}$ goes to S in the subset automaton.
Our aim is to describe an sv-fooling set for L^* of size $2^{n-1} - 1$. To this aim, for
every non-empty subset S of $\{0, 1, \ldots, n - 2\}$, define the string $v_S = v_0 v_1 \cdots v_{n-2}$
of length $n - 1$ over $\{c, d\}$ as follows:

$$v_{n-2-i} = \begin{cases} c, & \text{if } i \in S, \\ d, & \text{if } i \notin S, \end{cases}$$

that is, the string v_S somehow describes the set S, however, in a reversed order:
we can assign the symbol $\sigma(i) = c$ to each state i in S and the symbol $\sigma(i) = d$ to

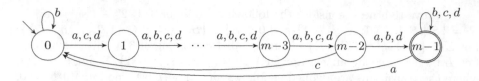

Fig. 7. The quaternary DFA of a language L with $\mathrm{svsc}(L^*) \geq 2^{n-1} - 1$.

each state i outside the set S, and then we have $v_S = \sigma(n-2)\sigma(n-3)\cdots\sigma(1)\sigma(0)$. Then, for every set S, the string v_S is accepted by B^* from every state outside the set S, while v_S is rejected by B^* from every state in S. It follows that $\{(u_S, v_S) \mid \emptyset \neq S \subseteq \{0, 1, \ldots, n-2\}\}$ is an sv-fooling set for L^*. □

3.4 Left and Right Quotients

The left quotient of a language L by a string w is $w\backslash L = \{x \mid wx \in L\}$, and the left quotient of a language L by a language K is the language $K\backslash L = \bigcup_{w \in K} w\backslash L$. The state complexity of the left quotient operation is $2^n - 1$ [23], and its nondeterministic state complexity is $n+1$ [10]. In both cases, the worst-case examples are defined over a binary alphabet.

The right quotient of a language L by a string w is $L/w = \{x \mid xw \in L\}$, and the right quotient of a language L by a language K is $L/K = \bigcup_{w \in K} L/w$. If a language L is accepted by an n-state DFA $A = (Q, \Sigma, \cdot, s, F)$, then the language L/K is accepted by a DFA that is exactly the same as the DFA A, except for the set of final states that consists of all the states q of A, such that there exists a string w in K with $q \cdot w \in F$ [23]. Thus $\mathrm{sc}(L/K) \leq n$. The tightness of this upper bound has been shown using binary languages in [23].

Here we show that the self-verifying complexity of the left quotient operation is $2^n - 1$. To prove tightness, we use an exponential alphabet. Then, using a four letter alphabet, we get a lower bound $2^{n-1} - 1$. Finally, we show that the self-verifying state complexity of right quotient is given by the function $g(n)$, where $g(n)$ is the tight upper bound for SVFA-to-DFA conversion given in (1) on page 3.

Theorem 14 ([9])**.** Let $K, L \subseteq \Sigma$, $\mathrm{svsc}(K) = m$, and $\mathrm{svsc}(L) = n$. Then
(i) $\mathrm{svsc}(K\backslash L) \leq 2^n - 1$, and the bound is tight if $|\Sigma| \geq 2^n + 1$;
(ii) the bound $2^{n-1} - 1$ can be met by quaternary languages.

Proof. (i) Let L be accepted by an SVFA $A = (Q, \Sigma, \delta, s, F^a, F^b)$. Then the language $K\backslash L$ is accepted by an NNFA $N = (Q, \Sigma, \delta, I, F^a)$, where a state q is in I if it can be reached from the initial state of A by a string in K. After applying the subset construction to the NNFA N, we get a DFA for $K\backslash L$, in which the empty set is unreachable. This gives the upper bound.

To prove tightness, consider the family $\mathcal{R}$ of all non-empty subsets of $\{0, 1, \ldots, n-1\}$. Let $\Sigma = \{a, b\} \cup \{c_S \mid S \in \mathcal{R}\}$ be an alphabet consisting of $2^n + 1$ symbols. Let $K = a^* \cup a^* b^{m-2}$ be a language over Σ. Then K is accepted by an m-state DFA, and the set $\{(b^i, b^{m-2-i}) \mid 0 \le i \le m-2\} \cup \{(b^{m-1}a, \varepsilon)\}$ is an sv-fooling set of size m for the language K. Hence $\mathrm{svsc}(K) = m$.

Let L be accepted by an n-state DFA $B = (\{0, 1, \ldots, n-1\}, \Sigma, \delta, 0, \{n-1\})$, where the transitions are defined as follows: $\delta(i, a) = (i+1) \bmod n$; $\delta(0, b) = \delta(1, b) = 0$, and $\delta(i, b) = i$ if $2 \le i \le n-1$; and for each subset S of $\{0, \ldots, n-1\}$, we have $\delta(i, c_S) = 0$ if $i \in S$, and $\delta(i, c_S) = n-1$ otherwise.

Construct an NNFA N for the language $K \backslash L$ from the DFA B by making all the states of B initial. Each subset S in $\mathcal{R}$ is reachable in the subset automaton of the NNFA N by a string u_S. Now, in the same way as in the proof of Theorem 13, we can prove that the set of pairs $\{(u_S, c_S) \mid S \in \mathcal{R}\}$ is an sv-fooling set of size $2^n - 1$ for the language $K \backslash L$.

(ii) The language K over $\{a, b, c, d\}$ is the same as in (i). The language L is accepted by the DFA B', in which the transitions on a and b are the same as in the DFA B above, and the transitions on c and d are the same as in Fig. 7. In a similar way as in the proof of Theorem 13 (ii), we can describe an sv-fooling set $\{(u_S, v_S) \mid \emptyset \ne S \subseteq \{0, 1, \ldots, n-2\}\}$ of size $2^{n-1} - 1$ for $K \backslash L$.   □

Theorem 15 ([9]). *Let $K, L \subseteq \Sigma$, $\mathrm{svsc}(K) = m$, and $\mathrm{svsc}(L) = n$. Then*
  (i) $\mathrm{svsc}(L/K) \le g(n)$, and the bound is tight if $|\Sigma| \ge g(n) + 2$;
  (ii) the bound $\Omega(2^{n/3})$ can be met by quaternary languages.   □

Proof. (i) Let a language L be accepted by an n-state SVFA. First, convert this SVFA to an equivalent minimal DFA. By Theorem 4, this DFA has at most $g(n)$ states. By making certain states final based on the language K, we get a DFA for L/K of at most $g(n)$ states.

For tightness, let $n = 1 + 3k$ and $k \ge 2$; the arguments can be extended to the other values of n in a straightforward way. Consider the grid $Q = \{(i, j) \mid 0 \le i \le 2 \text{ and } 1 \le j \le k\}$ of $3k$ nodes. Let $\mathcal{R}$ be the following family of 3^k subsets $\mathcal{R} = \{\{(i_1, 1), (i_2, 2), \ldots, (i_k, k)\} \mid i_1, i_2, \ldots, i_k \in \{0, 1, 2\}\}$, that is, each subset in $\mathcal{R}$ corresponds to a choice of one element in each column of the grid Q. Let $\Sigma = \{a, b, c\} \cup \{d_S \mid S \in \mathcal{R}\}$ be an alphabet consisting of $3 + 3^k$ symbols.

Let $K = \{c^\ell \mid \ell \ge m-2\}$ be the language over Σ that contains all the strings in c^* of length at least $m-2$. We have $\mathrm{svsc}(K) = m$. Let L be accepted by a $(3k+1)$-state SVFA B, in which the transitions on a, b are the same as in the binary witness for SVFA-to-DFA conversion in Theorem 4. Next, symbol c performs the cyclic permutation on each row of the grid Q, and maps the initial state to each state in the first row. Finally, for each set S in $\mathcal{R}$, symbol d_S maps every state (i, j) of S to the state $(1, j)$, and it maps every state (i, j) outside S to $(0, j)$. Then we can show that $\mathrm{svsc}(L/K) = g(n)$.

(ii) Let $\Sigma = \{a, b, c, d\}$. Let $K = \{c^\ell \mid \ell \ge m-2\}$ be a language over Σ with $\mathrm{svsc}(K) = m$. Let L be accepted by an n-state SVFA B' in which the transitions on a, b are the same as in the SVFA B in case (i). By c and d, the state q_0 goes to $\{(0, 1), \ldots, (0, k)\}$, and each state (i, j) with $j \le k-1$ goes to $\{(i, j+1)\}$. The state $(0, k)$ goes to $\{(1, 1)\}$ on both c, d. The state $(1, k)$ goes to $\{(0, 1)\}$ on c, and it goes to $\{(2, 1)\}$ on d. The state $(2, k)$ goes to $\{(2, 1)\}$ on both c, d. Then we get $\mathrm{svsc}(L/K) \in \Omega(2^{n/3})$.   □

3.5   Concatenation

The state complexity of concatenation is $m2^n - 2^{n-1}$, and its nondeterministic state complexity is $m + n$. In both cases, the worst-case examples can be defined over a binary alphabet [7,10,15,23]. The aim of this subsection is to get asymptotically tight bound $\Theta(3^{m/3} \cdot 2^n)$ on the self-verifying state complexity of the concatenation operation. Recall that $g(n)$ is the tight upper bound for SVFA-to-DFA conversion given in (1) on page 3.

Theorem 16 ([9]). *Let* $K, L \subseteq \Sigma$, $\mathrm{svsc}(K) = m$, *and* $\mathrm{svsc}(L) = n$. *Then*
  (i) $\mathrm{svsc}(KL) \leq g(m) \cdot 2^n$;
  (ii) *the bound* $1/2 \cdot g(m) \cdot 2^n$ *can be met if* $|\Sigma| \geq g(m) + 2^n + 4$;
  (iii) *the bound* $\Omega(2^{m/3} \cdot 2^n)$ *can be met if* $|\Sigma| \geq 8$.

Proof. (i) Let K and L be accepted by SVFAs A and B, respectively. First, convert the SVFA A to a minimal DFA A'. Then, construct an NNFA N for the language KL from automata A' and B in a usual way. Next, apply the subset construction to N. In the subset automaton of N, every reachable subset can be expressed as $\{q\} \cup T$, where q is a state of A' and T is a subset of the state set of B. Since A is an SVFA, the DFA A' has at most $g(m)$ states by Theorem 4. Thus the subset automaton of N has at most $g(m) \cdot 2^n$ reachable states.

(ii) For the sake of simplicity, we consider the case of $m = 1 + 3k$ a $k \geq 2$. Consider the grid $Q = \{(i, j) \mid 0 \leq i \leq 2 \text{ and } 1 \leq j \leq k\}$ of $3k$ nodes. Let $\mathcal{R} = \{\{(i_1, 1), (i_2, 2), \ldots, (i_k, k)\} \mid i_1, i_2, \ldots, i_k \in \{0, 1, 2\}\}$. Let $\Sigma = \{a, b, c, d, e\} \cup \{f_S \mid S \in \mathcal{R}\} \cup \{g_T \mid T \subseteq \{0, 1, \ldots, n - 1\}\}$ be an alphabet consisting of $5 + 3^{\frac{m-1}{3}} + 2^n$ symbols. Let K be the language over Σ accepted by m-state SVFA $A = (Q \cup \{q_0\}, \Sigma, \delta, q_0, F^a, F^r)$, where the transitions on a, b, c are the same as in the case of right quotient, the symbols d, e, g_T are ignored, and transitions on f_S are defined by

$$\delta((i, j), f_S) = \begin{cases} \{(1, j)\}, & \text{if } (i, j) \in S, \\ \{(0, j)\}, & \text{if } (i, j) \notin S; \end{cases}$$

Let L be the language accepted by DFA $B = (\{0, 1, \ldots, n - 1\}, \Sigma, \cdot, 0, \{0\})$, where $i \cdot a = i \cdot b = i \cdot c = i \cdot f_S = i$; $i \cdot d = (i + 1) \bmod n$; $0 \cdot e = 0$, $i \cdot b = i + 1$ if $1 \leq i \leq n - 2$, and $(n - 1) \cdot b = 1$;

$$i \cdot g_T = \begin{cases} n - 1, & \text{if } i \in T, \\ 0, & \text{if } i \notin T. \end{cases}$$

Construct an NFA N for KL and show that each set in the family $\mathcal{R}_N = \{S \cup T \mid S \in \mathcal{R} \text{ with } (0, k) \notin S, \text{ and } T \subseteq \{0, 1, \ldots, n - 1\}\}$ is reachable in the corresponding subset automaton. Then prove that $\mathcal{F} = \{(u_{S \cup T}, g_T \cdot f_S \cdot c^k) \mid S \cup T \in \mathcal{R}_N\}$ is an sv-fooling set for the language KL.

(iii) The idea of the proof is to define strings v_S and v_T over an eight-letter alphabet for some sets S in $\mathcal{R}$, namely, for those that consist only of the states in the first and second row of the grid Q, and for each subset T of $\{1, \ldots, n-2\}$. As a result, we get an sv-fooling set $\{(u_{S \cup T}, v_T \cdot v_S \cdot c^{2k}) \mid S \in \mathcal{R}', T \subseteq \{1, \ldots, n-2\}\}$, where $\mathcal{R}'$ contains all the sets in $\mathcal{R}$ which only have states in the the first or second row of the grid Q. This gives a lower bound in $\Omega(2^{m/3} \cdot 2^n)$. $\square$

Table 1. The state complexity, nondeterministic, and self-verifying state complexity of basic regular operations.

	DFAs	NFAs	SVFAs	$|\Sigma|$
complement	n	2^n	n	1
intersection	mn	mn	mn	2
union	mn	$m+n+1$	mn	2
difference	mn	?	mn	2
symmetric difference	mn	?	mn	2
reversal	2^n	$n+1$	$2n+1$	2
star	$3/4 \cdot 2^n$	$n+1$	$3/4 \cdot 2^n$	$3/4 \cdot 2^n + 1$
left quotient	$2^n - 1$	$n+1$	$2^n - 1$	$2^n + 1$
right quotient	n	n	$g(n)$	$g(n) + 2$
concatenation	$(m - \frac{1}{2}) \cdot 2^n$	$m+n$	$\Theta(3^{m/3} \cdot 2^n)$	$g(m) + 2^n + 4$

4   Conclusions

Table 1 summarizes the results on the self-verifying state complexity of considered operations, and compares them to the known results on their state complexity and nondeterministic state complexity. The last column of the table displays the size of an alphabet which was used to define witness languages. For star and quotients, an exponential lower bound can be obtained by using a four-letter alphabet. In the case of concatenation, a lower bound in $\Omega(2^{m/3}2^n)$ is met by languages defined over an eight-letter alphabet. The tight upper bound for the concatenation operation remains open even in the case of a growing alphabet.

References

1. Assent, I., Seibert, S.: An upper bound for transforming self-verifying automata into deterministic ones. Theor. Inf. Appl. **41**, 261–265 (2007)
2. Birget, J.C.: Partial orders on words, minimal elements of regular languages, and state complexity. Theoret. Comput. Sci. **119**, 267–291 (1993)
3. Brzozowski, J.A.: Quotient complexity of regular languages. J. Autom. Lang. Comb. **15**, 71–89 (2010)
4. Ďuriš, P., Hromkovič, J., Rolim, J., Schnitger, G.: Las Vegas versus determinism for one-way communication complexity, finite automata, and polynomial-time computations. In: Reischuk, R., Morvan, M. (eds.) STACS 1997. LNCS, vol. 1200, pp. 117–128. Springer, Heidelberg (1997)
5. Ershov, U.L.: On a conjecture of W.U. Uspenskii. Algebra i Logika (Seminar) **1**, 45–48 (1962). (in Russian)
6. Glaister, I., Shallit, J.: A lower bound technique for the size of nondeterministic finite automata. Inf. Process. Lett. **59**, 75–77 (1996)
7. Holzer, M., Kutrib, M.: Nondeterministic descriptional complexity of regular languages. Int. J. Found. Comput. Sci. **14**, 1087–1102 (2003)

8. Hromkovič, J., Schnitger, G.: On the power of Las Vegas for one-way communication complexity, OBDDs, and finite automata. Inf. Comput. **169**, 284–296 (2001)
9. Jirásek, J.Š., Jirásková, G., Szabari, A.: Operations on self-verifying finite automata. In: Beklemishev, L.D. (ed.) CSR 2015. LNCS, vol. 9139, pp. 231–261. Springer, Heidelberg (2015)
10. Jirásková, G.: State complexity of some operations on binary regular languages. Theoret. Comput. Sci. **330**, 287–298 (2005)
11. Jirásková, G., Pighizzini, G.: Optimal simulation of self-verifying automata by deterministic automata. Inf. Comput. **209**, 528–535 (2011)
12. Jirásková, G., Šebej, J.: Reversal of binary regular languages. Theoret. Comput. Sci. **449**, 85–92 (2012)
13. Leiss, E.: Succinct representation of regular languages by Boolean automata. Theoret. Comput. Sci. **13**, 323–330 (1981)
14. Lupanov, O.B.: A comparison of two types of finite automata. Problemy Kibernetiki **9**, 321–326 (1963). (in Russian)
15. Maslov, A.N.: Estimates of the number of states of finite automata. Soviet Math. Doklady **11**, 1373–1375 (1970)
16. Mirkin, B.G.: On dual automata. Kibernetika (Kiev) **2**, 7–10 (1966). (in Russian)
17. Moon, J.W., Moser, L.: On cliques in graphs. Israel J. Math. **3**, 23–28 (1965)
18. Moore, F.: On the bounds for state-set size in the proofs of equivalence between deterministic, nondeterministic, and two-way finite automata. IEEE Trans. Comput. **C–20**, 1211–1214 (1971)
19. Rabin, M., Scott, D.: Finite automata and their decision problems. IBM J. Res. Develop. **3**, 114–125 (1959)
20. Sakoda, W.J., Sipser, M.: Nondeterminism and the size of two-way finite automata. In: Proceedings of the 10th Annual ACM STOC, pp. 275–286 (1978)
21. Sipser, M.: Introduction to the Theory of Computation. PWS Publishing Company, Boston (1997)
22. Yu, S.: Chapter 2: Regular languages. In: Rozenberg, G., Salomaa, A. (eds.) Handbook of Formal Languages, vol. I, pp. 41–110. Springer, Heidelberg (1997)
23. Yu, S., Zhuang, Q., Salomaa, K.: The state complexity of some basic operations on regular languages. Theoret. Comput. Sci. **125**, 315–328 (1994)
24. Čevorová, K.: Kleene star on unary regular languages. In: Jurgensen, H., Reis, R. (eds.) DCFS 2013. LNCS, vol. 8031, pp. 277–288. Springer, Heidelberg (2013)

On the State Complexity of Partial Derivative Automata For Regular Expressions with Intersection

Rafaela Bastos, Sabine Broda, António Machiavelo, Nelma Moreira$^{(\boxtimes)}$, and Rogério Reis

CMUP and Faculdade de Ciências da Universidade do Porto, Porto, Portugal
{rrbastos,sbb,nam,rvr}@dcc.fc.up.pt, ajmachia@fc.up.pt

Abstract. Extended regular expressions (with complement and intersection) are used in many applications due to their succinctness. In particular, regular expressions extended with intersection only (also called semi-extended) can already be exponentially smaller than standard regular expressions or equivalent nondeterministic finite automata (NFA). For practical purposes it is important to study the average behaviour of conversions between these models. In this paper, we focus on the conversion of regular expressions with intersection to nondeterministic finite automata, using partial derivatives and the notion of support. First, we give a tight upper bound of $2^{O(n)}$ for the worst-case number of states of the resulting partial derivative automaton, where n is the size of the expression. Using the framework of analytic combinatorics, we then establish an upper bound of $(1.056 + o(1))^n$ for its asymptotic average-state complexity, which is significantly smaller than the one for the worst case.

1 Introduction

Regular expressions with additional operators are used in applications such as programming languages [12], XML processing [23], or runtime verification [22]. Most of these operators do not increase their language expressive power but lead to gains in the succinctness of the representation. This is the case for intersection. For regular expressions with intersection ($\mathsf{RE}_\cap$) (or semi-extended), several computational complexity decision problems, such as membership, equivalence and emptiness, were studied by various authors. Petersen [21] has shown that the membership problem is LOGCFL-complete, while for standard regular expressions (RE) it is NL-complete [19]. Fürer [14] has proved that inequivalence and non-empty complement are EXPSPACE-complete, which contrasts with the PSPACE-completeness

This work was partially supported by CMUP (UID/MAT/00144/2013), which is funded by FCT (Portugal) with national (MEC) and european structural funds through the programs FEDER, under the partnership agreement PT2020.

of these problems for RE. The complexity of the conversions from regular expressions with intersection to standard regular expressions, and to finite automata, were recently studied by Gelade and Neven [16], Gruber and Holzer [18], and Gelade [15]. The conversion from $RE_\cap$ to RE or to nondeterministic finite automata (NFA) is exponential and it is double exponential to deterministic finite automata (DFA). The conversion from $\alpha \in RE_\cap$ to a DFA can be accomplished using Brzozowski's derivatives [8]. From RE to NFA a standard algorithm is the partial derivative automaton construction ($\mathcal{A}_{pd}$) introduced by Antimirov [1], which coincides with the resolution of systems of equations by Mirkin [20]. The average complexity of these conversions was recently studied using the framework of analytic combinatorics [4,5], and also their extension to regular expressions with shuffle [7]. For these studies, Mirkin's construction is essential as it provides inductive definitions that can be used to obtain generating functions.

Caron et al. [9] extended the $\mathcal{A}_{pd}$ to regular expressions with both intersection and complement (extended regular expressions)[1]. In their approach a partial derivative is a set of sets of expressions (akin a disjunctive normal form), whereas here it is simply a set of expressions. In the worst-case, their approach also leads to NFAs that can be exponentially larger than the original expressions. Moreover, considering sets of sets of expressions would turn the analytic combinatoric analysis much harder.

In this paper we show that for $RE_\cap$, Mirkin's construction can lead to automata not initially connected and thus larger than the ones built by Antimirov's construction. However, the two constructions can produce identical NFAs. We present an exponential worst-case upper bound which is tight for both. Using the framework of analytic combinatorics, we give an upper bound for the asymptotic average-state complexity for the Mirkin's construction, which turns out to be much smaller than the worst-case bound. This also means that Antimirov's construction is asymptotically and on average much smaller than the worst-case upper bound.

2   Regular Expressions with Intersection

Let $\Sigma = \{a_1, \ldots, a_k\}$ be an *alphabet* of size k. A *word* over Σ is a finite sequence of symbols of Σ. The *empty word* is denoted by ε. The set $\Sigma^\star$ is the set of all words over Σ. A *language* over Σ is a subset of $\Sigma^\star$. The set $RE_\cap$ of *regular expressions with intersection* over Σ contains the expression $\emptyset$ and all terms generated by the following grammar:

$$\alpha \rightarrow \varepsilon \mid a \mid (\alpha + \alpha) \mid (\alpha \cdot \alpha) \mid (\alpha \cap \alpha) \mid (\alpha^\star) \qquad (a \in \Sigma), \tag{1}$$

where the operator $\cdot$ (concatenation) is often omitted. Parenthesis can also be omitted considering the following precedences for the operators: $\star > \cdot > \cap > +$. The size of a regular expression $\alpha \in RE_\cap$ is denoted by $\|\alpha\|$ and defined as the number of occurrences of symbols (parenthesis not counted) in α.

---

[1] And a more general framework is also reported in [10].

Similarly, $|\alpha|_\Sigma$ denotes the number of occurrences of alphabet symbols in α, and $|\alpha|_\cap$ the number of occurrences of the binary operator $\cap$. The language $\mathcal{L}(\alpha)$ for $\alpha \in \mathsf{RE}_\cap$ is defined as usual, with $\mathcal{L}(\alpha \cap \beta) = \mathcal{L}(\alpha) \cap \mathcal{L}(\beta)$. We say that two regular expressions $\alpha, \beta \in \mathsf{RE}_\cap$ are *equivalent*, if $\mathcal{L}(\alpha) = \mathcal{L}(\beta)$, and write $\alpha \doteq \beta$ in this case. For a set $S \subseteq \mathsf{RE}_\cap$, the language of S is defined as $\mathcal{L}(S) = \bigcup_{\alpha \in S} \mathcal{L}(\alpha)$. The notion of equivalence extends naturally to sets of regular expressions. The *left-quotient* of a language $\mathcal{L}$ w.r.t. a word $w \in \Sigma^\star$ is defined as $w^{-1}\mathcal{L} = \{\, x \mid wx \in \mathcal{L} \,\}$. The algebraic structure $(\mathsf{RE}_\cap, +, \cdot, \emptyset, \varepsilon)$ constitutes an idempotent semiring, that with the unary operator $\star$ is a Kleene algebra. Antimirov and Mosses [2] presented a complete and sound axiomatization for $\mathsf{RE}_\cap$, where the binary operator $\cap$ is idempotent, commutative, associative, distributes over $+$, and also satisfies the following axioms, where $a_i, a_j \in \Sigma$:

$$(\varepsilon \cap \beta) \doteq \emptyset \wedge (\alpha \doteq \beta\alpha + \gamma) \Rightarrow \alpha \doteq \beta^\star\gamma, \qquad \varepsilon \cap \alpha^\star \doteq \varepsilon,$$
$$\varepsilon \cap (\alpha\beta) \doteq (\varepsilon \cap \alpha) \cap \beta, \qquad \varepsilon \cap a_i \doteq \emptyset \cap \alpha \doteq \emptyset,$$
$$(a_i\alpha) \cap (a_j\beta) \doteq (a_i \cap a_j)(\alpha \cap \beta), \qquad a_i \cap a_j \doteq \emptyset \quad (a_i \neq a_j),$$
$$(\alpha a_i) \cap (\beta a_j) \doteq (\alpha \cap \beta)(a_i \cap a_j), \qquad \alpha + (\alpha \cap \beta) \doteq \alpha.$$

With the usual abuse of notation, define the function $\varepsilon : \mathsf{RE}_\cap \to \{\emptyset, \varepsilon\}$ by $\varepsilon(\alpha) = \varepsilon$ if $\varepsilon \in \mathcal{L}(\alpha)$, and $\varepsilon(\alpha) = \emptyset$ otherwise. The methods developed in Sects. 3 and 4 are syntactical and aim at building automata equivalent to a given regular expression. To ensure the finiteness of the constructions it is not necessary to consider regular expressions modulo any of the above properties[2]. However, in some examples, for the sake of succinctness, we also consider regular expressions modulo the identities of $\cdot$ and $+$. Note that this does not affect the upper bounds of the number of states, both in the worst and in the average case.

3   Automata and Systems of Equations

We first recall the definition of a nondeterministic finite automaton (NFA) as a tuple $\mathcal{A} = \langle S, \Sigma, S_0, \delta, F \rangle$, where S is a finite set of states, Σ is a finite alphabet, $S_0 \subseteq S$ a set of initial states, $\delta : S \times \Sigma \to 2^S$ the transition function, and $F \subseteq S$ a set of final states. The *language* of $\mathcal{A}$ is $\mathcal{L}(\mathcal{A}) = \{w \in \Sigma^\star \mid \delta(S_0, w) \cap F \neq \emptyset\}$. The *right language* of a state s, denoted by $\mathcal{L}_s$, is the language accepted by $\mathcal{A}$ if we take $S_0 = \{s\}$. It is well known that, for each n-state NFA $\mathcal{A}$, over $\Sigma = \{a_1, \ldots, a_k\}$, having right languages $\mathcal{L}_1, \ldots, \mathcal{L}_n$, it is possible to associate a system of linear language equations

$$\mathcal{L}_i = a_1\mathcal{L}_{1i} \cup \cdots \cup a_k\mathcal{L}_{ki} \cup \varepsilon(\mathcal{L}_i), \text{ for } i \in [1, n],$$

where $\mathcal{L}_{ji} = \bigcup_{l \in \delta(i, a_j)} \mathcal{L}_l$ and $\mathcal{L}(\mathcal{A}) = \bigcup_{i \in S_0} \mathcal{L}_i$. In the same way, it is possible to associate to each regular expression a system of equations. We here extend Mirkin's contruction to regular expressions with intersection.

Definition 1. *Consider $\alpha_0 \in \mathsf{RE}_\cap$ over $\Sigma = \{a_1, \ldots, a_k\}$. A support of α_0 is a set $\{\alpha_1, \ldots, \alpha_n\}$ of regular expressions with intersection that satisfies a system of equations*

---

[2] As is the case, for instance, for Brzozowski DFA or Caron et al. approach.

$$\alpha_i \doteq a_1 \alpha_{1i} + \cdots + a_k \alpha_{ki} + \varepsilon(\alpha_i) \qquad i \in [0, n], \tag{2}$$

for some $\alpha_{1i}, \ldots, \alpha_{ki}$, where each $\alpha_{j,i}$ is a (possibly empty) sum of elements in $\{\alpha_1, \ldots, \alpha_n\}$.

It is clear that the existence of a support of α implies the existence of an NFA that accepts the language of α.

A support for a regular expression $\alpha \in \mathsf{RE}_\cap$ can be computed using the function $\pi : \mathsf{RE}_\cap \to 2^{\mathsf{RE}}_\cap$ defined below. First, we define some operations on sets of regular expressions. Given $S, T \subseteq \mathsf{RE}_\cap$ and $\beta \in \mathsf{RE}_\cap$, $S\beta = \{\alpha\beta \mid \alpha \in S\}$ and $S \cap T = \{\alpha \cap \beta \mid \alpha \in S, \beta \in T\}$. Note, in particular, that $\mathcal{L}(S \cap T) = \mathcal{L}(S) \cap \mathcal{L}(T)$.

Definition 2. Given $\alpha \in \mathsf{RE}_\cap$, the set $\pi(\alpha)$ is inductively defined by:

$$\pi(\emptyset) = \pi(\varepsilon) = \emptyset, \qquad \qquad \pi(\alpha + \beta) = \pi(\alpha) \cup \pi(\beta),$$
$$\pi(a) = \{\varepsilon\} \quad (a \in \Sigma), \qquad \pi(\alpha\beta) = \pi(\alpha)\beta \cup \pi(\beta),$$
$$\pi(\alpha^\star) = \pi(\alpha)\alpha^\star, \qquad \qquad \pi(\alpha \cap \beta) = \pi(\alpha) \cap \pi(\beta).$$

Proposition 3. If $\alpha \in \mathsf{RE}_\cap$, then $\pi(\alpha)$ is a support of α.

Proof. We will proceed by induction on the structure of α. The proof for all cases, excluding $\alpha \cap \beta$, can be found in [4, 11, 20]. Let $\pi(\alpha_0) = \{\alpha_1, \ldots, \alpha_n\}$ and $\pi(\beta_0) = \{\beta_1, \ldots, \beta_m\}$ be a support of α_0 and β_0, respectively. Thus,

$$\alpha_i \doteq a_1 \alpha_{1i} + \cdots + a_k \alpha_{ki} + \varepsilon(\alpha_i), \text{ for } i = 0, \ldots, n$$

and

$$\beta_j \doteq a_1 \beta_{1j} + \cdots + a_k \beta_{kj} + \varepsilon(\beta_j), \text{ for } j = 1, \ldots, m,$$

where, for all $l = 1, \ldots, k$, α_{li} and β_{lj} are linear combinations of elements of $\pi(\alpha_0)$ and $\pi(\beta_0)$, respectively. We want to prove that $\pi(\alpha_0 \cap \beta_0)$ is a support for $\alpha_0 \cap \beta_0$. For $i = 0, \ldots, n$ and $j = 0, \ldots, m$, and using the axioms for $\cap$, we have

$$\begin{aligned}
\alpha_i \cap \beta_j \doteq & (a_1\alpha_{1i} + \cdots + a_k\alpha_{ki} + \varepsilon(\alpha_i)) \cap (a_1\beta_{1j} + \cdots + a_k\beta_{kj} + \varepsilon(\beta_j)) \\
\doteq & (a_1\alpha_{1i} \cap a_1\beta_{1j}) + \cdots + (a_1\alpha_{1i} \cap a_k\beta_{kj}) + (a_1\alpha_{1i} \cap \varepsilon(\beta_j)) + \\
& \ldots + (a_k\alpha_{ki} \cap a_1\beta_{1j}) + \cdots + (a_k\alpha_{ki} \cap a_k\beta_{kj}) + (a_k\alpha_{ki} \cap \varepsilon(\beta_j)) + \\
& \ldots + (\varepsilon(\alpha_i) \cap a_1\beta_{1j}) + \cdots + (\varepsilon(\alpha_i) \cap a_k\beta_{kj}) + (\varepsilon(\alpha_i) \cap \varepsilon(\beta_j)) \\
\doteq & (a_1 \cap a_1)(\alpha_{1i} \cap \beta_{1j}) + \cdots + (a_k \cap a_k)(\alpha_{ki} \cap \beta_{kj}) + (\varepsilon(\alpha_i) \cap \varepsilon(\beta_j)) \\
\doteq & a_1(\alpha_{1i} \cap \beta_{1j}) + \cdots + a_k(\alpha_{ki} \cap \beta_{kj}) + \varepsilon(\alpha_i \cap \beta_j).
\end{aligned}$$

For each $l = 1, \ldots, k$, we know that $\alpha_{li} = \sum_{i' \in I_{li}} \alpha_{i'}$ and $\beta_{lj} = \sum_{j' \in J_{lj}} \beta_{j'}$, for $I_{li} \subseteq \{1, \ldots, n\}$ and $J_{lj} \subseteq \{1, \ldots, m\}$. And, since

$$\alpha_{li} \cap \beta_{lj} \doteq \sum_{i' \in I_{li}} \alpha_{i'} \cap \sum_{j' \in J_{lj}} \beta_{j'} \doteq \sum_{i' \in I_{li}, j' \in J_{lj}} (\alpha_{i'} \cap \beta_{j'}),$$

we conclude that $\pi(\alpha_0) \cap \pi(\beta_0) = \{\alpha_1 \cap \beta_1, \ldots, \alpha_1 \cap \beta_m, \ldots, \alpha_n \cap \beta_m\}$ is a support for $\alpha_0 \cap \beta_0$. □

Example 4. Given the regular expression $\alpha_1 = (b + ab + aab + abab) \cap (ab)^*$, $\pi(\alpha_1) = \{bab \cap b(ab)^*, \ ab \cap b(ab)^*, \ b \cap b(ab)^*, \ \varepsilon \cap b(ab)^*, \ bab \cap (ab)^*, \ ab \cap (ab)^*, \ b \cap (ab)^*, \ \varepsilon \cap (ab)^*\}$.

The next proposition provides an upper bound on the cardinality of the support of a regular expression.

Proposition 5. *For all $\alpha \in RE_\cap$, the inequality $|\pi(\alpha)| \leq 2^{|\alpha|_\Sigma - |\alpha|_\cap - 1}$ holds.*

Proof. We proceed by induction on the structure of the regular expression α. It is easily proved that the statement holds for the base cases ε, $\emptyset$ and $a \in \Sigma$. Assume that the result holds for some $\alpha, \beta \in RE_\cap$. We will make use of the fact that $2^m + 2^n \leq 2^{m+n+1}$, for any $m, n \geq 0$. For $\alpha + \beta$, one has

$$|\pi(\alpha + \beta)| = |\pi(\alpha) \cup \pi(\beta)| \leq |\pi(\alpha)| + |\pi(\beta)| \leq$$
$$\leq 2^{|\alpha|_\Sigma - |\alpha|_\cap - 1} + 2^{|\beta|_\Sigma - |\beta|_\cap - 1} \leq$$
$$\leq 2^{|\alpha|_\Sigma - |\alpha|_\cap - 1 + |\beta|_\Sigma - |\beta|_\cap - 1 + 1} = 2^{|\alpha+\beta|_\Sigma - |\alpha+\beta|_\cap - 1}.$$

The case for $\alpha\beta$ is analogous. For α^*, one has

$$|\pi(\alpha^*)| = |\pi(\alpha)\alpha^*| = |\pi(\alpha)| \leq 2^{|\alpha|_\Sigma - |\alpha|_\cap - 1} = 2^{|\alpha^*|_\Sigma - |\alpha^*|_\cap - 1}.$$

Finally, for $\alpha \cap \beta$, one has

$$|\pi(\alpha \cap \beta)| = |\pi(\alpha) \cap \pi(\beta)| \leq$$
$$\leq |\pi(\alpha)| \cdot |\pi(\beta)| \leq 2^{|\alpha|_\Sigma - |\alpha|_\cap - 1} \cdot 2^{|\beta|_\Sigma - |\beta|_\cap - 1} =$$
$$= 2^{|\alpha|_\Sigma - |\alpha|_\cap - 1 + |\beta|_\Sigma - |\beta|_\cap - 1} = 2^{|\alpha\cap\beta|_\Sigma - (|\alpha\cap\beta|_\cap - 1) - 2} =$$
$$= 2^{|\alpha\cap\beta|_\Sigma - |\alpha\cap\beta|_\cap - 1}.$$

□

The next examples present families of regular expressions that witnesses the tightness of the upper bound established in Proposition 5.

Example 6. Let the regular expression $r_n \in RE_\cap$ over $\Sigma = \{a, b\}$ be inductively defined by $r_0 = a^*b^*$, $r_1 = b^*a$ and $r_n = r_{n-2} \cap r_{n-1}^*$, for $n \geq 2$. Using the definition of support it is straightforward that $|\pi(r_0)| = |\{a^*b^*, b^*\}| = 2^1$, $|\pi(r_1)| = |\{b^*a, \varepsilon\}| = 2^1$, and $|\pi(r_n)| = |\pi(r_{n-2})| \cdot |\pi(r_{n-1})|$, for $n \geq 2$. Thus, we obtain $|\pi(r_n)| = 2^{\text{fib}(n)}$, for $n \geq 0$, and where fib(n) is the Fibonacci sequence. Also, $|r_0|_\Sigma - |r_0|_\cap - 1 = 2 - 0 - 1 = 1$, $|r_1|_\Sigma - |r_1|_\cap - 1 = 2 - 0 - 1 = 1$, and $|r_n|_\Sigma - |r_n|_\cap - 1 = |r_{n-2}|_\Sigma + |r_{n-1}|_\Sigma - |r_{n-2}|_\cap - |r_{n-1}|_\Sigma - 1 - 1 = (|r_{n-2}|_\Sigma - |r_{n-2}|_\cap - 1) + (|r_{n-1}|_\Sigma - |r_{n-1}|_\cap - 1)$, for $n \geq 2$. Consequently, $|r_n|_\Sigma - |r_n|_\cap - 1 = \text{fib}(n)$, for $n \geq 0$. We conclude that $|\pi(r_n)| = 2^{|r_n|_\Sigma - |r_n|_\cap - 1}$, for $n \geq 0$.

Example 7. Let the regular expression $r_n \in \mathsf{RE}_\cap$ over $\{a\}$, be defined inductively by $r_0 = a^\star a$ and $r_n = r_{n-1} \cap a^\star a$, for $n \geq 1$. We have $\pi(r_0) = \pi(a^\star a) = \{a^\star a, \varepsilon\}$, and for $n \geq 1$,

$$\pi(r_n) = \underbrace{\{a^\star a, \varepsilon\} \cap \cdots \cap \{a^\star a, \varepsilon\}}_{n+1}.$$

Thus $|\pi(r_0)| = 2$ and $|\pi(r_n)| = |\pi(r_0)|^{n+1} = 2^{n+1}$. Note that $|r_n|_\Sigma = 2n + 2$ and $|r_n|_\cap = n$. Therefore $|\pi(r_n)| = 2^{n+1} = 2^{2n+2-n-1} = 2^{|r_n|_\Sigma - |r_n|_\cap - 1}$.

4   Partial Derivatives

The notions of partial derivatives and partial derivative automata were introduced by Antimirov [1] for standard regular expressions. We now consider the Antimirov construction from $\mathsf{RE}_\cap$ expressions to NFAs.

Definition 8. *For a regular expression $\alpha \in \mathsf{RE}_\cap$ and a symbol $a \in \Sigma$, the set $\partial_a(\alpha)$ of partial derivatives of α w.r.t. a is defined by:*

$$\partial_a(\emptyset) = \emptyset, \qquad\qquad \partial_a(\alpha\beta) = \begin{cases} \partial_a(\alpha)\beta \cup \partial_a(\beta), & \text{if } \varepsilon(\alpha) = \varepsilon \\ \partial_a(\alpha)\beta & \text{otherwise,} \end{cases}$$

$$\partial_a(\varepsilon) = \emptyset,$$

$$\partial_a(b) = \begin{cases} \{\varepsilon\}, & \text{if } a = b \\ \emptyset & \text{otherwise,} \end{cases} \qquad \begin{aligned} \partial_a(\alpha + \beta) &= \partial_a(\alpha) \cup \partial_a(\beta), \\ \partial_a(\alpha \cap \beta) &= \partial_a(\alpha) \cap \partial_a(\beta), \\ \partial_a(\alpha^\star) &= \partial_a(\alpha)\alpha^\star. \end{aligned}$$

This definition is extended to words $w \in \Sigma^\star$ by $\partial_\varepsilon(\alpha) = \{\alpha\}$, $\partial_{wa}(\alpha) = \bigcup_{\alpha_i \in \partial_w(\alpha)} \partial_a(\alpha_i)$, and $\partial_w(R) = \bigcup_{\alpha_i \in R} \partial_w(\alpha_i)$, where $R \subseteq \mathsf{RE}_\cap$. It follows easily that $\mathcal{L}(\partial_w(\alpha)) = w^{-1}\mathcal{L}(\alpha)$. The set of partial derivatives of an expression α is $\partial(\alpha) = \bigcup_{w \in \Sigma^\star} \partial_w(\alpha)$. We also define $\partial^+(\alpha) = \bigcup_{w \in \Sigma^+} \partial_w(\alpha)$.

As for standard regular expressions, the partial derivative automaton of an expression $\alpha \in \mathsf{RE}_\cap$ is defined by $\mathcal{A}_{pd}(\alpha) = \langle \partial(\alpha), \Sigma, \{\alpha\}, \delta_\alpha, F_\alpha \rangle$, where $F_\alpha = \{\, \gamma \in \partial(\alpha) \mid \varepsilon(\gamma) = \varepsilon \,\}$ and $\delta_\alpha(\gamma, a) = \partial_a(\gamma)$. It follows that $\mathcal{L}(\mathcal{A}_{pd}(\alpha))$ is exactly $\mathcal{L}(\alpha)$. Mirkin's and Antimirov's constructions coincide for standard regular expressions. We will see that this is not true for regular expressions with intersection.

The following lemmas present some properties of the function ∂_w, used to prove Proposition 11 and are easy to prove.

Lemma 9. *For all $S, S' \subseteq \mathsf{RE}_\cap$ and $a \in \Sigma$, the following property holds*

$$\partial_a(S \cap S') = \partial_a(S) \cap \partial_a(S').$$

Let $\mathrm{suff}(w)$ be the set of all non-empty suffixes of w, being defined as $\mathrm{suff}(w) = \{\, v \in \Sigma^+ \mid \exists u \in \Sigma^\star : uv = w \,\}$. Except for the second case, the following lemma was shown by Antimirov.

Lemma 10. *For every regular expressions $\alpha, \beta \in RE_\cap$ and word $w \in \Sigma^+$, ∂_w satisfies the following:*

$$\partial_w(\alpha + \beta) = \partial_w(\alpha) \cup \partial_w(\beta), \tag{3}$$

$$\partial_w(\alpha \cap \beta) = \partial_w(\alpha) \cap \partial_w(\beta), \tag{4}$$

$$\partial_w(\alpha\beta) \subseteq \partial_w(\alpha)\beta \cup \bigcup_{v \in \text{suff}(w)} \partial_v(\beta), \tag{5}$$

$$\partial_w(\alpha^\star) \subseteq \bigcup_{v \in \text{suff}(w)} \partial_v(\alpha)\alpha^\star. \tag{6}$$

Proposition 11. *For every regular expressions $\alpha, \beta \in RE_\cap$, the following holds.*

$$\partial^+(\alpha + \beta) \subseteq \partial^+(\alpha) \cup \partial^+(\beta), \qquad \partial^+(\alpha \cap \beta) \subseteq \partial^+(\alpha) \cap \partial^+(\beta),$$
$$\partial^+(\alpha\beta) \quad \subseteq \partial^+(\alpha)\beta \cup \partial^+(\beta), \qquad \partial^+(\alpha^\star) \quad \subseteq \partial^+(\alpha)\alpha^\star.$$

Proof. First note that, given a set $E \subseteq RE_\cap$ and a regular expression $\alpha \in RE_\cap$, if, for all $w \in \Sigma^+$, we have that $\partial_w(\alpha) \subseteq E$, then we have $\bigcup_{w \in \Sigma^+} \partial_w(\alpha) \subseteq E$ and thus $\partial^+(\alpha) \subseteq E$. Moreover, we know that for every $w \in \Sigma^+$, $\partial_w(\alpha) \subseteq \partial^+(\alpha)$, since $\partial^+(\alpha) = \bigcup_{w \in \Sigma^+} \partial_w(\alpha)$. Let $\alpha, \beta \in RE_\cap$ be regular expressions over Σ. In order to prove the inclusions above, the facts mentioned above are used. The proof of each inclusion is given, respectively, by the following four proofs:

1. From Eq. (3), for all $w \in \Sigma^+$, the following holds:

$$\partial_w(\alpha + \beta) = \partial_w(\alpha) \cup \partial_w(\beta) \subseteq \partial^+(\alpha) \cup \partial^+(\beta).$$

   And thus, we can conclude that $\partial^+(\alpha + \beta) \subseteq \partial^+(\alpha) \cup \partial^+(\beta)$.
2. In the same way, from Eq. (4), for all $w \in \Sigma^+$, the following holds:

$$\partial_w(\alpha \cap \beta) \subseteq \partial_w(\alpha) \cap \partial_w(\beta) \subseteq \partial^+(\alpha) \cap \partial^+(\beta).$$

   And then, $\partial^+(\alpha \cap \beta) \subseteq \partial^+(\alpha) \cap \partial^+(\beta)$.
3. From Eq. (5), for all $w \in \Sigma^+$, the following holds:

$$\partial_w(\alpha\beta) \subseteq \partial_w(\alpha)\beta \cup \bigcup_{v \in \text{suff}(w)} \partial_v(\beta) \subseteq \partial^+(\alpha)\beta \cup \partial^+(\beta).$$

   Thus, $\partial^+(\alpha\beta) \subseteq \partial^+(\alpha)\beta \cup \partial^+(\beta)$.
4. Finally, from Eq. (6), for all $w \in \Sigma^+$, the following holds:

$$\partial_w(\alpha^\star) \subseteq \bigcup_{v \in \text{suff}(w)} \partial_v(\alpha)\alpha^\star \subseteq \partial^+(\alpha)\alpha^\star.$$

   Therefore, we have that $\partial^+(\alpha) \subseteq \partial^+(\alpha)\alpha^\star$. $\qquad\square$

Example 12. Consider again $\alpha_1 = (b + ab + aab + abab) \cap (ab)^\star$. We have $\partial^+(\alpha_1) = \{bab \cap b(ab)^\star, \ ab \cap b(ab)^\star, \ b \cap b(ab)^\star, \ ab \cap (ab)^\star, \ \varepsilon \cap (ab)^\star\}$. Now, with $\beta = (b + ab + aab + abab)$, one has

$$\partial^+(\beta) \cap \partial^+((ab)^\star) = \{bab \cap b(ab)^\star, ab \cap b(ab)^\star, \ b \cap b(ab)^\star,$$
$$\varepsilon \cap b(ab)^\star, \ bab \cap (ab)^\star, \ ab \cap (ab)^\star, \ b \cap (ab)^\star, \ \varepsilon \cap (ab)^\star\}.$$

Thus, we conclude that $\partial^+(\alpha_1) \subset \partial^+(b + ab + aab + abab) \cap \partial^+((ab)^\star)$.

The following proposition relates the function ∂^+ and the support π.

Proposition 13. *Given* $\alpha \in RE_\cap$, $\partial^+(\alpha) \subseteq \pi(\alpha)$.

Proof. The proof proceeds by induction on the structure of α. It is trivial that $\partial^+(\emptyset) = \pi(\emptyset)$, $\partial^+(\varepsilon) = \pi(\varepsilon)$ and $\partial^+(a) = \pi(a)$, for a symbol $a \in \Sigma$. Assume that $\partial^+(\alpha) \subseteq \pi(\alpha)$ and $\partial^+(\beta) \subseteq \pi(\beta)$ holds, for $\alpha, \beta \in RE_\cap$. For $\alpha + \beta$, we have $\partial^+(\alpha+\beta) \subseteq \partial^+(\alpha) \cup \partial^+(\beta) \subseteq \pi(\alpha) \cup \pi(\beta)$. For $\alpha \cap \beta$, there is $\partial^+(\alpha \cap \beta) \subseteq \partial^+(\alpha) \cap \partial^+(\beta) \subseteq \pi(\alpha) \cap \pi(\beta)$. For $\alpha\beta$, we have $\partial^+(\alpha\beta) \subseteq \partial^+(\alpha)\beta \cup \partial^+(\beta) \subseteq \pi(\alpha)\beta \cup \pi(\beta)$. Finally, for $\alpha^\star$, $\partial^+(\alpha^\star) \subseteq \partial^+(\alpha)\alpha^\star \subseteq \pi(\alpha)\alpha^\star$. $\square$

Since, for every regular expression $\alpha \in RE_\cap$, the set $\pi(\alpha)$ is finite, Proposition 13 also proves that the set $\partial^+(\alpha)$ is finite. For regular expressions without intersection it is known that π and ∂^+ coincide [11]. Examples 4 and 12 show that there exists $\alpha \in RE_\cap$ such that $\pi(\alpha) \neq \partial^+(\alpha)$. The following lemmas establish some conditions for the equality of $\pi(\alpha \cap \beta)$ and $\partial^+(\alpha \cap \beta)$ to hold for $\alpha, \beta \in RE_\cap$, and will be used in Proposition 16.

Lemma 14. *Given* $\alpha, \beta \in RE_\cap$, *one has* $\pi(\alpha \cap \beta) = \partial^+(\alpha \cap \beta)$ *if and only if* $\pi(\alpha) = \partial^+(\alpha)$, $\pi(\beta) = \partial^+(\beta)$ *and* $\partial^+(\alpha \cap \beta) = \partial^+(\alpha) \cap \partial^+(\beta)$.

Proof. ($\Rightarrow$) We have that $\pi(\alpha \cap \beta) = \partial^+(\alpha \cap \beta) \subseteq \partial^+(\alpha) \cap \partial^+(\beta)$. From Proposition 13 follows that $\partial^+(\alpha) \subseteq \pi(\alpha)$ and $\partial^+(\beta) \subseteq \pi(\beta)$. Suppose by contradiction that $\partial^+(\alpha) \subset \pi(\alpha)$ or $\partial^+(\beta) \subset \pi(\beta)$. Then $\partial^+(\alpha \cap \beta) \subseteq \partial^+(\alpha) \cap \partial^+(\beta) \subset \pi(\alpha) \cap \pi(\beta) = \pi(\alpha \cap \beta)$, a contradiction since $\pi(\alpha \cap \beta) = \partial^+(\alpha \cap \beta)$. Thus, we conclude that $\pi(\alpha) = \partial^+(\alpha)$ and $\pi(\beta) = \partial^+(\beta)$. Consequently, $\pi(\alpha \cap \beta) = \pi(\alpha) \cap \pi(\beta) = \partial^+(\alpha \cap \beta)$.

($\Leftarrow$) This follows trivially from the definition of support, i.e., $\pi(\alpha \cap \beta) = \pi(\alpha) \cap \pi(\beta)$, since $\pi(\alpha) = \partial^+(\alpha)$ and $\pi(\beta) = \partial^+(\beta)$. $\square$

Lemma 15. *Given* $\alpha, \beta \in RE_\cap$, *such that* $\partial_w(\alpha) = \pi(\alpha)$ *or* $\partial_w(\beta) = \pi(\beta)$ *holds for all* $w \in \Sigma^+$, *then* $\partial^+(\alpha \cap \beta) = \partial^+(\alpha) \cap \partial^+(\beta)$.

Proof. First, note that if $\gamma \in RE_\cap$ and $\partial_w(\gamma) = \pi(\gamma)$ for every $w \in \Sigma^+$, then $\partial^+(\gamma) = \bigcup_{w \in \Sigma^+} \partial_w(\gamma) = \pi(\gamma)$. Given $\alpha, \beta \in RE_\cap$, there are three possible cases to prove. First, suppose that, for all $w \in \Sigma^+$, we have $\partial_w(\alpha) = \pi(\alpha)$ and $\partial_w(\beta) = \pi(\beta)$. Then

$$\partial^+(\alpha \cap \beta) = \bigcup_{w \in \Sigma^+} (\partial_w(\alpha) \cap \partial_w(\beta)) = \pi(\alpha) \cap \pi(\beta) = \partial^+(\alpha) \cap \partial^+(\beta).$$

It remains to prove the cases that either $\partial_w(\alpha) = \pi(\alpha)$ or $\partial_w(\beta) = \pi(\beta)$, for all $w \in \Sigma^+$. The proof is the same for both cases. So, we will only present the proof for the first case. Suppose that, for all $w \in \Sigma^+$, $\partial_w(\alpha) = \pi(\alpha)$, it holds that

$$
\begin{aligned}
\partial^+(\alpha \cap \beta) &= \bigcup_{w \in \Sigma^+} (\partial_w(\alpha) \cap \partial_w(\beta)) = \bigcup_{w \in \Sigma^+} (\pi(\alpha) \cap \partial_w(\beta)) \\
&= \bigcup_{w \in \Sigma^+} \{\alpha_i \cap \beta_j \mid \alpha_i \in \pi(\alpha),\ \beta_j \in \partial_w(\beta)\} \\
&= \left\{ \alpha_i \cap \beta_j \,\middle|\, \alpha_i \in \pi(\alpha),\ \beta_j \in \bigcup_{w \in \Sigma^+} \partial_w(\beta) \right\} \\
&= \{\alpha_i \cap \beta_j \mid \alpha_i \in \pi(\alpha),\ \beta_j \in \partial^+(\beta)\} \\
&= \pi(\alpha) \cap \partial^+(\beta) = \partial^+(\alpha) \cap \partial^+(\beta).
\end{aligned}
$$

$\square$

By Proposition 13, $|\pi(\alpha)|$ is an upper bound for the cardinality of $\partial^+(\alpha)$. This upper bound can be achieved, as shown by the following proposition.

Proposition 16. *For any $n \in \mathbb{N}$ there exists a regular expression $r_n \in \mathsf{RE}_\cap$ of size $O(n)$ such that $|\partial^+(r_n)| = 2^{|r_n|_\Sigma - |r_n|_\cap - 1}$.*

Proof. Consider the regular expressions $r_n \in \mathsf{RE}_\cap$ from Example 7. We prove that $\pi(r_n) = \partial^+(r_n)$. The proof proceeds by induction on n. For $n = 0$ and for all $w \in \Sigma^+$, we have $\partial_w(a^\star a) = \{a^\star a, \epsilon\} = \partial^+(a^\star a) = \pi(a^\star a)$. Let us assume, by induction, that $\pi(r_n) = \partial^+(r_n)$, for $n \geq 1$. It follows from Lemma 15 that $\partial^+(r_{n+1}) = \partial^+(r_n \cap a^\star a) = \partial^+(r_n) \cap \partial^+(a^\star a)$. Since $\pi(a^\star a) = \partial^+(a^\star a)$, $\pi(r_n) = \partial^+(r_n)$, and $\partial^+(r_n \cap a^\star a) = \partial^+(r_n) \cap \partial^+(r_n)$, we conclude, from Lemma 14, that $\pi(r_{n+1}) = \pi(r_n \cap a^\star a) = \partial^+(r_n \cap a^\star a) = \partial^+(r_{n+1})$. $\square$

The next example provides another non-trivial family of regular expressions for which the set of partial derivatives and the support coincide.

Example 17. For $n \geq 0$ let the regular expression $s_n \in \mathsf{RE}_\cap$ be inductively defined by $s_0 = (a+b)^\star b(a+b)^\star$ and $s_n = ((a+b)s_{n-1}(a+b)) \cap ((a+b)^\star(a+b))$, for $n \geq 1$. The alphabetic length of s_n is $|s_n|_\Sigma = 5 + 8n$ and $|s_n|_\cap = n$. The cardinality of the support of s_n is given by: $|\pi(s_0)| = 2$, $|\pi(s_1)| = 6$ and $|\pi(s_n)| = \sum_{i=2}^{n} 2^i + 3 \cdot 2^n$, for $n \geq 2$. Thus, for $n \geq 2$ we have $|\pi(s_n)| = O(2^n)$. Let $m = |s_n|_\Sigma - |s_n|_\cap - 1 = 5 + 7n - 1$, i.e. $n = (m-4)/7$. Then, $|\pi(s_n)| = O(2^{\frac{1}{7}m}) = O(1.105^m)$, which is much smaller than the upper bound 2^m. For all $n \geq 0$, $\pi(s_n) = \partial^+(s_n)$.

5 Average Complexity Results

We know that the number of states in the partial derivative automaton of an expression α has $|\pi(\alpha)|$ as its tight upper bound. In this section we estimate an upper bound for the asymptotic average size of $\pi(\alpha)$. This is done using standard methods of analytic combinatorics as expounded by Flajolet and Sedgewick [13], which apply to generating functions $f(z) = \sum_n a_n z^n$ associated with combinatorial classes. Given some measure of the objects of a combinatorial class $\mathcal{A}$, the coefficient a_n represents the sum of the values of this measure for all objects of size n. We will use the notation $[z^n]f(z)$ for a_n. For an introduction to this approach applied to formal languages, we refer to Broda *et al.* [6].

Although the methods here used are the standard ones from the Analytic Combinatorics (and Complex Analysis), each application of these techniques is always a challenge, as one cannot foresee the analytic difficulties that one can incur into when conducting the study of the generation function. The generating function f can be seen as a complex analytic function, and the study of its behaviour near its dominant singularity η (in case there is only one, as it happens with the functions here considered) gives us access to the asymptotic form of its coefficients. In particular, if $f(z)$ is analytic in some appropriate neighbourhood of 0 containing η, then one has the following [6,13]:

Proposition 18. *If $f(z) = a - b\sqrt{1 - z/\rho} + o\left(\sqrt{1 - z/\rho}\right)$, with $a, b \in \mathbb{R}$, $b \neq 0$, then*

$$[z^n]f(z) \sim \frac{b}{2\sqrt{\pi}}\, \rho^{-n} n^{-3/2}.$$

If $f(z) = \frac{a}{\sqrt{1-z/\rho}} + o\left(\frac{1}{\sqrt{1-z/\rho}}\right)$, with $a \in \mathbb{R}$, and $a \neq 0$, then

$$[z^n]f(z) \sim \frac{a}{\sqrt{\pi}}\, \rho^{-n} n^{-1/2}.$$

5.1   Number of Expressions and Letters and $\cap$ Symbols

The study of the combinatorial behaviour of the $\mathsf{RE}_\cap$-expressions, both in terms of the number of expressions and the number of letters in them, is identical to the study of any other regular expressions with 3 binary operators and a single unary operator. Thus the results presented in Broda et al. [7] are valid for the case here studied. Denoting by $R_k(z)$ the generating function for the number of $\mathsf{RE}_\cap$-expressions without $\emptyset$ over a k letters alphabet, and by $L_k(z)$ the generating function for the number of letters in the expressions, one has:

$$[z^n]R_k(z) \sim c_k \rho_k^{-n-\frac{1}{2}} n^{-\frac{3}{2}}, \tag{7}$$

$$[z^n]L_k(z) \sim \frac{k}{12\pi c_k} \rho_k^{-n+\frac{1}{2}} n^{-\frac{1}{2}}, \tag{8}$$

where $c_k = \frac{\sqrt[4]{3+3k}}{6\sqrt{\pi}}$ and $\rho_k = \frac{-1+2\sqrt{3+3k}}{11+12k}$.

The average number of letters in an expression of size n is given by

$$\frac{[z^n]L_k(z)}{[z^n]R_k(z)}.$$

Using Eqs. (7) and (8), one obtains, asymptotically,

$$|\alpha|_\Sigma \sim \frac{3k\rho_k}{\sqrt{3+3k}}\,\|\alpha\| \xrightarrow[k\to\infty]{} \frac{1}{2}\|\alpha\|. \tag{9}$$

The number of intersections in the $\mathrm{RE}_\cap$-expressions under consideration can be computed as follows. Consider the bivariate generating function

$$\mathcal{I}_k(u,z) = \sum_{m,n} \iota_{mn} u^m z^n,$$

where ι_{mn} is the number of $\mathrm{RE}_\cap$-expressions with m intersection symbols and size n. From (1), and using the symbolic method, we can write

$$\mathcal{I}_k(u,z) = (k+1)z + 2z\mathcal{I}_k(u,z)^2 + uz\mathcal{I}_k(u,z)^2 + z\mathcal{I}_k(u,z).$$

Solving this for $\mathcal{I}_k(u,z)$, differentiating the result w.r.t. u, and making $u = 1$, we obtain an expression for the generating function for the cumulative number of intersection symbols in all $\mathrm{RE}_\cap$-expressions of size n:

$$I_k(z) = \frac{1}{18z}\sqrt{q_k(z)} + \frac{(k+1)z}{3\sqrt{q_k(z)}} + \frac{z-1}{18z}, \tag{10}$$

where $q_k(z) = 1 - 2z - (11 + 12k)z^2$, from which one obtains, using the same methods,

$$[z^n]I_k(z) \sim \frac{1}{6\sqrt{\pi}}\left(\frac{(k+1)\sqrt{\rho_k}}{\sqrt[4]{3+3k}\sqrt{n}} - \frac{\sqrt[4]{3+3k}}{3\sqrt{\rho_k}\,n^{3/2}}\right)\rho_k^{-n}. \tag{11}$$

The average number of symbols $\cap$ in an expression of size n is given by

$$\frac{[z^n]I_k(z)}{[z^n]R_k(z)}.$$

Using Eqs. (7) and (11), one obtains, asymptotically,

$$|\alpha|_\cap \sim \frac{(k+1)\rho_k}{\sqrt{3+3k}}\,\|\alpha\| \xrightarrow[k\to\infty]{} \frac{1}{6}\|\alpha\|. \tag{12}$$

5.2   Average Size of π

Let $P_k(z)$ denote the generating function for the size of $\pi(\alpha)$ for expressions without $\emptyset$. From Definition 2 it follows that, given an expression α, an upper bound, $p(\alpha)$, for the number of elements[3] in the set $\pi(\alpha)$ satisfies:

$$p(\varepsilon) = 0, \qquad\qquad p(\alpha + \beta) = p(\alpha) + p(\beta),$$
$$p(a) = 1, \;\; \text{for } a \in \Sigma, \qquad p(\alpha\beta) = p(\alpha) + p(\beta),$$
$$p(\alpha^\ast) = p(\alpha), \qquad\qquad p(\alpha \cap \beta) = p(\alpha)p(\beta).$$

---

[3] This upper bound corresponds to the case where all unions in $\pi(\alpha)$ are disjoint.

From this, we directly get

$$P_k(z) = kz + 4zP_k(z)R_k(z) + zP_k(z) + zP_k(z)^2,$$

from which we obtain the following closed expression

$$P_k(z) = \frac{1 - z + 2\sqrt{q_k(z)} - \sqrt{p_k(z) + 4(1-z)\sqrt{q_k(z)}}}{6z}, \tag{13}$$

where

$$p_k(z) = 5 - 10z - (43 + 84k)z^2. \tag{14}$$

One now needs to determine the dominant singularity of $P_k(z)$ which can either be a root of $q_k(z)$ or a root of $r_k(z) = p_k(z) + 4(1-z)\sqrt{q_k(z)}$. We need to know which of the two expressions $r_k(z)$ or $q_k(z)$ has the smallest positive zero. Because this is not trivial (note that one needs to decide this for all k), one will do it indirectly using the method expounded in the following paragraphs.

Observing that $r_k(0) = 9$ is positive and

$$r_k(\rho_k) = \frac{12\left(13 - 14k - 24k^2 + (8k - 4)\sqrt{3 + 3k}\right)}{(11 + 12k)^2} < 0,$$

by Bolzano theorem, $r_k(z)$ must have a positive zero smaller than ρ_k. This conclusion could be achieved, directly, from the fact that the absolute value of the negative zero of $q_k(z)$ is smaller than its positive zero, and thus, by Pringsheim theorem [13], another smaller positive singularity of $P_k(z)$ necessarily exists that can only be due to $r_k(z)$. Letting

$$\bar{\rho}_k = \frac{-1 - 2\sqrt{3 + 3k}}{11 + 12k},$$

and observing that

$$r_k(\bar{\rho}_k) = -\frac{12\left(-13 + 14k + 24k^2 + (8k - 4)\sqrt{3 + 3k}\right)}{(11 + 12k)^2} < 0,$$

one concludes that $r_k(z)$ has necessarily two real zeros in its domain, $[\bar{\rho}_k, \rho_k]$. Analogously, $s_k(z) = p_k(z) - 4(1-z)\sqrt{q_k(z)}$ has also two real zeros in the same interval, and since $r_k(z)s_k(z)$ is a fourth degree polynomial, it follows that $r_k(z)$ has exactly two zeros, η_k and η_k', which are real. Since $s_k(0) = 1 < r_k(0) = 9$, and $r_k(x) = s_k(x)$ only at the end points of $[\bar{\rho}_k, \rho_k]$ it follows that $s_k(x) < r_k(x)$ in $]\bar{\rho}_k, \rho_k[$. Considering the four real zeros of the polynomial $r_k(z)s_k(z)$, given what we just said, we conclude that the two more distant zeros from the origin are the roots of $r_k(z)$. In fact, we can obtain an explicit expression for the zeros of $r_k(z)s_k(z)$ by noticing that

$$p_k(z) \pm 4(1-z)\sqrt{q_k(z)} = \left(1 - z \pm 2\sqrt{q_k(z)}\right)^2 - 36kz^2$$

$$= \left(1 - z \pm 2\sqrt{q_k(z)} - 6\sqrt{k}z\right)\left(1 - z \pm 2\sqrt{q_k(z)} + 6\sqrt{k}z\right),$$

and thus, solving the equations resulting of nulling those factors, we obtain the four zeros of $r_k(z)s_k(z)$:

$$\eta_k = \frac{4\sqrt{2k+1}+2\sqrt{k}-1}{28k+4\sqrt{k}+15}, \qquad \eta'_k = -\frac{4\sqrt{2k+1}+2\sqrt{k}+1}{28k-4\sqrt{k}+15},$$

$$\eta''_k = \frac{4\sqrt{2k+1}-2\sqrt{k}-1}{28k-4\sqrt{k}+15}, \qquad \eta'''_k = -\frac{4\sqrt{2k+1}-2\sqrt{k}+1}{28k+4\sqrt{k}+15}. \qquad (15)$$

It is possible to verify that η_k and η'_k are the roots of $r_k(z)$ and the other two the roots from $s_k(z)$. Therefore, one has

$$r_k(z)s_k(z) = (7056k^2 + 7416k + 2025)(z - \eta_k)(z - \eta'_k)(z - \eta''_k)(z - \eta'''_k). \quad (16)$$

From (13) one has

$$6zP_k(z) = 1 - z - \sqrt{r_k(z)} + 2\sqrt{q_k(z)}, \qquad (17)$$

and we split the study of the coefficients of the series of $P_k(z)$ into the study of the coefficients of $1 - z - \sqrt{r_k(z)}$ and of $2\sqrt{q_k(z)}$. For the first one, we use that

$$r_k(z) = \frac{7056k^2 + 7416k + 2025}{s_k(z)}\eta_k(\eta'_k - z)(\eta''_k - z)(\eta'''_k - z)\left(1 - \frac{z}{\eta_k}\right),$$

and the fact that given a complex function f, defined in a neighbourhood of η such that $\lim_{z\to\eta} f(z) = a$, one has, for all $r \in \mathbb{R}$, $f(z)(1-z/\eta)^r = a(1-z/\eta)^r + o((1-z/\eta)^r)$, together with Proposition 18, to obtain

$$[z^n]\left(1 - z - \sqrt{r_k(z)}\right) \sim \lambda_k \eta_k^{-n} n^{-\frac{3}{2}},$$

where

$$\lambda_k = \left(\frac{(7056k^2 + 7416k + 2025)(\eta'_k - \eta_k)(\eta''_k - \eta_k)(\eta'''_k - \eta_k)\eta_k}{2\pi s_k(\eta_k)}\right)^{\frac{1}{2}}. \qquad (18)$$

For the last summand one has, similarly,

$$2\sqrt{q_k(z)} = 4\sqrt[4]{3+3k}\,\rho_k^{\frac{1}{2}}(\rho_k - \bar{\rho}_k)^{\frac{1}{2}}(1 - z/\rho_k)^{\frac{1}{2}} + o\left((1 - z/\rho_k)^{\frac{1}{2}}\right),$$

from which it follows, $[z^n]2\sqrt{q_k(z)} \sim -\mu_k \rho_k^{-n} n^{-\frac{3}{2}}$, where

$$\mu_k = 2\pi^{-\frac{1}{2}}\rho_k^{\frac{1}{2}}\sqrt[4]{3+3k}. \qquad (19)$$

Summing up, we get that

$$[z^n]P_k(z) \sim \frac{1}{6}\left(\lambda_k \eta_k^{-(n+1)} - \mu_k \rho_k^{-(n+1)}\right)n^{-\frac{3}{2}}. \qquad (20)$$

In order to see what this result entails for the average case when compared with the worst case result, expressed in Proposition 5, attend to the following.

$$\left(\frac{[z^n]P_k(z)}{[z^n]R_k(z)} \right)^{\frac{1}{n}} \sim \left(\frac{\frac{1}{6}\lambda_k \eta_k^{-(n+1)} n^{-\frac{3}{2}}}{c_k \rho_k^{-n-\frac{1}{2}} (n+1)^{-\frac{3}{2}}} \right)^{\frac{1}{n}} \xrightarrow[n\to\infty]{} \frac{\rho_k}{\eta_k}.$$

Setting $\gamma_k = \frac{\rho_k}{\eta_k}$, this means that, on average,

$$|\pi(\alpha)| \sim \gamma_k^{\|\alpha\|}.$$

One has $\gamma_2 \sim 1.01655$, $\gamma_{10} \sim 1.04137$, $\gamma_{100} \sim 1.05294$, and

$$\lim_{k\to\infty} \gamma_k = \frac{7\sqrt{3}}{6\sqrt{2}+3} \sim 1.05564.$$

Proposition 19. *For large values of k and n an upper bound for the average number of states of $\mathcal{A}_{pd}$ is $(1.056 + o(1))^n$.*

Considering the estimates given in (9) and (12), the worst-case upper bound $2^{|\alpha|_\Sigma - |\alpha|_\cap - 1}$ from Proposition 5 leads to an upper bound for the average case roughly of $\sqrt[3]{2}^{\|\alpha\|}$, for α large enough. As $\sqrt[3]{2} \sim 1.25992$, the result just obtained shows that the upper bound for the average complexity is significantly smaller than the one for the worst case.

6   Conclusions

The conversion of a regular expression with intersection α to NFA is in the worst-case $2^{\Omega(\|\alpha\|)}$ [15,17,18]. This fact leads to the assumption that, although succinct, these expressions are not useful in practical applications. Here we show that, asymptotically, an upper bound for the average-state complexity of $\mathcal{A}_{pd}(\alpha)$ is exponential but with a base only slightly above 1. Actually, experimental results using a uniform distribution suggest that the average-state complexity of $\mathcal{A}_{pd}(\alpha)$ may even be polynomial [3].

References

1. Antimirov, V.: Partial derivatives of regular expressions and finite automaton constructions. Theoret. Comput. Sci. **155**(2), 291–319 (1996)
2. Antimirov, V.M., Mosses, P.D.: Rewriting extended regular expressions. In: Rozenberg, G., Salomaa, A. (eds.) 1st DLT. pp. 195–209. World Scientific (1994)
3. Bastos, R.: Manipulation of Extended Regular Expressions with Derivatives. Master's thesis, Faculdade de Ciências da Universidade do Porto (2015)
4. Broda, S., Machiavelo, A., Moreira, N., Reis, R.: On the average state complexity of partial derivative automata. Int. J. Found. Comput. Sci. **22**(7), 1593–1606 (2011)
5. Broda, S., Machiavelo, A., Moreira, N., Reis, R.: On the average size of Glushkov and partial derivative automata. Int. J. Found. Comput. Sci. **23**(5), 969–984 (2012)

6. Broda, S., Machiavelo, A., Moreira, N., Reis, R.: A Hitchhiker's guide to descriptional complexity through analytic combinatorics. Theoret. Comput. Sci. **528**, 85–100 (2014)
7. Broda, S., Machiavelo, A., Moreira, N., Reis, R.: Partial derivative automaton for regular expressions with shuffle. In: Shallit, J., Okhotin, A. (eds.) DCFS 2015. LNCS, vol. 9118, pp. 21–32. Springer, Heidelberg (2015)
8. Brzozowski, J.A.: Derivatives of regular expressions. JACM **11**(4), 481–494 (1964)
9. Caron, P., Champarnaud, J.-M., Mignot, L.: Partial derivatives of an extended regular expression. In: Dediu, A.-H., Inenaga, S., Martín-Vide, C. (eds.) LATA 2011. LNCS, vol. 6638, pp. 179–191. Springer, Heidelberg (2011)
10. Caron, P., Champarnaud, J., Mignot, L.: A general framework for the derivation of regular expressions. RAIRO - Theor. Inf. Appl. **48**(3), 281–305 (2014)
11. Champarnaud, J.M., Ziadi, D.: From Mirkin's prebases to Antimirov's word partial derivatives. Fundam. Inform. **45**(3), 195–205 (2001)
12. Christiansen, T., Foy, B.D., Wall, L., Orwant, J.: Programming Perl, 4th edn. O'Reilly Media, Sebastopol (2012)
13. Flajolet, P., Sedgewick, R.: Analytic Combinatorics. CUP, Cambridge (2008)
14. Fürer, M.: The complexity of the inequivalence problem for regular expressions with intersection. In: de Bakker, J.W., van Leeuwen, J. (eds.) ICALP 1980. LNCS, vol. 85, pp. 234–245. Springer, Heidelberg (1980)
15. Gelade, W.: Succinctness of regular expressions with interleaving, intersection and counting. Theoret. Comput. Sci. **411**(31–33), 2987–2998 (2010)
16. Gelade, W., Neven, F.: Succinctness of the complement and intersection of regular expressions. In: Albers, S., Weil, P. (eds.) 25th STACS. LIPIcs, vol. 1, pp. 325–336. Schloss Dagstuhl - Leibniz-Zentrum fuer Informatik, Germany (2008)
17. Gruber, H.: On the descriptional and algorithmic complexity of regular languages. Ph.D. thesis, Justus Liebig University Giessen (2010)
18. Gruber, H., Holzer, M.: Finite automata, digraph connectivity, and regular expression size. In: Aceto, L., Damgård, I., Goldberg, L.A., Halldórsson, M.M., Ingólfsdóttir, A., Walukiewicz, I. (eds.) ICALP 2008, Part II. LNCS, vol. 5126, pp. 39–50. Springer, Heidelberg (2008)
19. Jiang, T., Ravikumar, B.: A note on the space complexity of some decision problems for finite automata. Inf. Process. Lett. **40**(1), 25–31 (1991)
20. Mirkin, B.G.: An algorithm for constructing a base in a language of regular expressions. Eng. Cybern. **5**, 51–57 (1966)
21. Petersen, H.: The membership problem for regular expressions with intersection is complete in LOGCFL. In: Alt, H., Ferreira, A. (eds.) STACS 2002. LNCS, vol. 2285, pp. 513–522. Springer, Heidelberg (2002)
22. Sen, K., Rosu, G.: Generating optimal monitors for extended regular expressions. Electr. Notes Theor. Comput. Sci. **89**, 231–250 (2003)
23. van der Vlist, E.: RELAX NG. O'Reilly Media, Cambridge (2003)

Unrestricted State Complexity of Binary Operations on Regular Languages

Janusz Brzozowski[✉]

David R. Cheriton School of Computer Science,
University of Waterloo, Waterloo, ON N2L 3G1, Canada
brzozo@uwaterloo.ca

Abstract. I study the state complexity of binary operations on regular languages over different alphabets. It is well known that if L'_m and L_n are languages restricted to be over the same alphabet, with m and n quotients, respectively, the state complexity of any binary boolean operation on L'_m and L_n is mn, and that of the product (concatenation) is $(m-1)2^n + 2^{n-1}$. In contrast to this, I show that if L'_m and L_n are over their own different alphabets, the state complexity of union and symmetric difference is $mn + m + n + 1$, that of intersection is $mn + 1$, that of difference is $mn + m + 1$, and that of the product is $m2^n + 2^{n-1}$.

Keywords: Boolean operation · Concatenation · Different alphabets · Most complex languages · Product · Quotient complexity · Regular language · State complexity · Stream · Unrestricted complexity

1 Motivation

Formal definitions are postponed until Sect. 2.

The first paper on state complexity was published by A. N. Maslov [9] in 1970, but this work was unknown in the West for many years. Maslov wrote:

> *An important measure of the complexity of [sets of words representable in finite automata] is the number of states in the minimal representing automaton. ... if $T(A) \cup T(B)$ are representable in automata A and B with m and n states respectively ..., then:*
> *1. $T(A) \cup T(B)$ is representable in an automaton with $m \cdot n$ states;*
> *2. $T(A).T(B)$ is representable in an automaton with $(m-1)2^n + 2^{n-1}$ states.*

The second paper on state complexity was published by S. Yu, Q. Zhuang and K. Salomaa [11] in 1994. Here the authors wrote:

This work was supported by the Natural Sciences and Engineering Research Council of Canada grant No. OGP0000871.

© IFIP International Federation for Information Processing 2016
Published by Springer International Publishing Switzerland 2016. All Rights Reserved
C. Câmpeanu et al. (Eds.): DCFS 2016, LNCS 9777, pp. 60–72, 2016.
DOI: 10.1007/978-3-319-41114-9_5

1. ... *for any pair of complete m-state DFA A and n-state DFA B defined on the same alphabet Σ, there exists a DFA with at most $m2^n - 2^{n-1}$ states which accepts $L(A)L(B)$.*
2. ... *$m \cdot n$ states are ... sufficient for a DFA to accept the intersection (union) of an m-state DFA language and an n-state DFA language.*

Here DFA stands for *deterministic finite automaton*, and *complete* means that there is a transition from every state under every input letter.

I will show that statements 1 and 2 of Maslov are incorrect without the restriction that the languages are over the same alphabet. In [11] the first statement includes that restriction, but the second omits it (presumably it's implied).

The same-alphabet restriction is unnecessary: There is no reason why we should not be able to find, for example, the union of languages $L' = \{a,b\}^*b$ and $L = \{a,c\}^*c$ accepted by the minimal complete two-state automata $\mathcal{D}'_2$ and $\mathcal{D}_2$ of Fig. 1, where an incoming arrow denotes the initial state and a double circle represents a final state.

Fig. 1. Two minimal complete DFAs $\mathcal{D}'_2$ and $\mathcal{D}_2$.

The union of L' and L is a language over three letters. To find the DFA for $L' \cup L$, we view $\mathcal{D}'_2$ and $\mathcal{D}_2$ as incomplete DFA's, the first missing all transitions under c, and the second under b. After adding the missing transitions we obtain DFAs $\mathcal{D}'_3$ and $\mathcal{D}_3$ of Fig. 2. Now we can proceed as is usually done in the same-alphabet approach, and take the direct product of $\mathcal{D}'_3$ and $\mathcal{D}_3$ to find $L' \cup L$. Here it turns out that six states are necessary to represent $L' \cup L$, but the state complexity of union is actually $(m + 1)(n + 1)$.

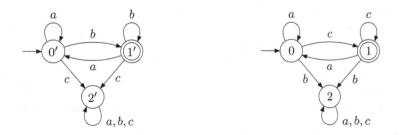

Fig. 2. DFAs $\mathcal{D}'_3$ and $\mathcal{D}_3$ over three letters.

In general, when calculating the result of a binary operation on regular languages with different alphabets, we deal with special incomplete DFAs that are only missing some letters and all the transitions caused by these letters. The complexity of incomplete DFAs has been studied previously by Gao, K. Salomaa, and Yu [6] and by Maia, Moreira and Reis [8]. However, the objects studied there are *arbitrary* incomplete DFAs, whereas we are interested only in *complete DFAs with some missing letters*. Secondly, we study *state* complexity, whereas the above-mentioned papers deal mainly with *transition* complexity. Nevertheless, there is some overlap. It was shown in [6, Corollary 3.2] that the incomplete state complexity of union is less than or equal to $mn + m + n$, and that this bound is tight in some special cases. In [8, Theorem 2], witnesses that work in all cases were found. These complexities correspond to my result for union in Theorem 1. Also in [8, Theorem 5], the incomplete state complexity of product is shown to be $m2^n + 2^{n-1} - 1$, and this corresponds to my result for product in Theorem 2.

In this paper I remove the restriction of equal alphabets of the two operands. I prove that the complexity of union and symmetric difference is $mn + m + n + 1$, that of intersection is $mn + 1$, that of difference is $mn + m - 1$, and that of the product is $m2^n + 2^{n-1}$, if each language's own alphabet is used. I exhibit a new most complex regular language that meets the complexity bounds for boolean operations, product, star, and reversal, has a maximal syntactic semigroup and most complex atoms. All the witnesses used here are derived from that one most complex language.

2   Terminology and Notation

A basic complexity measure of a regular language L over an alphabet Σ is the number n of distinct (left) quotients of L, where a *(left) quotient* of L by a word $w \in \Sigma^*$ is $w^{-1}L = \{x \mid wx \in L\}$. The number of quotients of L is its *quotient complexity* [2], $\kappa(L)$. A concept equivalent to quotient complexity is the *state complexity* [11] of L, which is the number of states in a complete minimal deterministic finite automaton (DFA) recognizing L. Since we do not use any other measures of complexity in this paper (with the exception of one mention of time and space complexity in the next paragraph), we refer to quotient/state complexity simply as *complexity*.

Let $L'_m \subseteq \Sigma'^*$ and $L_n \subseteq \Sigma^*$ be regular languages of complexities m and n, respectively. The *complexity of a binary operation* $\circ$ on L'_m and L_n is the maximal value of $\kappa(L'_m \circ L_n)$ as a function $f(m, n)$, as L'_m and L_n range over all regular languages of complexity m and n, respectively. The complexity of an operation gives a worst-case lower bound on the time and space complexity of the operation. For this reason it has been studied extensively; see [2,3,10,11] for additional references.

A *deterministic finite automaton (DFA)* is a quintuple $\mathcal{D} = (Q, \Sigma, \delta, q_0, F)$, where Q is a finite non-empty set of *states*, Σ is a finite non-empty *alphabet*, $\delta: Q \times \Sigma \to Q$ is the *transition function*, $q_0 \in Q$ is the *initial* state, and $F \subseteq Q$ is the set of *final* states. We extend δ to a function $\delta: Q \times \Sigma^* \to Q$ as usual. A DFA $\mathcal{D}$ accepts a word $w \in \Sigma^*$ if $\delta(q_0, w) \in F$. The language accepted by $\mathcal{D}$ is denoted by $L(\mathcal{D})$. If q is a state of $\mathcal{D}$, then the language L^q of q is the language accepted by the DFA $(Q, \Sigma, \delta, q, F)$. A state is *empty* (or *dead* or a *sink state*) if its language is empty. Two states p and q of $\mathcal{D}$ are *equivalent* if $L^p = L^q$. A state q is *reachable* if there exists $w \in \Sigma^*$ such that $\delta(q_0, w) = q$. A DFA is *minimal* if all of its states are reachable and no two states are equivalent. Usually DFAs are used to establish upper bounds on the complexity of operations, and also as witnesses that meet these bounds.

If $\delta(q, a) = p$ for a state $q \in Q$ and a letter $a \in \Sigma$, we say there is a *transition* under a from q to p in $\mathcal{D}$. The DFAs defined above are *complete* in the sense that there is *exactly one* transition for each state $q \in Q$ and each letter $a \in \Sigma$. If there is *at most one* transition for each state of Q and letter of Σ, the automaton is an *incomplete* DFA.

A *nondeterministic finite automaton (NFA)* is a 5-tuple $\mathcal{D} = (Q, \Sigma, \delta, I, F)$, where Q, Σ and F are defined as in a DFA, $\delta \colon Q \times \Sigma \to 2^Q$ is the *transition function*, and $I \subseteq Q$ is the *set of initial states*. An ε-*NFA* is an NFA in which transitions under the empty word ε are also permitted.

To simplify the notation, without loss of generality we use $Q_n = \{0, \ldots, n-1\}$ as the set of states of every DFA with n states. A *transformation* of Q_n is a mapping $t \colon Q_n \to Q_n$. The *image* of $q \in Q_n$ under t is denoted by qt. For $k \geqslant 2$, a transformation (permutation) t of a set $P = \{q_0, q_1, \ldots, q_{k-1}\} \subseteq Q$ is a k-*cycle* if $q_0 t = q_1, q_1 t = q_2, \ldots, q_{k-2} t = q_{k-1}, q_{k-1} t = q_0$. This k-cycle is denoted by $(q_0, q_1, \ldots, q_{k-1})$, and acts as the identity on the states in $Q_n \setminus P$. A 2-cycle (q_0, q_1) is called a *transposition*. A transformation that changes only one state p to a state $q \neq p$ and acts as the identity for the other states is denoted by $(p \to q)$. The identity transformation is denoted by $\mathbf{1}$.

In any DFA, each $a \in \Sigma$ induces a transformation δ_a of the set Q_n defined by $q\delta_a = \delta(q, a)$; we denote this by $a \colon \delta_a$. For example, when defining the transition function of a DFA, we write $a \colon (0, 1)$ to mean that $\delta(q, a) = q(0, 1)$, where the transformation $(0, 1)$ acts on state q as follows: if q is 0 it maps it to 1, if q is 1 it maps it to 0, and it acts as the identity on the remaining states.

By a slight abuse of notation we use the letter a to denote the transformation it induces; thus we write qa instead of $q\delta_a$. We extend the notation to sets of states: if $P \subseteq Q_n$, then $Pa = \{pa \mid p \in P\}$. We also find it convenient to write $P \xrightarrow{a} Pa$ to indicate that the image of P under a is Pa. If s, t are transformations of Q, their composition is denoted by $s * t$ and defined by $q(s * t) = (qs)t$; the $*$ is usually omitted. Let $\mathcal{T}_{Q_n}$ be the set of all n^n transformations of Q_n; then $\mathcal{T}_{Q_n}$ is a monoid under composition.

A sequence $(L_n, n \geqslant k) = (L_k, L_{k+1}, \ldots)$, of regular languages is called a *stream*; here k is usually some small integer, and the languages in the stream usually have the same form and differ only in the parameter n. For example, $(\{a, b\}^* a^n \{a, b\}^* \mid n \geqslant 2)$ is a stream. To find the complexity of a binary operation $\circ$ we need to find an upper bound on this complexity and two streams $(L'_m, m \geqslant h)$ and $(L_n, n \geqslant k)$ of languages meeting this bound. In general, the two streams are different, but there are many examples where L'_n "differs only slightly" from L_n; such a language L'_n is called a *dialect* [3] of L_n.

Let $\Sigma = \{a_1, \ldots, a_k\}$ be an alphabet; we assume that its elements are ordered as shown. Let π be a *partial permutation* of Σ, that is, a partial function $\pi\colon \Sigma \to \Gamma$ where $\Gamma \subseteq \Sigma$, for which there exists $\Delta \subseteq \Sigma$ such that π is bijective when restricted to Δ and undefined on $\Sigma \setminus \Delta$. We denote undefined values of π by "$-$", that is, we write $\pi(a) = -$, if π is undefined at a.

If $L \subseteq \Sigma^*$, we denote it by $L(a_1, \ldots, a_k)$ to stress its dependence on Σ. If π is a partial permutation, let $s_\pi(L(a_1, \ldots, a_k))$ be the language obtained from $L(a_1, \ldots, a_k)$ by the substitution s_π defined as follows: for $a \in \Sigma$, $a \mapsto \{\pi(a)\}$ if $\pi(a)$ is defined, and $a \mapsto \emptyset$ otherwise. The *permutational dialect*, or simply *dialect*, of $L(a_1, \ldots, a_k)$ defined by π is the language $L(\pi(a_1), \ldots, \pi(a_k)) = s_\pi(L(a_1, \ldots, a_k))$.

Similarly, let $\mathcal{D} = (Q, \Sigma, \delta, q_0, F)$ be a DFA; we denote it by $\mathcal{D}(a_1, \ldots, a_k)$ to stress its dependence on Σ. If π is a partial permutation, then the *permutational dialect*, or simply *dialect*, $\mathcal{D}(\pi(a_1), \ldots, \pi(a_k))$ of $\mathcal{D}(a_1, \ldots, a_k)$ is obtained by changing the alphabet of $\mathcal{D}$ from Σ to $\pi(\Sigma)$, and modifying δ so that in the modified DFA $\pi(a_i)$ induces the transformation induced by a_i in the original DFA. One verifies that if the language $L(a_1, \ldots, a_k)$ is accepted by DFA $\mathcal{D}(a_1, \ldots, a_k)$, then $L(\pi(a_1), \ldots, \pi(a_k))$ is accepted by $\mathcal{D}(\pi(a_1), \ldots, \pi(a_k))$.

If the letters for which π is undefined are at the end of the alphabet Σ, then they are omitted. For example, if $\Sigma = \{a, b, c, d\}$ and $\pi(a) = b$, $\pi(b) = a$, and $\pi(c) = \pi(d) = -$, then we write $L_n(b, a)$ for $L_n(b, a, -, -)$, etc.

3 Boolean Operations

A binary boolean operation is *proper* if it is not a constant and does not depend on only one variable. We study the complexities of four proper boolean operations only: union ($\cup$), symmetric difference ($\oplus$), difference ($\setminus$), and intersection ($\cap$); the complexity of any other proper operation can be deduced from these four. For example, $\kappa(\overline{L'} \cup L) = \kappa\left(\overline{\overline{L'} \cup L}\right) = \kappa(L' \cap \overline{L}) = \kappa(L' \setminus L)$, where we have used the well-known fact that $\kappa(\overline{L}) = \kappa(L)$, for any L.

The DFA of Definition 1 is required for the next theorem; this DFA is the 4-input "universal witness" called $\mathcal{U}_n(a, b, c, d)$ in [3].

Definition 1. *For $n \geqslant 3$, let $\mathcal{D}_n = \mathcal{D}_n(a, b, c, d) = (Q_n, \Sigma, \delta_n, 0, \{n-1\})$, where $\Sigma = \{a, b, c, d\}$, and δ_n is defined by the transformations $a\colon (0, \ldots, n-1)$, $b\colon (0, 1)$, $c\colon (n-1 \to 0)$, and $d\colon \mathbf{1}$. Let $L_n = L_n(a, b, c, d)$ be the language accepted by $\mathcal{D}_n$. The structure of $\mathcal{D}_n(a, b, c, d)$ is shown in Fig. 3.*

Theorem 1. *For $m, n \geqslant 3$, let L'_m (respectively, L_n) be a regular language with m (respectively, n) quotients over an alphabet Σ', (respectively, Σ). Then $\kappa(L'_m \cup L_n) = \kappa(L'_m \oplus L_n) = mn + m + n + 1$, $\kappa(L'_m \setminus L_n) = mn + m + 1$, $\kappa(L'_m \cap L_n) = mn + 1$.*

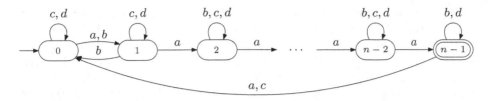

Fig. 3. DFA of Definition 1.

Proof. Let $\mathcal{D}'_m = (Q'_m, \Sigma', \delta', 0', F')$ and $\mathcal{D}_n = (Q_n, \Sigma, \delta, 0, F)$ be minimal DFAs for L'_m and L_n, respectively. To calculate an upper bound for the boolean operations assume that $\Sigma' \setminus \Sigma$ and $\Sigma \setminus \Sigma'$ are non-empty. We add an empty state to $\mathcal{D}'_m$ to send all transitions under the letters from $\Sigma \setminus \Sigma'$ to that state; thus we get an $(m+1)$-state DFA $\mathcal{D}'_{m,\emptyset}$. Similarly, we add an empty state to $\mathcal{D}_n$ to get $\mathcal{D}_{n,\emptyset}$. Now we have two DFAs over the same alphabet, and an ordinary problem of finding an upper bound for the boolean operations on two languages over the same alphabet, *except that these languages both contain empty quotients*. It is clear that $(m+1)(n+1)$ is an upper bound for all four operations; however, this bound can be improved for difference and intersection. Consider the direct product $\mathcal{P}_{m,n}$ of $\mathcal{D}'_{m,\emptyset}$ and $\mathcal{D}_{n,\emptyset}$. For difference, all $n+1$ states of $\mathcal{P}_{m,n}$ that have the form $(\emptyset, q)$, where $q \in Q_n$ are empty. Hence the bound can be reduced by n states to $mn + m + 1$. For intersection, all n states $(\emptyset, q)$, $q \in Q_n$, and all m states $(p', \emptyset)$, $p' \in Q'_m$, are equivalent to the empty state $(\emptyset, \emptyset)$, thus reducing the upper bound to $mn + 1$.

To prove that the bounds are tight, we start with $\mathcal{D}_n(a, b, c, d)$ of Definition 1. For $m, n \geqslant 3$, let $\mathcal{D}'_m(a, b, -, c)$ be the dialect of $\mathcal{D}'_m(a, b, c, d)$ where c plays the role of d and the alphabet is restricted to $\{a, b, c\}$, and let $\mathcal{D}_n(b, a, -, d)$ be the dialect of $\mathcal{D}_n(a, b, c, d)$ in which a and b are permuted, and the alphabet is restricted to $\{a, b, d\}$; see Fig. 4.

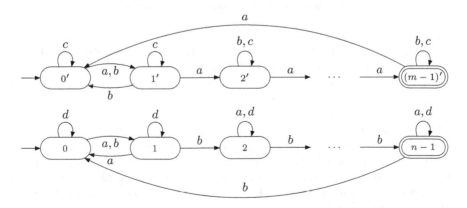

Fig. 4. Witnesses $\mathcal{D}'_m(a, b, -, c)$ and $\mathcal{D}_n(b, a, -, d)$ for boolean operations.

To finish the proof, we complete the two DFAs by adding empty states, and construct their direct product as illustrated in Fig. 5. If we restrict both DFAs to the alphabet $\{a, b\}$, we have the usual problem of determining the complexity of two DFAs over the same alphabet. By [1, Theorem 1], all mn states of the form $\{p', q\}$, $p' \in Q'_m$, $q \in Q_n$, are reachable and pairwise distinguishable by words in $\{a, b\}^*$ for all proper boolean operations if $(m, n) \notin \{(3, 4), (4, 3), (4, 4)\}$. For our application, the three exceptional cases were verified by computation.

To prove that the remaining states are reachable, observe that $(0', 0) \xrightarrow{d} (\emptyset', 0)$ and $(\emptyset', 0) \xrightarrow{b^q} (\emptyset', q)$, for $q \in Q_n$. Symmetrically, $(0', 0) \xrightarrow{c} (0', \emptyset)$ and $(0', \emptyset) \xrightarrow{a^p} (p', \emptyset)$, for $p' \in Q'_m$. Finally, $(\emptyset', n-1) \xrightarrow{c} (\emptyset', \emptyset)$, and all $(m+1)(n+1)$ states of the direct product are reachable.

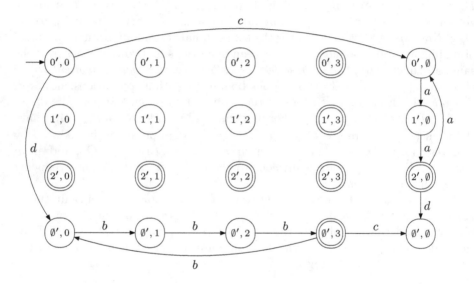

Fig. 5. Direct product for union shown partially.

It remains to verify that the appropriate states are pairwise distinguishable. From [1, Theorem 1], we know that all states in $Q'_m \times Q_n$ are distinguishable. Let $H = \{(\emptyset', q) \mid q \in Q_n\}$, and $V = \{(p', \emptyset) \mid p' \in Q'_m\}$. For the operations consider four cases:

Union: The final states of $\mathcal{P}_{m,n}$ are $\{((m-1)', q) \mid q \in Q_n \cup \{\emptyset\}\}$, and $\{(p', n-1) \mid p' \in Q'_m \cup \{\emptyset'\}\}$. Every state in V accepts a word with a c, whereas no state in H accepts such words. Similarly, every state in H accepts a word with a d, whereas no state in V accepts such words. Every state in $Q'_m \times Q_n$ accepts a word with a c and a word with a d. State $(\emptyset', \emptyset)$ accepts no words at all. Hence any two states chosen from different sets (the sets being $Q'_m \times Q_n$, H, V, and $\{(\emptyset', \emptyset)\}$) are distinguishable. States in H are distinguishable by words in b^* and those in V, by words in a^*. Therefore all $mn + m + n + 1$ states are pairwise distinguishable.

Symmetric Difference: The final states here are all the final states for union except $((m-1)', n-1)$. The rest of the argument is the same as for union.

Difference: The final states now are $\{((m-1)', q) \mid q \neq n-1\}$. The n states of the form $(\emptyset', q)$, $q \in Q_n$, are now equivalent to the empty state $\{(\emptyset', \emptyset)\}$. The remaining states are pairwise distinguishable by the arguments used for union. Hence we have $mn + m + 1$ distinguishable states.

Intersection: Here only $((m-1)', n-1)$ is final and all states $(p', \emptyset)$, $p' \in Q_m'$, and $(\emptyset', q)$, $q \in Q_n$ are equivalent to $\{(\emptyset', \emptyset)\}$, leaving $mn + 1$ distinguishable states. □

Remark 1 (Marek Szykuła, personal communication). In the case of intersection the alphabet of one of the witnesses can be binary: $L_m'(a, b, -, c)$ and $L_n(b, a)$ meet the bound $mn + 1$. Reachability and distinguishability of all mn states of the form $\{p', q\}$, $p' \in Q_m'$, $q \in Q_n$, is the same as above. State $(p', \emptyset)$ can be reached from $(p', 0)$ by c, and is equivalent to the empty state, thus giving $mn + 1$ states in the intersection.

4   Product

Theorem 2. *For $m, n \geqslant 3$, let L_m' (respectively, L_n) be a regular language with m (respectively, n) quotients over an alphabet Σ', (respectively, Σ). Then $\kappa(L_m' L_n) = m2^n + 2^{n-1}$.*

Proof. First we derive the upper bound. Let $\mathcal{D}_m' = (Q_m', \Sigma', \delta', 0', F')$ and $\mathcal{D}_n = (Q_n, \Sigma, \delta, 0, F)$ be minimal DFAs of L_m' and L_n, respectively. We use the normal construction of an ε-NFA $\mathcal{N}$ to recognize $L_m' L_n$, by introducing an ε-transition from each final state of $\mathcal{D}_m'$ to the initial state of $\mathcal{D}_n$, and changing all final states of $\mathcal{D}_m'$ to non-final. This is illustrated in Fig. 6, where $(m-1)'$ is the only final state of $\mathcal{D}_m'$. We then determinize $\mathcal{N}$ using the subset construction to get the DFA $\mathcal{D}$ for $L_m' L_n$.

Suppose $\mathcal{D}_m'$ has k final states, where $1 \leqslant k \leqslant m-1$. I will show that $\mathcal{D}$ can have only the following types of states: (a) at most $(m-k)2^n$ states $\{p'\} \cup S$, where $p' \in Q_m' \setminus F'$, and $S \subseteq Q_n$, (b) at most $k2^{n-1}$ states $\{p', 0\} \cup S$, where $p' \in F'$ and $S \subseteq Q_n \setminus \{0\}$, and (c) at most 2^n states $S \subseteq Q_n$. Because $\mathcal{D}_m'$ is deterministic, there can be at most one state p' of Q_m' in any reachable subset. If $p' \notin F'$, it may be possible to reach any subset of states of Q_n along with p', and this accounts for (a). If $p' \in F'$, then the set must contain 0 and possibly any subset of $Q_n \setminus \{0\}$, giving (b). It may also be possible to have any subset S of Q_n by applying an input that is not in Σ' to $\{0'\} \cup S$ to get S, and so we have (c). Altogether, there are at most $(m-k)2^n + k2^{n-1} + 2^n = (2m-k)2^{n-1} + 2^n$ reachable subsets. This expression reaches its maximum when $k = 1$, and hence we have at most $m2^n + 2^{n-1}$ states in $\mathcal{D}$.

To prove that the bound is tight, we use the same witnesses as for boolean operations; see Fig. 6. If $S = \{q_1, \ldots, q_k\} \subseteq Q_n$ then $S + i = \{q_1 + i, \ldots, q_k + i\}$

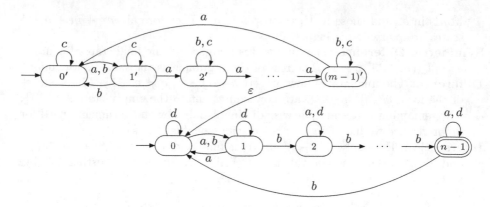

Fig. 6. An NFA for the product of $L'_m(a, b, -, c)$ and $L_n(b, a, -, d)$.

and $S - i = \{q_1 - i, \ldots, q_k - i\}$, where addition and subtraction are modulo n. Note that b^2 and a^m (a^2 and b^n) act as the identity on Q'_m (Q_n). If $p < m - 1$, then $\{p'\} \cup S \xrightarrow{b^2} \{p'\} \cup (S+2)$, for all $S \subseteq Q_n$. If n is odd, then $(b^2)^{(n-1)/2} = b^{n-1}$ and $\{p'\} \cup S \xrightarrow{b^{n-1}} \{p'\} \cup (S - 1)$, for all $q \in Q_n$. If $0, 1 \notin S$ or $\{0, 1\} \subseteq S$, then a acts as the identity on S.

Remark 2. If $1 \notin S$ and $\{(m - 2)'\} \cup S$ is reachable, then $\{0', 1\} \cup S$ is reachable for all $S \subseteq Q_n \setminus \{1\}$.

Proof. If $0 \in S$, then $\{(m - 2)', 0\} \cup S \setminus \{0\} \xrightarrow{a} \{(m - 1)', 0, 1\} \cup S \setminus \{0\} \xrightarrow{a} \{0', 0, 1\} \cup S \setminus \{0\} = \{0', 1\} \cup S$. If $0 \notin S$, then $\{(m - 2)'\} \cup S \xrightarrow{a} \{(m - 1)', 0\} \cup S \xrightarrow{a} \{0', 1\} \cup S$. □

We now prove that the languages of Fig. 6 meet the upper bound.

Claim 1: *All sets of the form* $\{p'\} \cup S$, *where* $p' \in Q'_{m-1}$ *and* $S \subseteq Q_n$, *are reachable.* We show this by induction on the size of S.

Basis: $|S| = 0$. The initial set is $\{0'\}$, and from $\{0'\}$ we reach $\{p'\}$, $p' \in Q'_{m-1}$, by a^p, without reaching any states of Q_n. Thus the claim holds if $|S| = 0$.

Induction Assumption: $\{p'\} \cup S$, where $p' \in Q'_{m-1}$ and $S \subseteq Q_n$, is reachable if $|S| \leqslant k$.

Induction Step: We prove that if $|S| = k + 1$, then $\{p'\} \cup S$ is reachable. Let $S = \{q_0, q_1, \ldots, q_k\}$, where $0 \leqslant q_0 < q_1 < \cdots < q_k \leqslant n - 1$. Suppose $q \in S$. By assumption, sets $\{p'\} \cup (S \setminus \{q\} - (q - 1))$ are reachable for all $p' \in Q'_{m-1}$.

- *All sets of the form* $\{0'\} \cup S$ *are reachable.*
Note that $1 \notin (S \setminus \{q\} - (q - 1))$. By assumption, $\{(m - 2)'\} \cup (S \setminus \{q\} - (q - 1))$ is reachable. By Remark 2, $\{0', 1\} \cup (S \setminus \{q\} - (q - 1))$ is reachable.

1. If there is an odd state q in S, then $\{0', 1\} \cup (S \setminus \{q\} - (q - 1)) \xrightarrow{b^{q-1}} \{0', q\} \cup (S \setminus \{q\}) = \{0'\} \cup S$.

2. If there is no odd state in S and n is odd, then $S \subseteq \{0, 2, \ldots, n - 1\}$. Pick $q \in S$. Then $\{0', 1\} \cup (S \setminus \{q\} - (q - 1)) \xrightarrow{b^{q}} \{0', q+1\} \cup (S \setminus \{q\} + 1) \xrightarrow{b^{n-1}} \{0', q\} \cup S \setminus \{q\} = \{0'\} \cup S$.

3. If there is no odd state and n is even, then $S \subseteq \{0, 2, \ldots, n-2\}$ (so $n-1 \notin S$).

   (a) If $0 \notin S$, then $0, 1 \notin S + 1$. By 1, $\{0'\} \cup (S + 1)$ is reachable, since $S + 1$ contains an odd state. Then $\{0'\} \cup (S+1) \xrightarrow{a} \{1'\} \cup (S+1) \xrightarrow{b^{n-1}} \{0'\} \cup S$.

   (b) If $2 \notin S$, then $0, 1 \notin S - 1$. By 1, $\{0'\} \cup (S - 1)$ is reachable, since $S - 1$ contains an odd state. Then $\{0'\} \cup (S-1) \xrightarrow{a} \{1'\} \cup (S-1) \xrightarrow{b^{n+1}} \{0'\} \cup S$.

   (c) If $\{0, 2\} \subseteq S$, then $0 \notin S - 1$, and $1, n - 1 \in S - 1$. By 1, $\{0'\} \cup (S - 1)$ is reachable, since $1 \in S - 1$. Note that aba sends 1 to 0, $n - 1$ to 1, and adds 1 to each state $q \geqslant 3$ of $S - 1$; thus $2 \notin (S - 1)aba$, and $\{0'\} \cup (S - 1) \xrightarrow{aba} \{1', 0, 1\} \cup S \setminus \{0, 2\}$. Next, b^{n-1} sends 0 to $n - 1$ and subtracts 1 from every other element of $S \setminus \{0, 2\}$. Hence $\{1', 0, 1\} \cup S \setminus \{0, 2\} \xrightarrow{b^{n-1}} \{0', n - 1, 0\} \cup (S \setminus \{0, 2\} - 1) \xrightarrow{ab} \{0', 0, 2\} \cup (S \setminus \{0, 2\}) = \{0'\} \cup S$.

- *All sets of the form $\{1'\} \cup S$ are reachable.*
If 0 and 1 are not in S or are both in S, then $\{0'\} \cup S \xrightarrow{a} \{1'\} \cup S$. If $0 \in S$ but $1 \notin S$, then $\{0', 1\} \cup S \setminus \{0\} \xrightarrow{a} \{1', 0\} \cup S \setminus \{0\} = \{1'\} \cup S$. If $1 \in S$ but $0 \notin S$, then $\{0', 0\} \cup S \setminus \{1\} \xrightarrow{a} \{1', 1\} \cup S \setminus \{1\} = \{1'\} \cup S$.

- *All sets of the form $\{p'\} \cup S$, where $2 \leqslant p \leqslant m - 2$, are reachable.*
If p is even, then $\{0'\} \cup S \xrightarrow{a^p} \{p'\} \cup S$.
If p is odd, then $\{1'\} \cup S \xrightarrow{a^{p-1}} \{p'\} \cup S$.

Claim 2: *All sets of the form $\{(m - 1)', 0\} \cup S$ are reachable.*

1. By Claim 1, $\{(m - 3)'\} \cup S$ is reachable. If $q_0 = 1$, then
   $\{(m - 3)', 1\} \cup S \setminus \{1\} \xrightarrow{a^2} \{(m - 1)', 0, 1\} \cup S \setminus \{1\} = \{(m - 1)', 0\} \cup S$.
2. By Claim 1, $\{(m - 2)'\} \cup S$ is reachable. If $q_0 \geqslant 2$, then
   $\{(m - 2)'\} \cup S \xrightarrow{a} \{(m - 1)', 0\} \cup S$.

Claim 3: *All sets of the form S are reachable.*

By Claim 1, $\{0'\} \cup S$ is reachable for every S, and $\{0'\} \cup S \xrightarrow{d} S$.

For distinguishability, note that only state q accepts $w_q = b^{n-1-q}$ in $\mathcal{D}_n$. Hence, if two states of the product have different sets S and S' and $q \in S \oplus S'$, then they can be distinguished by w_q. State $\{p'\} \cup S$ is distinguished from S by $ca^{m-1-p}b^{n-1}$. If $p < q$, states $\{p'\} \cup S$ and $\{q'\} \cup S$ are distinguished as follows. Use ca^{m-1-q} to reach $\{(p + m - 1 - q)'\}$ from p' and $\{(m - 1)'\} \cup \{0\}$ from q'. The reached states are distinguishable since they differ in their subsets of Q_n. $\square$

5   Most Complex Regular Languages

A *most complex* regular language stream is one that, together with some dialects, meets the complexity bounds for all boolean operations, product, star, and reversal, and has the largest syntactic semigroup and most complex atoms [3]. A most complex stream should have the smallest possible alphabet sufficient to meet all the bounds. Most complex streams are useful in systems dealing with regular languages and finite automata. One would like to know the maximal sizes of automata that can be handled by the system. In view of the existence of most complex streams, one stream can be used to test all the operations. Here we present a stream similar to that of [3] but with one added input letter that induces the identity transformation, as shown in Fig. 3.

Theorem 3 (Most Complex Regular Languages). *For each $n \geqslant 3$, the DFA of Definition 1 is minimal and its language $L_n(a, b, c, d)$ has complexity n. The stream $(L_m(a, b, c, d) \mid m \geqslant 3)$ with dialect streams $(L_n(a, b, -, c) \mid n \geqslant 3)$ and $(L_n(b, a, -, d) \mid n \geqslant 3)$ is most complex in the class of regular languages. In particular, it meets all the complexity bounds below, which are maximal for regular languages. In several cases the bounds can be met with a restricted alphabet.*

1. *The syntactic semigroup of $L_n(a, b, c)$ has cardinality n^n.*
2. *Each quotient of $L_n(a)$ has complexity n.*
3. *The reverse of $L_n(a, b, c)$ has complexity 2^n, and $L_n(a, b, c)$ has 2^n atoms[1].*
4. *For each atom A_S of $L_n(a, b, c)$, the complexity $\kappa(A_S)$ satisfies:*

$$\kappa(A_S) = \begin{cases} 2^n - 1, & \text{if } S \in \{\emptyset, Q_n\}; \\ 1 + \sum_{x=1}^{|S|} \sum_{y=1}^{n-|S|} \binom{n}{x}\binom{n-x}{y}, & \text{if } \emptyset \subsetneq S \subsetneq Q_n. \end{cases}$$

5. *The star of $L_n(a, b)$ has complexity $2^{n-1} + 2^{n-2}$.*
6. *The product $L'_m(a, b, -, c)L_n(b, a, -, d)$ has complexity $m2^n + 2^{n-1}$.*
7. *The complexity of $L'_m(a, b, -, c) \circ L_n(b, a, -, d)$ is $mn + m + n + 1$ if $\circ \in \{\cup, \oplus\}$, $mn + m + 1$ if $\circ = \setminus$, and $mn + 1$ if $\circ = \cap$.*

Proof. The proofs of 1–5 can be found in [3], and Claims 6 and 7 are proved in the present paper, Theorems 1 and 2.   □

Proposition 1 (Marek Szykuła, personal communication). *At least four inputs are required for a most complex regular language. In particular, four inputs are needed for union: two inputs are needed to reach all pairs of states in $Q'_m \times Q_n$, one input in $\Sigma' \setminus \Sigma$ for pairs $(p', \emptyset)$ with $p' \in Q'_m$, and one in $\Sigma \setminus \Sigma'$ for pairs $(\emptyset', q)$ with $q \in Q_n$.*

---

[1] The *atom congruence* is a left congruence defined as follows: two words x and y are equivalent if $ux \in L$ if and only if $uy \in L$ for all $u \in \Sigma^*$. Thus x and y are equivalent if $x \in u^{-1}L$ if and only if $y \in u^{-1}L$. An equivalence class of this relation is called an *atom* of L [5,7]. It follows that an atom is a non-empty intersection of complemented and uncomplemented quotients of L. The number of atoms and their quotient complexities are possible measures of complexity of regular languages [3]. For more information about atoms and their complexity, see [4,5,7].

6   Conclusions

Two complete DFAs over different alphabets Σ' and Σ are incomplete DFAs over $\Sigma' \cup \Sigma$. Each DFA can be completed by adding an empty state and sending all transitions induced by letters not in the DFA's alphabet to that state. This results in an $(m + 1)$-state DFA and an $(n + 1)$-state DFA. From the theory about DFAs over the same alphabet we know that $(m + 1)(n + 1)$ is an upper bound for all boolean operations on the original DFAs, and that $m2^{n+1} + 2^n$ is an upper bound for product. We have shown that the tight bounds for boolean operations are $(m + 1)(n + 1)$ for union and symmetric difference, $mn + m + 1$ for difference, and $mn + 1$ for intersection, while the tight bound for product is $m2^n + 2^{n-1}$. In the same-alphabet case the tight bound is mn for all boolean operations and it is $(m - 1)2^n + 2^{n-1}$ for product. In summary, the restriction of identical alphabets is unnecessary and leads to incorrect results.

It should be noted that if the two languages in question already have empty quotients, then making the alphabets the same does not require the addition of any states, and the traditional same-alphabet methods are correct. This is the case, for example, for prefix-free, suffix-free and finite languages.

Acknowledgment. I am very grateful to Sylvie Davies, Bo Yang Victor Liu and Corwin Sinnamon for careful proofreading and constructive comments. I thank Marek Szykuła for contributing the important Remark 1 and Proposition 1.

References

1. Bell, J., Brzozowski, J., Moreira, N., Reis, R.: Symmetric groups and quotient complexity of boolean operations. In: Esparza, J., Fraigniaud, P., Husfeldt, T., Koutsoupias, E. (eds.) ICALP 2014, Part II. LNCS, vol. 8573, pp. 1–12. Springer, Heidelberg (2014)
2. Brzozowski, J.: Quotient complexity of regular languages. J. Autom. Lang. Comb. **15**(1/2), 71–89 (2010)
3. Brzozowski, J.: In search of the most complex regular languages. Int. J. Found. Comput. Sci. **24**(6), 691–708 (2013)
4. Brzozowski, J., Tamm, H.: Complexity of atoms of regular languages. Int. J. Found. Comput. Sci. **24**(7), 1009–1027 (2013)
5. Brzozowski, J., Tamm, H.: Theory of átomata. Theoret. Comput. Sci. **539**, 13–27 (2014)
6. Gao, Y., Salomaa, K., Yu, S.: Transition complexity of incomplete DFAs. Fund. Inform. **110**, 143–158 (2011)
7. Iván, S.: Complexity of atoms, combinatorially. Inform. Process. Lett. **116**(5), 356–360 (2016)
8. Maia, E., Moreira, N., Reis, R.: Incomplete operational transition complexity of regular languages. Inform. Comput. **244**, 1–22 (2015)

9. Maslov, A.N.: Estimates of the number of states of finite automata. Dokl. Akad. Nauk SSSR **194**, 1266–1268 (1970) (Russian). English translation: Soviet Math. Dokl. **11**, 1373–1375 (1970)
10. Yu, S.: State complexity of regular languages. J. Autom. Lang. Comb. **6**, 221–234 (2001)
11. Yu, S., Zhuang, Q., Salomaa, K.: The state complexities of some basic operations on regular languages. Theoret. Comput. Sci. **125**, 315–328 (1994)

On the State Complexity of the Shuffle of Regular Languages

Janusz Brzozowski[1], Galina Jirásková[2], Bo Liu[1], Aayush Rajasekaran[1], and Marek Szykuła[3(✉)]

[1] David R. Cheriton School of Computer Science, University of Waterloo, Waterloo, ON N2L 3G1, Canada
{brzozo,b23liu,arajasek}@uwaterloo.ca
[2] Mathematical Institute, Slovak Academy of Sciences, Grešákova 6, 040 01 Košice, Slovakia
jiraskov@saske.sk
[3] Institute of Computer Science, University of Wrocław, Joliot-Curie 15, 50-383 Wrocław, Poland
msz@cs.uni.wroc.pl

Abstract. We investigate the shuffle operation on regular languages represented by complete deterministic finite automata. We prove that $f(m,n) = 2^{mn-1} + 2^{(m-1)(n-1)}(2^{m-1}-1)(2^{n-1}-1)$ is an upper bound on the state complexity of the shuffle of two regular languages having state complexities m and n, respectively. We also state partial results about the tightness of this bound. We show that there exist witness languages meeting the bound if $2 \leqslant m \leqslant 5$ and $n \geqslant 2$, and also if $m = n = 6$. Moreover, we prove that in the subset automaton of the NFA accepting the shuffle, all 2^{mn} states can be distinguishable, and an alphabet of size three suffices for that. It follows that the bound can be met if all $f(m,n)$ states are reachable. We know that an alphabet of size at least mn is required provided that $m, n \geqslant 2$. The question of reachability, and hence also of the tightness of the bound $f(m,n)$ in general, remains open.

Keywords: Regular language · Shuffle · State complexity · Upper bound

1 An Upper Bound for the Shuffle Operation

The *state complexity of a regular language* L [6] is the number of states in a complete minimal deterministic finite automaton (DFA) recognizing the language;

This work was supported by the Natural Sciences and Engineering Research Council of Canada under grant No. OGP0000871, by VEGA grant 2/0084/15, and by the National Science Centre, Poland under project number 2014/15/B/ST6/00615.
B. Liu—Present address: Google Inc., 1600 Amphitheatre Parkway, Mountain View, CA 94043, USA.

C. Câmpeanu et al. (Eds.): DCFS 2016, LNCS 9777, pp. 73–86, 2016.
DOI: 10.1007/978-3-319-41114-9_6

it will be denoted by $\kappa(L)$. The *state complexity of an operation* on regular languages is the maximal state complexity of the result of the operation expressed as a function of the state complexities of the operands.

Let Σ be a finite non-empty alphabet. The *shuffle* $u \sqcup\!\sqcup v$ of words $u, v \in \Sigma^*$ is defined as follows:

$$u \sqcup\!\sqcup v = \{u_1 v_1 \cdots u_k v_k \mid u = u_1 \cdots u_k, v = v_1 \cdots v_k, u_1, \ldots, u_k, v_1, \ldots, v_k \in \Sigma^*\}.$$

The shuffle of two languages K and L over Σ is defined by

$$K \sqcup\!\sqcup L = \bigcup_{u \in K, v \in L} u \sqcup\!\sqcup v.$$

Note that the shuffle operation is commutative on both words and languages.

The state complexity of the shuffle operation was first studied by Câmpeanu et al. [2], but they considered only bounds for incomplete deterministic automata. In particular, they proved that $2^{mn} - 1$ is a tight upper bound for that case. Since we can convert an incomplete deterministic automaton into complete one by adding the empty state, it follows that $2^{(m-1)(n-1)} - 1$ is a lower bound for the case of complete deterministic automata. Here we show that this lower bound can be improved, and we derive an upper bound for two regular languages represented by complete deterministic automata, but the question whether this bound is tight remains open.

A *nondeterministic finite automaton* (NFA) is a quintuple $\mathcal{A} = (Q, \Sigma, \delta, s, F)$, where Q is a finite non-empty set of states, Σ is a finite alphabet of input symbols, $\delta \colon Q \times \Sigma \to 2^Q$ is the transition function which is extended to the domain $2^Q \times \Sigma^*$ in the natural way, $s \in Q$ is the initial state, and $F \subseteq Q$ is the set of final states. The language accepted by NFA $\mathcal{A}$ is the set of words $L(\mathcal{A}) = \{w \in \Sigma^* \mid \delta(s, w) \cap F \neq \emptyset\}$.

An NFA $\mathcal{A}$ is *deterministic and complete* (DFA) if $|\delta(q, a)| = 1$ for each q in Q and each a in Σ. In such a case, we write $\delta(q, a) = q'$ instead of $\delta(q, a) = \{q'\}$. A DFA is *minimal* (with respect to the number of states) if all its states are reachable, and no two distinct states are equivalent.

Every NFA $\mathcal{A} = (Q, \Sigma, \delta, s, F)$ can be converted to an equivalent DFA $\mathcal{A}' = (2^Q, \Sigma, \delta, \{s\}, F')$, where $F' = \{R \in 2^Q \mid R \cap F \neq \emptyset\}$. The DFA $\mathcal{A}'$ is called the *subset automaton* of NFA $\mathcal{A}$. The subset automaton may not be minimal since some of its states may be unreachable or equivalent to other states.

Let K and L be regular languages over an alphabet Σ recognized by deterministic finite automata $\mathcal{K} = (Q_K, \Sigma, \delta_K, q_K, F_K)$ and $\mathcal{L} = (Q_L, \Sigma, \delta_L, q_L, F_L)$, respectively. Then $K \sqcup\!\sqcup L$ is accepted by the nondeterministic finite automaton

$$\mathcal{N} = (Q_K \times Q_L, \Sigma, \delta, (q_K, q_L), F_K \times F_L),$$

where

$$\delta((p, q), a) = \{(\delta_K(p, a), q), (p, \delta_L(q, a))\}.$$

Let $\mathcal{D} = (2^{Q_K \times Q_L}, \Sigma, \delta', \{(q_K, q_L)\}, F')$ be the subset automaton of $\mathcal{N}$. If $|Q_K| = m$ and $|Q_L| = n$, then NFA $\mathcal{N}$ has mn states. It follows that DFA $\mathcal{D}$ has at most 2^{mn} reachable and pairwise distinguishable states. However, this upper bound cannot be met, as we will show.

In the sequel, we assume that $Q_K = \{1, 2, \ldots, m\}$, $q_K = 1$, $Q_L = \{1, 2, \ldots, n\}$, and $q_L = 1$. We say that a state (p, q) of NFA $\mathcal{N}$ is in row i if $p = i$, and it is in column j if $q = j$.

Proposition 1. *Let $a \in \Sigma$. Let S be a state of $\mathcal{D}$. Let $\pi_{\mathrm{col}}(S) = \{p \mid (p, q) \in S \text{ for some } q\}$, and $\pi_{\mathrm{row}}(S) = \{p \mid (p, q) \in S \text{ for some } p\}$. Then $\pi_x(S) \subseteq \pi_x(S \cdot a)$ for $x \in \{\mathrm{col}, \mathrm{row}\}$.*

Proof. Let $p \in \pi_{\mathrm{col}}(S)$; then we have $(p, q) \in S$ for some q. Since $\delta((p, q), a) = \{(\delta_K(p, a), q), (p, \delta_L(q, a))\}$, we have $(p, \delta_L(q, a)) \in \delta(S, a)$, so $p \in \pi_{\mathrm{col}}(\delta(S, a))$. By symmetry, the same claim holds for π_{row}.   □

We claim that in the subset automaton $\mathcal{D}$, every reachable subset S of $Q_K \times Q_L$ must contain a state in column 1 and a state in row 1, that is, it must satisfy the following condition.

Condition (C): There exist states $(s, 1)$ and $(1, t)$ in S for some $s \in Q_K$ and $t \in Q_L$.

Lemma 2. *Every reachable subset S of subset automaton $\mathcal{D}$ satisfies Condition (C).*

Proof. The initial subset of $\mathcal{D}$ is $\{(1, 1)\}$, and it satisfies Condition (C). By Proposition 1, for every $a \in \Sigma$ we get that $1 \in \pi_{\mathrm{col}}(\delta(S, a))$ and $1 \in \pi_{\mathrm{row}}(\delta(S, a))$, so $\delta(S, a)$ satisfies Condition (C). By induction, all reachable subsets satisfy Condition (C).   □

Theorem 3 (Shuffle: Upper Bound). *Let $\kappa(K) = m$ and $\kappa(L) = n$. Then the state complexity of the shuffle of K and L is at most*

$$f(m, n) = 2^{mn-1} + 2^{(m-1)(n-1)}(2^{m-1} - 1)(2^{n-1} - 1). \tag{1}$$

Proof. By Lemma 2, every reachable subset of $\mathcal{D}$ must contain a state in row 1 and a state in column 1. There are 2^{mn-1} subsets containing state $(1, 1)$, and $2^{(m-1)(n-1)}(2^{m-1} - 1)(2^{n-1} - 1)$ subsets not containing $(1, 1)$ but containing $(s, 1)$ for some $s \in \{2, 3, \ldots, m\}$ and $(1, t)$ for some $t \in \{2, 3, \ldots, n\}$. This gives $f(m, n)$.   □

Let K and L be two regular languages over Σ. If $\kappa(K) = \kappa(L) = 1$, then each of K, L, and $K \amalg L$ is either $\emptyset$ or Σ^*, and $\kappa(K \amalg L) = 1$; hence the bound $f(1, 1) = 1$ is tight.

Now suppose that $\kappa(K) = 1$; here we have two possible choices for K, the empty language or Σ^*. The first choice leads to $\kappa(K \amalg L) = 1$. Hence only the second choice is of interest, where the language $K \amalg L = \Sigma^* \amalg L$ is the all-sided ideal [1] generated by L. If $\kappa(L) = 2$, the upper bound $f(1, 2) = 2$ is met by the unary language $L = aa^*$. Hence assume that $\kappa(K) = 1$ and $\kappa(L) \geqslant 3$. The next observation shows that in such a case, the tight bound is less than $f(1, n) = 2^{n-1}$.

Proposition 4 (Okhotin [4]). *If $\kappa(L) \geqslant 3$, then the state complexity of $\Sigma^* \sqcup L$ is at most $2^{n-2} + 1$, and this bound can be reached only if $|\Sigma| \geqslant n - 2$.*

Okhotin showed that the language $L = (a_1 \Sigma^* a_1 \cup \cdots \cup a_{n-2} \Sigma^* a_{n-2}) \Sigma^*$, where $\Sigma = \{a_1, \ldots, a_{n-2}\}$, meets this bound [4]. This takes care of the case $\kappa(K) = 1$ and, by symmetry, of the case $\kappa(L) = 1$.

In what follows we assume that $m \geqslant 2$ and $n \geqslant 2$. First, let us show that the upper bound $f(m, n)$ cannot be met by regular languages defined over a fixed alphabet.

Proposition 5. *Let K and L be regular languages over Σ with $\kappa(K) = m$ and $\kappa(L) = n$, where $m, n \geqslant 2$. If $\kappa(K \sqcup L) = f(m, n)$, then $|\Sigma| \geqslant mn - 1$.*

Proof. For $s = 2, 3, \ldots, m$ and $t = 2, 3, \ldots, n$ denote

$$A_s = \{(1,1), (s,1)\},$$
$$B_t = \{(1,1), (1,t)\},$$
$$C_{st} = \{(s,1), (1,t)\}.$$

If all the subsets satisfying Condition (C) are reachable, then, in particular, all the subsets A_s, B_t, and C_{st} must be reachable. Let us show that all these subsets must be reached from some subsets containing state $(1,1)$ by distinct symbols.

Suppose that a set A_s is reached from a reachable set S with $S \neq A_s$ by a symbol a, that is, we have $A_s = \delta(S, a)$ and $S \neq A_s$. The set A_s contains only states in column 1 and rows 1 or s. By Proposition 1, the set S may only contain states in column 1 and in rows 1 or s, that is, we have $S \subseteq \{(1,1), (s,1)\}$. Since $S \neq A_s$, we must have $S = \{(1,1)\}$.

By symmetry, each B_t can only be reached from $\{(1,1)\}$.

Suppose that a set C_{st} is reached from a reachable set S with $S \neq C_{st}$ by a symbol a. By Proposition 1, we must have $S \subseteq \{(1,1), (s,1), (1,t), (s,t)\}$. Let us show that $(1,1) \in S$. Suppose for a contradiction that $(1,1) \notin S$. Then, since S is reachable, it must contain a state in column 1 and a state in row 1, that is, we must have $\{(s,1), (1,t)\} \subseteq S$. But then $(s,t) \in S$ since $S \neq C_{st}$. However, then $\delta_K(s, a) = 1$ and $\delta_L(t, a) = 1$ which implies that $(1,1) \in \delta((s,1), a)$, and so $(1,1) \in C_{st}$. This is a contradiction. Therefore C_{st} is reached from a set containing $(1,1)$.

Thus each A_s is reached from $\{(1,1)\}$ by a symbol a_s, each B_t is reached from $\{(1,1)\}$ by a symbol b_t, each C_{st} is reached from a set containing $(1,1)$ by a symbol c_{st}, and we must have

$$\delta_K(1, a_s) = s \text{ and } \delta_L(1, a_s) = 1,$$
$$\delta_K(1, b_t) = 1 \text{ and } \delta_L(1, b_t) = t,$$
$$\delta_K(1, c_{st}) = s \text{ and } \delta_L(1, c_{st}) = t.$$

It follows that all the symbols a_s, b_t, and c_{st} must be pairwise distinct. Therefore we have $|\Sigma| \geqslant m - 1 + n - 1 + (m-1)(n-1) = mn - 1$. $\qquad \square$

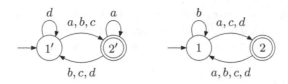

Fig. 1. Witness DFAs $\mathcal{K}$ and $\mathcal{L}$ for shuffle with $|Q_K| = 2$, $|Q_L| = 2$.

Unfortunately, this lower bound on the size of the alphabet is not tight, as is demonstrated by the following example:

Example 6. If t is a transformation of the set $\{1, 2, \ldots, n\}$ and $q \in \{1, 2, \ldots, n\}$, let qt be the image of q under t. Transformation t can now be denoted by $[1t, 2t, \ldots, nt]$.

(1) If $m = n = 2$, we have $f(2, 2) = 10$. Let $\Sigma = \{a, b, c, d\}$, and let the DFAs $\mathcal{K}$ and $\mathcal{L}$ be as shown in Fig. 1, and let K and L be their languages. Then $\kappa(K \sqcup\!\sqcup L) = 10$. We have used GAP [3] to show that the bound cannot be reached with a smaller alphabet, and that the DFAs of Fig. 1 are unique up to isomorphism.

(2) For $m = 2$ and $n = 3$, the minimal size of the alphabet of a witness pair is 6. We have verified this by a dedicated algorithm enumerating all pairs of non-isomorphic DFAs with 2 and 3 states. In contrast to the previous case, over a minimal alphabet there are more than 60 non-isomorphic DFAs of L – even if we do not distinguish them by sets of final states – that meet the bound with some K. One of the witness pairs is described below.

Let $\Sigma = \{a, b, c, d, e, f\}$. Let $\mathcal{K} = (\{1, 2\}, \Sigma, \delta_K, 1, \{2\})$, and let $a = [1, 2]$, $b = c = [2, 1]$, $d = [1, 1]$, $e = [2, 2]$, and $f = [2, 1]$. Let $\mathcal{L} = (\{1, 2, 3\}, \Sigma, \delta_L, 1, \{1\})$, and let $a = [2, 2, 3]$, $b = [2, 1, 3]$, $c = [1, 1, 1]$, $d = e = [3, 1, 2]$, $f = [3, 1, 1]$. Then $\kappa(K \sqcup\!\sqcup L) = 44 = f(2, 3)$.

The bound $mn - 1$ on the size of the alphabet is not tight for $m = n = 2$, where an alphabet of size four is required. For any $m, n \geqslant 2$ the subsets of $\{1, 2\} \times \{1, 2\}$ satisfying (C) must be also reachable, and to reach them we can use only transformations mapping 1 to either 1 or 2. There are only three such transformations counted in Proposition 5; thus we need one more letter.

2   Partial Results About Tightness

To prove that the upper bound $f(m, n)$ of Eq. (1) is tight, we must exhibit two languages K and L with state complexities m and n, respectively, such that $\kappa(K \sqcup\!\sqcup L) = f(m, n)$. As usual, we use DFAs to represent the languages: Let $\mathcal{K}$ and $\mathcal{L}$ be minimal *complete* DFAs for K and L. We first construct the NFA $\mathcal{N}$ as defined in Sect. 1, and we consider the subset automaton $\mathcal{D}$ of NFA $\mathcal{N}$. We must then show that $\mathcal{D}$ has $f(m, n)$ states reachable from the initial state

$\{(1,1)\}$, and that these states are pairwise distinguishable. We were unable to prove this for all m and n, but we have some partial results about reachability in Subsect. 2.1, and we deal with distinguishability in Subsect. 2.2.

2.1 Reachability

We performed computations verifying reachability of the upper bound for small values of m and n. These results are summarized in Table 1.

The computation in the hardest case with $m = n = 6$ took about 48 days on a computer with AMD Opteron(tm) Processor 6380 (2500 MHz) and 64 GB of RAM. Moreover, we verified that in all these cases, every subset of size at least 3 is directly reachable from some smaller subset. We also verified that for reachability in case of $m = n = 3$ an alphabet of size 12 is sufficient, and in case of $m = n = 4$ an alphabet of size 50 is sufficient. Using these results, we are going to prove reachability for all m, n with $2 \leqslant m \leqslant 5$ and $n \geqslant 2$.

Table 1. Computational verification of reachability of the bound. The fields with $\checkmark^*$ follow from the proofs of Subsect. 2.1.

$m\backslash n$	2	3	4	5	6	7	$\geqslant 8$
2	$\checkmark$	$\checkmark$	$\checkmark$	$\checkmark$	$\checkmark$	$\checkmark$	$\checkmark^*$
3		$\checkmark$	$\checkmark$	$\checkmark$	$\checkmark$	$\checkmark$	$\checkmark^*$
4			$\checkmark$	$\checkmark$	$\checkmark$	$\checkmark$	$\checkmark^*$
5				$\checkmark$	$\checkmark$	$\checkmark$	$\checkmark^*$
6					$\checkmark$	?	?
7						?	?
$\geqslant 8$							?

Without loss of generality, the set of states of any n-state DFA is denoted by $Q_n = \{1, 2, \ldots, n\}$. Let $\mathcal{T}_n$ be the monoid of all transformations of the set Q_n. Let $p, q \in Q_n$ and $P \subseteq Q_n$. Let $\mathbf{1}$ denote the identity transformation. Let $(p \to q)$ denote the transformation that maps state p to state q and acts as the identity on all the other states. Let (p, q) denote the transformation that transposes p and q.

Here we deal only with reachability, so final states do not matter. We assume that the sets of final states are empty in this subsection.

Let $\Sigma_{m,n} = \{a_{s,t} \mid s \in \mathcal{T}_m \text{ and } t \in \mathcal{T}_n\}$ be an alphabet consisting of $m^m n^n$ symbols. If an input a induces transformations s in $\mathcal{T}_m$ and t in $\mathcal{T}_n$, this will be indicated by $a\colon s; t$.

Define DFAs $\mathcal{K}_{m,n} = (Q_m, \Sigma_{m,n}, \delta_m, 1, \emptyset)$ and $\mathcal{L}_{m,n} = (Q_n, \Sigma_{m,n}, \delta_n, 1, \emptyset)$, where $\delta_m(p, a_{s,t}) = ps$ if $p \in Q_m$ and $\delta_n(q, a_{s,t}) = qt$ if $q \in Q_n$. Let $\mathcal{N}_{m,n}$ be the NFA for the shuffle of languages recognized by DFAs $\mathcal{K}_{m,n}$ and $\mathcal{L}_{m,n}$ as described in Sect. 1, and let $\mathcal{D}_{m,n}$ be the subset automaton of $\mathcal{N}_{m,n}$. The NFA $\mathcal{N}_{m,n}$ has alphabet $\Sigma_{m,n}$, and so has an input letter for every pair of transformations in $\mathcal{T}_m \times \mathcal{T}_n$. Therefore the addition of another input letter to the DFAs $\mathcal{K}_{m,n}$ and $\mathcal{L}_{m,n}$ cannot add any new set of states of $\mathcal{N}_{m,n}$ that would be reachable from $\{(1,1)\}$ in $\mathcal{D}_{m,n}$.

Let $m' \leqslant m$ and $n' \leqslant n$. Then DFA $\mathcal{K}_{m',n'} = (Q_{m'}, \Sigma_{m',n'}, \delta_{m'}, 1, \emptyset)$ (respectively, the DFA $\mathcal{L}_{m',n'} = (Q_{n'}, \Sigma_{m',n'}, \delta_{n'}, 1, \emptyset)$) is a sub-DFA of $\mathcal{K}_{m,n}$ (respectively, of $\mathcal{L}_{m,n}$), in the sense that $Q_{m'} \subseteq Q_m$, $\Sigma_{m',n'} \subseteq \Sigma_{m,n}$, and $\delta_{m'} \subseteq \delta_m$. As well, NFA $\mathcal{N}_{m',n'}$ is a sub-NFA of $\mathcal{N}_{m,n}$. Note that $\mathcal{D}_{m,n}$ is extremal for the shuffle: every language $K \sqcup\!\sqcup L$, where K and L are languages with state complexities m and n respectively, is recognized by some sub-DFA of $\mathcal{D}(m,n)$ after possibly renaming some letters.

For the next lemma it is convenient to consider a subset S of states (p,q) of $\mathcal{N}_{m,n}$ as an $m \times n$ matrix, where the entry in row p and column q is (p,q) if $(p,q) \in S$, and it is empty otherwise. We first introduce the following notions.

Definition 7. *Let $i, i' \in Q_m$, $i \neq i'$, and $j, j' \in Q_n$, $j \neq j'$.*
(a) A row i' contains row i, if $(i,j) \in S$ implies $(i',j) \in S$ for all $j \in Q_n$.
(b) A column j' contains column j if $(i,j) \in S$ implies $(i,j') \in S$ for all $i \in Q_m$.
(c) A subset of $Q_m \times Q_n$ is valid if it satisfies Condition (C) from Lemma 2, that is, if it contains a state in row 1 and a state in column 1.

Lemma 8. *Let S be a valid subset of $Q_m \times Q_n$ with the property that there are distinct i, i' or j, j' such that either row i' contains row i or column j' contains column j. Assume that every valid subset S' of $Q_{m'} \times Q_{n'}$, where $m' < m$, or $n' < n$, or $|S'| < |S|$, is reachable in DFA $\mathcal{D}_{m',n'}$. Then S is reachable in $\mathcal{D}_{m,n}$.*

Proof. If S contains an empty row or column, then without loss of generality we can renumber the n states of $\mathcal{L}_{m,n}$ in such a way that column n is the empty column in S. By the inductive assumption we know that S is reachable in $\mathcal{D}_{m,n-1}$ by some word w. Since $\mathcal{N}_{m,n-1}$ is a sub-NFA of $\mathcal{N}_{m,n}$, S is reachable in $\mathcal{D}_{m,n}$ as well by the same word. Suppose that S has neither an empty row nor an empty column. By symmetry, it is sufficient to consider the case with distinct i and i' such that row i' contains row i. Let $S' = S \backslash \{(i',j) \mid (i,j) \in S \text{ for } j \in \{1, \ldots, n\}\}$. Since $|S'| < |S|$, the set S' is reachable by assumption. To obtain S, we apply the letter that induces the transformation $i \to i'; 1$. $\square$

Lemma 9. *Let S be a valid subset of $Q_m \times Q_n$ such that there is a column or a row with exactly one element. Assume that every valid subset S' of $Q_{m'} \times Q_{n'}$, where $m' < m$, or $n' < n$, or $|S'| < |S|$, is reachable in $\mathcal{D}_{m',n'}$. Then S is reachable in $\mathcal{D}_{m,n}$.*

Proof. Recall that we can assume $m \geqslant 2$ and $n \geqslant 2$. We may assume that there is neither an empty row nor an empty column in S; otherwise S is reachable by Lemma 8. It is sufficient to consider the case involving a column, since the case involving a row follows by symmetric arguments. Let (p,q) be the only element in column q. If there are more elements in row p, then column q is contained in another column and by Lemma 8, the set S is reachable.

Let S' be the subset of $Q_{m-1} \times Q_{n-1}$ obtained by removing row p and column q, and renumbering the states to $Q_{m-1} \times Q_{n-1}$ in the way such that $i \in Q_m$ becomes $i - 1$ if $i > p$ and otherwise remains the same, and $j \in Q_n$ becomes $j - 1$ if $j > q$ and otherwise remains the same. We have that S' is a valid subset, and by the inductive assumption it is reachable in $\mathcal{D}_{m-1,n-1}$ by some word u'; let u be the word corresponding to u' in the original numbering of the states. We consider four cases.

Case $p \neq 1$ and $q \neq 1$: State $\{(1,1),(p,q)\}$ is reachable in $\mathcal{D}_{m,n}$ by word a^2, where $a \colon (1,p);(1,q)$. Then S is reachable by $a^2 u$.

Case $p = 1$ and $q \neq 1$: State $\{(2,1),(1,q)\}$ is reachable in $\mathcal{D}_{m,n}$ by word a^2, where $a \colon (1,2);(1,q)$. Then state $(2,1)$ corresponds to state $(1,1)$ after the renumbering, and S is reachable by $a^2 u$.

Case $p \neq 1$ and $q = 1$: This is symmetrical to the previous case.

Case $p = 1$ and $q = 1$: State $\{(1,1),(2,2)\}$ is reachable in $\mathcal{D}_{m,n}$ by word a^2, where $a \colon (1,2);(1,2)$. Then state $(2,2)$ corresponds to state $(1,1)$ after the renumbering, and S is reachable by $a^2 u$.   □

Theorem 10. *If for some h every valid subset can be reached in $\mathcal{D}_{h,\binom{h}{\lfloor h/2 \rfloor}}$ then for every $m \leqslant h$ and every n, every valid subset can be reached in $\mathcal{D}_{m,n}$.*

Proof. This follows by induction on m, n, and $|S|$.

For $m = 1$ this follows by induction on n: if $n = 1$ then $\mathcal{D}_{1,1}$ consists of a single valid subset $\{(1,1)\}$, and if $n > 1$, then we apply Lemma 8. For $m \leqslant h$ and $n \leqslant \binom{h}{\lfloor h/2 \rfloor}$ this holds by assumption, since $\mathcal{N}_{m,n}$ is a sub-NFA of $\mathcal{N}_{h,\binom{h}{\lfloor h/2 \rfloor}}$. If $|S| = 1$, then $\{(1,1)\}$ is the only valid subset, and it is reachable since it is the initial subset of $\mathcal{D}_{m,n}$.

Let S be a valid subset of $Q_m \times Q_n$, where $m \leqslant h$ and $n > \binom{h}{\lfloor h/2 \rfloor}$, and assume that every valid subset S' of $Q_{m'} \times Q_{n'}$ is reachable if $m' < m$, or $n' < n$, or $|S'| < |S|$. By Sperner's theorem [5], the maximal number of subsets of an m-element set such that none of them contains any other subset is $\binom{m}{\lfloor m/2 \rfloor}$. This is not larger than $\binom{h}{\lfloor h/2 \rfloor}$; hence, there exist some columns j, j' with $j \neq j'$ such that the j-th column is contained in j'-th column. By Lemma 8, the subset S is reachable.   □

Corollary 11. *Let $1 \leqslant m \leqslant 4$ and $n \geqslant 1$. Then every valid subset can be reached in $\mathcal{D}_{m,n}$.*

Proof. Since we have verified the reachability of all valid subsets for $m = 4$ and $n = 6 = \binom{4}{2}$, Theorem 10 applies with $h = 4$.   □

To strengthen this result and show reachability for $m \leqslant 5$, we need to introduce another concept with permutations. Let φ be any permutation of m rows. We split subsets of Q_m (subsets of rows) into equivalence classes under φ. For $U \subseteq Q_m$, $[U]_\varphi = \{V \subseteq Q_m \mid V = \varphi^i(U) \text{ for some } i \geqslant 0\}$ denotes the equivalence class of U. See Tables 2, 3, 4 for examples of subsets whose columns U are partitioned into equivalence classes under some φ.

For a subset S of $Q_m \times Q_n$, by $\text{col}(S, i)$ we denote the subset of Q_m contained in the i-th column. Then $\text{cols}(S) = \bigcup_{1 \leqslant i \leqslant n} \text{col}(S, i)$ is the set of the subsets in the columns of S.

The following lemma assures reachability (under an inductive assumption) of a special kind of subsets whose columns form only full and empty equivalence classes under some permutation φ.

Lemma 12. *Let φ be a permutation of m rows. Let S be a valid subset of $Q_m \times Q_n$ such that $[U]_\varphi \subseteq \text{cols}(S)$ for every $U \in \text{cols}(S)$, and there is a column $V \in \text{cols}(S)$ such that $|[V]_\varphi| \geqslant 2$. Assume that every valid subset S' of $Q_{m'} \times Q_{n'}$, where $m' < m$, or $n' < n$, or $|S'| < |S|$, is reachable in $\mathcal{D}_{m',n'}$. Then S is reachable in $\mathcal{D}_{m,n}$.*

Proof. We can assume that no two columns contain the same subset of rows, no column is empty, and the first row contains at least two elements; otherwise S is reachable by Lemma 8 or by Lemma 9.

Let $S_j = \text{col}(S, j)$ be the j-th column of a valid subset S. Thus we have $S = \{(i, j) \mid 1 \leqslant j \leqslant n \text{ and } i \in S_j\}$. Since $|[V]_\varphi| \geqslant 2$, we can always choose V so that $\varphi^{-1}(V)$ is in a k-th column S_k with $k \neq 1$. Let S' be the set obtained from S by omitting the states in the k-th column and by taking the pre-image of S_j under φ in any other column, that is,

$$S' = \{(i, j) \mid 1 \leqslant j \leqslant n, j \neq k, \text{ and } i \in \varphi^{-1}(S_j)\}.$$

Since $k \neq 1$ and the first row of S contains at least two elements, the set S' is valid. Since V is non-empty, we have $|S'| < |S|$. Let ψ be a permutation that maps a column j to the column containing $\varphi^{-1}(S_j)$, that is, we have $S_{\psi(j)} = \varphi^{-1}(S_j)$. Let t be the transformation given by $a_{\varphi,\psi}$. Let us show that $S't = S$.

Let $(i, j) \in S'$. Then $i \in \varphi^{-1}(S_j)$, so $\varphi(i) \in S_j$, and we have $(i, j)t = \{(\varphi(i), j), (i, \psi(j))\} \subseteq S$. Hence $S't \subseteq S$.

Now let $(i, j) \in S$. First let $j \neq k$. Then $i \in S_j$, so $\varphi^{-1}(i) \in \varphi^{-1}(S_j)$. Therefore $(\varphi^{-1}(i), j) \in S'$. Since $(i, j) \in (\varphi^{-1}(i), j)t$, we have $(i, j) \in S't$. Now let $j = k$. Then $i \in \varphi^{-1}(V)$ and $S_{\psi^{-1}(k)} = V$. Thus $(i, \psi^{-1}(k)) \in S'$, and we have $(i, k) \in (i, \psi^{-1}(k))t$. Hence $S \subseteq S't$. Our proof is complete. $\qquad\square$

Corollary 13. *Let $1 \leqslant m \leqslant 5$ and $n \geqslant 1$. Then every valid subset can be reached in $\mathcal{D}_{m,n}$.*

Proof. The proof follows by analysis of valid subsets $S \subseteq Q_5 \times Q_n$, with the aid of Corollary 11, Lemmas 8 and 12, and the results from Table 1.

Suppose that there is a valid subset $S \subseteq Q_5 \times Q_n$ that is not reachable; let S be chosen so that n is the smallest number and S is a smallest non-reachable subset of $Q_5 \times Q_n$.

By Corollary 11 and the choice of n, every valid subset $S' \subset Q_{m'} \times Q_{n'}$, where $m' < 5$, or $n' < n$, or $|S'| < |S|$, is reachable. Hence, S has no column containing another column; otherwise, we can apply Lemma 8. Since we have verified the reachability of all valid subsets for $m = 5$ and $n \leqslant 7$ (Table 1), we must have

Table 2. A subset and the equivalence classes of columns under $\varphi = [2,3,1,4,5]$.

	1	2	3	4	5	6	7	8	9
1	o		o	o			o		
2	o	o			o			o	
3		o	o			o			o
4				o	o	o			
5							o	o	o
eq	A	A	A	B	B	B	C	C	C

Table 3. A subset and the equivalence classes of columns under $\varphi = [1,2,3,5,4]$.

	1	2	3	4	5	6	7	8
1	o	o			o			
2	o		o			o		
3	o			o			o	
4		o	o	o				o
5					o	o	o	o
eq	A	B	C	D	B	C	D	E

Table 4. A subset and the equivalence classes of columns under $\varphi = [2,3,4,1,5]$.

	1	2	3	4	5	6	7	8
1		o	o	o	o			
2	o		o	o		o		
3	o	o		o			o	
4	o	o	o					o
5					o	o	o	o
eq	A	A	A	A	B	B	B	B

$n \geqslant 8$ and so S has at least 8 distinct columns. Obviously there is neither an empty nor a full column. If there is a column U with $|U| = 1$ or $|U| = 4$, then by Sperner's theorem if $n > \binom{4}{2} = 6$, then S has a column containing another column; hence S can have only columns U with $|U| = 3$ or $|U| = 2$.

Let C_3 be the number of 3-element columns ($|U| = 3$), and C_2 be the number of 2-element columns ($|U| = 2$). We are searching for possible subsets S that do not have a column containing another column, and with $C_3 + C_2 \geqslant 8$. We consider the following six cases.

(1) Let $C_3 = 0$. If $C_2 = 10$, which implies that S contains all possible 2-element subsets, then under $\varphi = [2,3,4,5,1]$ we have two full and non-trivial equivalence classes. Hence S is reachable from a smaller subset by Lemma 12.

If $C_2 = 9$, then without loss of generality let the missing 2-element subset be $\{4, 5\}$; see Table 2. Under $\varphi = [2, 3, 1, 4, 5]$ we have three full and non-trivial equivalence classes, and S is reachable by Lemma 12. Finally, if $C_2 = 8$, then we have two subcases. If the two missing 2-element subsets have a common element, then without loss of generality let them be $\{2, 3\}$ and $\{4, 5\}$. Under $\varphi = [1, 4, 5, 2, 3]$ we have four full and non-trivial equivalence classes, and S is reachable by Lemma 12. If they have a common element, then without loss of generality let them be $\{3, 4\}$ and $\{4, 5\}$. Under $\varphi = [1, 2, 5, 4, 3]$ we have six full equivalence classes and two of them are non-trivial. Thus S is reachable by Lemma 12.

(2) Let $C_3 = 1$. The only possible subset, up to permutation of columns and rows, is shown in Table 3. It has all columns with two elements that are not contained in the 3-element column. By Lemma 12 with $\varphi = [1, 2, 3, 5, 4]$, it is reachable.

(3) Let $C_3 = 2$. A simple analysis reveals that if the 3-element columns have only one common element, then C_2 is at most 4. If they have two common elements, then C_2 is at most 5. Thus in this case, we have $C_2 + C_3 \leqslant 7$.

(4) Let $C_3 = 3$. Here C_2 is at most 4.

(5) Let $C_3 = 4$. The only possible subset, up to permutation of columns and rows, is shown in Table 4. By Lemma 12 with $\varphi = [2, 3, 4, 1, 5]$, it is reachable.

(6) Let $C_3 \geqslant 5$. These cases are symmetrical to those with $C_3 \leqslant 3$; it is sufficient to consider the complement of S.

Since these cover all the possibilities for set S, this set is reachable. $\square$

2.2   Proof of Distinguishability

The aim of this section is to show that there are regular languages defined over a three-letter alphabet such that the subset automaton of the NFA for their shuffle does not have equivalent states.

To this aim let $\mathcal{A} = (Q, \Sigma, \delta, s, F)$ be an NFA. We say that a state q in Q is *uniquely distinguishable* if there is a word w in Σ^* which is accepted by $\mathcal{A}$ from and only from the state q, that is, if there is a word w such that $\delta(p, w) \in F$ if and only if $p = q$. First, let us prove the following two observations.

Proposition 14. *If each state of an NFA $\mathcal{A}$ is uniquely distinguishable, then the subset automaton of $\mathcal{A}$ does not have equivalent states.*

Proof. Let S and T be two distinct subsets in 2^Q. Then, without loss of generality, there is a state q in Q with $q \in S \setminus T$. Since q is uniquely distinguishable, there is a word w which is accepted by $\mathcal{A}$ from and only from q. Therefore, the subset automaton of $\mathcal{A}$ accepts w from S and it rejects w from T. Hence w distinguishes S and T. $\square$

Proposition 15. *Let a state q of an NFA $\mathcal{A} = (Q, \Sigma, \delta, s, F)$ be uniquely distinguishable. Assume that there is a symbol a in Σ and exactly one state p in Q that goes to q on a, that is, (p, a, q) is a unique in-transition on a going to q. Then the state p is uniquely distinguishable as well.*

Proof. Let w be a word which is accepted by $\mathcal{A}$ from and only from q. The word aw is accepted from p since $q \in \delta(p, a)$ and w is accepted from q. Let $r \neq p$. Then $q \notin \delta(r, a)$ since (p, a, q) is a unique in-transition on a going to q. It follows that the word w is not accepted from any state in $\delta(r, a)$. Thus $\mathcal{A}$ rejects aw from r, so p is uniquely distinguishable. $\square$

Now we can prove the following result.

Theorem 16. *Let $m, n \geqslant 2$. There exist ternary languages K and L with $\kappa(K) = m$ and $\kappa(L) = n$ such that the subset automaton of the NFA accepting $K \sqcup L$ does not have equivalent states.*

Proof. Let m and n be arbitrary but fixed integers with $m, n \geqslant 2$. Let K be accepted by the DFA $\mathcal{K} = (\{1, 2, \ldots, m\}, \{a, b, c\}, \delta_K, 1, \{m\})$, where for each i in $\{1, 2, \ldots, m\}$,

$\delta_K(i, a) = i + 1$ if $i \leqslant m - 1$ and $\delta_K(m, a) = 1$;
$\delta_K(i, b) = 1$;
$\delta_K(1, c) = 2$ and $\delta_K(i, c) = 1$ if $i \geqslant 2$.

Let L be accepted by the DFA $\mathcal{L} = (\{1, 2, \ldots, n\}, \{a, b, c\}, \delta_L, 1, \{n\})$, where for each j in $\{1, 2, \ldots, n\}$,

$\delta_L(j, a) = 1$;
$\delta_L(j, b) = j + 1$ if $j \leqslant n - 1$ and $\delta_L(n, b) = 1$;
$\delta_L(j, c) = n$.

The DFAs $\mathcal{K}$ and $\mathcal{L}$ are shown in Fig. 2.

Construct the NFA $\mathcal{N}$ for $K \sqcup L$ as described in Sect. 1 on page 2. The transitions on a, b, c in $\mathcal{N}$ for $m = 4$ and $n = 5$ are shown in Fig. 3. Notice that each state (i, j) with $2 \leqslant i \leqslant m$ and $2 \leqslant j \leqslant n$ has a unique in-transition on symbol a and this transition goes from state $(i - 1, j)$; see the dashed transitions in Fig. 3 (top-left). Next, each state (m, j) with $2 \leqslant j \leqslant n$ has a unique in-transition on b which goes from $(m, j - 1)$, and each state $(i, 2)$ with $2 \leqslant i \leqslant m$

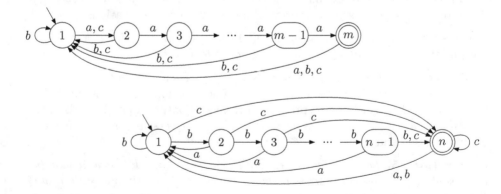

Fig. 2. The DFAs $\mathcal{K}$ and $\mathcal{L}$.

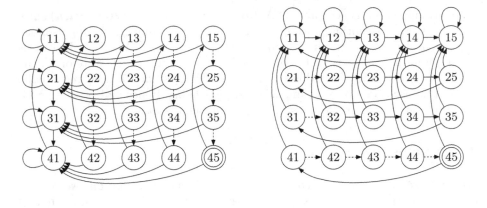

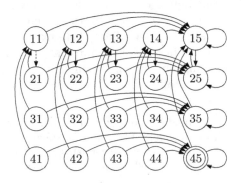

Fig. 3. NFA $\mathcal{N}$ for $m = 4$ and $n = 5$; the transitions on a (top-left), b (top-right), and c (bottom).

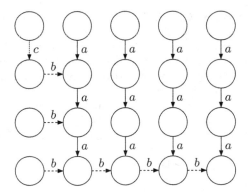

Fig. 4. The subgraph of unique in-transitions in NFA $\mathcal{N}$; $m = 4$ and $n = 5$.

has a unique in-transition on b going from $(i, 1)$; see the dashed transitions in Fig. 3 (top-right). Finally, the state $(2, 1)$ has a unique in-transition on c going from $(1, 1)$; see the dashed transition in Fig. 3 (bottom).

The empty word is accepted by $\mathcal{N}$ from and only from the state (m, n) since this is a unique accepting state of $\mathcal{N}$. Thus (m, n) is uniquely distinguishable. Next, consider the subgraph of unique in-transitions in $\mathcal{N}$. Figure 4 shows this subgraph in the case of $m = 4$ and $n = 5$. Notice that from each state of $\mathcal{N}$, the state (m, n) is reachable in this subgraph. By Proposition 15, used repeatedly, we get that each state of $\mathcal{N}$ is uniquely distinguishable. Hence by Proposition 14, the subset automaton of $\mathcal{N}$ does not have equivalent states.   □

3   Conclusions

We have examined the state complexity of the shuffle operation on two regular languages of state complexities m and n, respectively, and found an upper bound for it. We know that this bound can be reached for any m with $1 \leqslant m \leqslant 5$ and any $n \geqslant 1$, and also for $m = n = 6$. For the remaining values of m and n, however, the problem remains open. Since there exist two languages K and L for which all pairs of states in the subset automaton of the NFA accepting the shuffle $K \sqcup L$ are distinguishable, the main difficulty consists of proving that all valid states in the subset automaton can be reached for the witness languages.

Acknowledgments. We would like to thank an anonymous referee for proposing the notions of a uniquely distinguishable state and of a subgraph of unique in-transitions which allow us to simplify the proof of distinguishability. We are also grateful for his comments and suggestions that helped us improve the presentation of the paper.

References

1. Brzozowski, J., Jirásková, G., Li, B.: Quotient complexity of ideal languages. Theoret. Comput. Sci. **470**, 36–52 (2013)
2. Câmpeanu, C., Salomaa, K., Yu, S.: Tight lower bound for the state complexity of shuffle of regular languages. J. Autom. Lang. Comb. **7**(3), 303–310 (2002)
3. The GAP Group: GAP – Groups, Algorithms, and Programming, Version 4.8.3 (2016). http://www.gap-system.org
4. Okhotin, A.: On the state complexity of scattered substrings and superstrings. Fund. Inform. **99**(3), 325–338 (2010)
5. Sperner, E.: Ein Satz über Untermengen einer endlichen Menge. Math. Z. **27**, 544–548 (1928)
6. Yu, S.: State complexity of regular languages. J. Autom. Lang. Comb. **6**, 221–234 (2001)

MSO-definable Properties of Muller Context-Free Languages Are Decidable

Zoltán Ésik and Szabolcs Iván[⊠]

University of Szeged, Szeged, Hungary
szabivan@inf.u-szeged.hu

Abstract. We show that it is decidable given an MSO-definable property P of countable words and a Muller context-free grammar G, whether every word in the language generated by G satisfies P.

1 Introduction

A word, called 'arrangement' in [9], is an isomorphism type of a countable labeled linear order. Such words form a generalization of the classic notions of finite and ω-words.

Finite automata on ω-words have by now a vast literature, see [13] for a comprehensive treatment. Finite automata acting on well-ordered words longer than ω have been investigated in [1,6,7,16,17], to mention a few references. In the last decade, the theory of automata on well-ordered words has been extended to automata on all countable words, including scattered and dense words. In [2,3,5], both operational and logical characterizations of the class of languages of countable words recognized by finite automata were obtained.

Context-free grammars generating ω-words were introduced in [8] and subsequently studied in [4,12]. Context-free grammars generating arbitrary countable words were defined in [10,11]. Actually, two types of grammars were defined, context-free grammars with Büchi acceptance condition (BCFG), and context-free grammars with Muller acceptance condition (MCFG). These grammars generate the Büchi and the Muller context-free languages of countable words, abbreviated as BCFLs and MCFLs. Every BCFL is clearly an MCFL, but there exists an MCFL of well-ordered words that is not a BCFL, for example the set of all countable well-ordered words over some alphabet. In contrast, the set of all countable words over an alphabet is a BCFL.

In [11], it was shown that it is decidable (in polynomial time) whether a given MCFG generates well-ordered (or scattered) words only. This result was obtained by analysing the structure of a finite graph canonically associated with the grammar. In this note we establish a generic decidability result to the effect that whenever P is some property of countable words definable in monadic second-order logic (MSO), e.g., being well-ordered or scattered, then it is decidable

This research was supported by NKFI grant number K108448.

C. Câmpeanu et al. (Eds.): DCFS 2016, LNCS 9777, pp. 87–97, 2016.
DOI: 10.1007/978-3-319-41114-9_7

whether an MCFG generates only words satisfying P. Of course for such a general setting one cannot hope for efficient algorithms since model checking MSO is nonelementary.

The main idea of the proof is that one can associate with each MCFG a regular tree in which every derivation tree of the grammar can be represented and moreover, the set of all derivation trees is MSO-definable.

2   Notation

Countable Words and Muller Context-Free Grammars. An *alphabet* is a finite nonempty set Σ of symbols, usually called *letters*. A *word* over Σ is a strict linear ordering $(I, <)$ equipped with a *labeling function* $\lambda : I \to \Sigma$. The *empty word*, denoted ε, is the unique word over the empty ordering. It is assumed that no alphabet contains ε. When Σ is an alphabet, Σ_ε stands for $\Sigma \cup \{\varepsilon\}$. An *embedding of words* is a mapping between the respective underlying linear orderings that preserves the order and the labeling; a surjective embedding is an *isomorphism*. We usually identify isomorphic words and denote by $\Sigma^\sharp$ the set of all countable words over the alphabet Σ. As usual, we denote the collections of finite and ω-words over Σ by Σ^* and Σ^ω, respectively. Sometimes we will also use the same notation for infinite sets.

Let $\mathbb{N}$ denote the set of positive integers. When $u \in \mathbb{N}^*$ and $i \in \mathbb{N}$, we usually write ui as $u \cdot i$. A *tree domain* D is a prefix- and left-sibling closed nonempty (but possibly infinite) subset of $\mathbb{N}^*$. Thus, whenever $u \cdot (i + 1)$ is in D, where $i \in \mathbb{N}$, then $u \cdot i$ is also in D, and $u \cdot i \in D$ implies $u \in D$ as well. Elements of a tree domain D are also called the *nodes* of D. When u and $u \cdot i$ are nodes of D, where $u \in \mathbb{N}^*$ and $i \in \mathbb{N}$, then $u \cdot i$ is called a *child* of u. A *descendant* of a node u is a node of the form uv, or $u \cdot v$, where $v \in \mathbb{N}^*$. Nodes of D having no child are the *leaves* of D. The leaves, equipped with the order inherited from the lexicographic ordering $\prec_\ell$ of $\mathbb{N}^*$ (that is, $u \prec_\ell v$ iff $u = wiw'$ and $v = wjw''$ for some $i < j \in \mathbb{N}$, $w, w', w'' \in \mathbb{N}^*$) form the *frontier* of D, denoted $\mathrm{fr}(D)$. An *inner node* of D is a non-leaf node. A *path* of a tree domain D is a (finite or infinite) prefix-closed subset π of D such that each node of π has at most one child in π. Given a tree domain D and some node $u \in D$, the *sub-tree domain* of D rooted at u is the tree domain $D|_u = \{v : uv \in D\}$.

A *tree* over some alphabet Δ, or a Δ-tree for short, is a mapping $t : \mathrm{dom}(t) \to \Delta_\varepsilon$, where $\mathrm{dom}(t)$ is a tree domain, such that inner vertices are mapped to letters in Δ. Notions such as nodes, paths etc. of tree domains are lifted to trees. When π is a path of the tree t, then $\mathrm{labels}(\pi) = \{t(u) : u \in \pi\}$ is the set of labels occurring on π and $\mathrm{inflabels}(\pi) = \bigcap_{u \in \pi} \{t(v) : uv \in \pi\} \subseteq \mathrm{labels}(\pi)$ is the set of labels occurring infinitely often. Given a tree t and some node $u \in \mathrm{dom}(t)$, the *subtree* of t rooted at u is the tree $t|_u$ with domain $\mathrm{dom}(t|_u) = \mathrm{dom}(t)|_u$ and labeling $t|_u(v) = t(uv)$. A tree is *regular* if it has finitely many subtrees.

The *frontier word* $\mathrm{lfr}(t)$ of a tree t is determined by the leaves *not* labeled by ε, which is equipped with the lexicographic ordering of $\mathbb{N}^*$ and the labeling function inherited from t. The *root symbol* of t is $t(\varepsilon)$.

A *Muller context-free grammar* [11], or MCFG for short, is a system $G = (V, \Sigma, R, S, \mathcal{F})$, where V and Σ are the pairwise disjoint alphabets of *nonterminals* and *terminals* respectively, R is the finite set of *productions* of the form $A \to \alpha$ with $A \in V$ and $\alpha \in (\Sigma \cup V)^*$, $S \in V$ is the *start symbol* and $\mathcal{F} \subseteq P(V)$ is the set of *accepting sets*.

A $(V \cup \Sigma)$-tree t is *locally consistent* with the above grammar G if it satisfies the following conditions:

1. The root symbol of t is S.
2. For each inner node u of t there exists a production $A \to X_1 \ldots X_n$ in R with $t(u) = A$, $X_i \in V \cup \Sigma$ such that:
    (a) either $n > 0$, the children of u are exactly $u \cdot 1, \ldots, u \cdot n$ and for each $1 \le i \le n$, $t(u \cdot i) = X_i$;
    (b) or $n = 0$ and u has a single child $u \cdot 1$ labeled ε.
3. The leaves of t are labeled in Σ_ε.

A *derivation tree* of the above grammar G is a locally consistent tree t satisfying the additional condition that for each infinite path π of t, infLabels(π) is an accepting set of G.

The language $L(G) \subseteq \Sigma^\sharp$ *generated* by G is the set of frontier words of derivation trees. A *Muller context-free language*, or MCFL for short, is a language generated by some MCFG.

Example 1. If $G = (\{S, I\}, \{a, b\}, R, S, \{\{I\}\})$, with

$$R = \{S \to a, S \to b, S \to \varepsilon, S \to I, I \to SI\},$$

then $L(G)$ consists of all the well-ordered words over $\{a, b\}$.

Indeed, assume $t_1, t_2, \ldots$ are derivation trees. Then so is the tree t depicted in Fig. 1 with frontier word lfr(t_1)lfr$(t_2) \ldots$. Thus, $L(G)$ contains the empty word (by $S \to \varepsilon$), the words of length 1 (by $S \to a$ and $S \to b$), and is closed under taking "ω-products". Since the least class of order types which contains $\mathbf{0}$, $\mathbf{1}$ and which is closed under ω-sums is the class of all countable ordinals (see e.g. [15]), $L(G)$ contains all the well-ordered words over $\{a, b\}$.

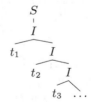

Fig. 1. Derivation tree corresponding to Example 1

For the other direction, assume t is a derivation tree having a frontier word containing an infinite descending chain $u_1 \succ_\ell u_2 \succ_\ell \ldots$. Then let us define the path $v_0, v_1, \ldots$ in t: $v_0 = \varepsilon$ and v_{i+1} is $v_i \cdot 1$ if this node is an ancestor of infinitely many u_j and $v_i \cdot 2$ otherwise (which happens if v_i corresponds to the production $I \to SI$ and the node $v_i \cdot 1$ (which is labeled S) has no descendant of the form u_j at all). Note that for each u_j there exists a unique v_{i_j} such that v_{i_j} is an ancestor of u_j and v_{i_j+1} is not, since the length of the words v_i grows without a bound. Now these nodes v_{i_j} correspond to the production $I \to SI$ and $v_{i_j+1} = v_{i_j} \cdot 1$, so that the successor of v_{i_j} along the path is labeled by S. Hence $v_0, v_1, \ldots$, is a path π in t such that $\mathrm{infLabels}(\pi)$ contains S, which is a contradiction since the only accepting set is $\{I\}$.

MSO on Trees and Words. Let $\mathcal{X}_1$ and $\mathcal{X}_2$ be fixed, countably infinite, disjoint sets of *first-order* and *second-order* variables, respectively. It is assumed that $\mathcal{X}_1$ and $\mathcal{X}_2$ are disjoint from alphabets, they do not contain ε, etc.

Given an alphabet Δ, the set of *monadic second-order*, or MSO-formulas (for *trees* over Δ) is the least set satisfying the following conditions:

1. When $\boldsymbol{x}$ is a first-order variable and $\delta \in \Delta_\varepsilon$ is a symbol, then $\delta(\boldsymbol{x})$ is an MSO-formula.
2. When $\boldsymbol{x}$ and $\boldsymbol{y}$ are first-order variables, then $\boldsymbol{y} = \boldsymbol{x} \cdot 1$ and $\mathrm{sibling}(\boldsymbol{x}, \boldsymbol{y})$ are MSO-formulas.
3. When $\boldsymbol{x}$ is a first-order and $\boldsymbol{X}$ is a second-order variable, then $\boldsymbol{X}(\boldsymbol{x})$, also written $\boldsymbol{x} \in \boldsymbol{X}$ is an MSO-formula.
4. When φ and ψ are MSO-formulas, then so are $(\varphi \vee \psi)$ and $(\neg\varphi)$.
5. When $\boldsymbol{x}$ ($\boldsymbol{X}$, resp.) is a first-order (second-order, resp.) variable and φ is an MSO-formula, then so is $(\exists \boldsymbol{x} \varphi)$ $((\exists \boldsymbol{X} \varphi)$, resp).

We also use the standard abbreviations of $\varphi \wedge \psi = \neg(\neg\varphi \vee \neg\psi)$, $\varphi \to \psi = (\neg\varphi) \vee \psi$, $\forall \boldsymbol{x} \varphi = \neg\exists \boldsymbol{x} \neg\varphi$ etc., and omit some parentheses for the sake of readability. Formulas over Δ are interpreted on Δ-trees in the expected way. A *structure* is a triple (t, Π_1, Π_2) where t is a Δ-tree, $\Pi_1 : \mathcal{X}_1 \to \mathrm{dom}(t)$ assigns a node of t to each first-order variable, and $\Pi_2 : \mathcal{X}_2 \to P(\mathrm{dom}(t))$ assigns a set of nodes of t to each second-order variable. Then, the above structure *satisfies* the formula φ, denoted $(t, \Pi_1, \Pi_2) \models \varphi$, if and only if one of the following conditions holds:

1. $\varphi = \delta(\boldsymbol{x})$ for $\delta \in \Delta_\varepsilon$ and $\boldsymbol{x} \in \mathcal{X}_1$, and $t(\Pi_1(\boldsymbol{x})) = \delta$.
2. $\varphi = (\boldsymbol{y} = \boldsymbol{x} \cdot 1)$ for $\boldsymbol{x}, \boldsymbol{y} \in \mathcal{X}_1$ and $\Pi_1(\boldsymbol{y}) = \Pi_1(\boldsymbol{x}) \cdot 1$.
3. $\varphi = \mathrm{sibling}(\boldsymbol{x}, \boldsymbol{y})$ for $\boldsymbol{x}, \boldsymbol{y} \in \mathcal{X}_1$ and there exist $u \in \mathbb{N}^*$, $i \in \mathbb{N}$ with $\Pi_1(\boldsymbol{x}) = u \cdot i$, $\Pi_1(\boldsymbol{y}) = u \cdot (i+1)$.
4. $\varphi = \boldsymbol{X}(\boldsymbol{x})$ for $\boldsymbol{x} \in \mathcal{X}_1$, $\boldsymbol{X} \in \mathcal{X}_2$ and $\Pi_1(\boldsymbol{x}) \in \Pi_2(\boldsymbol{X})$.
5. $\varphi = (\varphi_1 \vee \varphi_2)$ and (t, Π_1, Π_2) satisfies φ_1 or φ_2 (or both).
6. $\varphi = (\neg\varphi_1)$ and it is not the case that the structure satisfies φ_1.
7. $\varphi = (\exists \boldsymbol{x} \varphi_1)$ and there is a structure (t, Π'_1, Π_2) satisfying φ_1 such that $\Pi_1(\boldsymbol{y}) = \Pi'_1(\boldsymbol{y})$ for each first-order variable $\boldsymbol{y} \neq \boldsymbol{x}$.
8. $\varphi = (\exists \boldsymbol{X} \varphi_1)$ and there is a structure (t, Π_1, Π'_2) satisfying φ_1 such that $\Pi_2(\boldsymbol{Y}) = \Pi'_2(\boldsymbol{Y})$ for each second-order variable $\boldsymbol{Y} \neq \boldsymbol{X}$.

It is clear that satisfaction depends on $\Pi_1(\boldsymbol{x})$ or $\Pi_2(\boldsymbol{X})$ only if the appropriate variable occurs *freely* in the formula (i.e. not within the scope of some $\exists$ quantifier). Hence when φ is a *sentence*, a formula without free variable occurrences, it makes sense to write $t \models \varphi$ instead of $(t, \Pi_1, \Pi_2) \models \varphi$.

In order to ease notation, when Π_1 and Π_2 are clear from the context, we write $\underline{\boldsymbol{x}}$ and $\underline{\boldsymbol{X}}$ for $\Pi_1(\boldsymbol{x})$ and $\Pi_2(\boldsymbol{X})$.

Example 2. One can define the i-th child relation $\boldsymbol{y} = \boldsymbol{x} \cdot i$ for $i \in \mathbb{N}$ inductively as $\exists \boldsymbol{z}(\boldsymbol{z} = \boldsymbol{x} \cdot (i-1) \ \wedge \ \mathsf{sibling}(\boldsymbol{z}, \boldsymbol{y}))$ (which is satisfied by a structure if and only if $\underline{\boldsymbol{y}} = \underline{\boldsymbol{x}} \cdot i$).

Consider the formula

$$\mathsf{child}(\boldsymbol{x}, \boldsymbol{y}) = \exists \boldsymbol{z}(\boldsymbol{z} = \boldsymbol{x} \cdot 1) \ \wedge$$
$$\forall \boldsymbol{X}\Big(\big(\forall \boldsymbol{z}((\boldsymbol{z} = \boldsymbol{x} \cdot 1) \to \boldsymbol{z} \in \boldsymbol{X})\big) \ \wedge$$
$$\forall \boldsymbol{z}\forall \boldsymbol{w}(\boldsymbol{z} \in \boldsymbol{X} \wedge \mathsf{sibling}(\boldsymbol{z}, \boldsymbol{w}) \to \boldsymbol{w} \in \boldsymbol{X}) \to \boldsymbol{y} \in \boldsymbol{X}\Big).$$

Then, $\mathsf{child}(\boldsymbol{x}, \boldsymbol{y})$ holds in the structure iff $\underline{\boldsymbol{x}}$ has a first child and if whenever a set $\underline{\boldsymbol{X}}$ contains the first child of $\underline{\boldsymbol{x}}$ and is closed under taking right siblings, then $\underline{\boldsymbol{X}}$ contains $\underline{\boldsymbol{y}}$ as well, that is, if and only if $\underline{\boldsymbol{y}} = \underline{\boldsymbol{x}} \cdot i$ for some i.

As another example, the formula

$$\exists \boldsymbol{x}(\boldsymbol{x} \in \boldsymbol{X}) \ \wedge \ \forall \boldsymbol{x}\forall \boldsymbol{y}(\boldsymbol{x} \in \boldsymbol{X} \wedge \mathsf{child}(\boldsymbol{y}, \boldsymbol{x}) \to \boldsymbol{y} \in \boldsymbol{X})$$

holds in a structure if $\underline{\boldsymbol{X}}$ is a nonempty, prefix-closed subset of the nodes.

It is well-known [14] that given any *regular* tree t and MSO sentence φ, it is decidable whether $t \models \varphi$ holds.

For countable Σ-words, the syntax and semantics of MSO are slightly changed due to the differing relational structure: the atomic formulas are of the form $a(\boldsymbol{x})$ for $a \in \Sigma$ and $\boldsymbol{x} \in \mathcal{X}_1$ and $\boldsymbol{x} < \boldsymbol{y}$ for $\boldsymbol{x}, \boldsymbol{y} \in \mathcal{X}_1$, interpreted in the expected way. A property P of countable Σ-words is called *MSO-definable* if there exists an MSO sentence φ_P which is satisfied exactly by those Σ-words having property P.

3 Result

Let us fix a Muller context-free grammar $G = (V, \Sigma, R, S, \mathcal{F})$ for this section, with R being disjoint from $V \cup \Sigma$. Without loss of generality we assume that each $A \in V$ is the left-hand side of at least one production. We define the *grammar tree* associated with G as the unique derivation tree $\mathcal{T}$ of the following grammar $G' = (V \cup R, \Sigma, R', S, P(V \cup R))$ with R' consisting of productions of the following form:

1. When $A \to \alpha_1, \ldots, A \to \alpha_k$ are all the productions of G having A on their left side in some fixed ordering of the productions, then $A \to (A \to \alpha_1)(A \to \alpha_2) \ldots (A \to \alpha_k)$ is a production of G' (the right-hand side of this single production is in R^* while the left-hand side is in V).

2. For each production $A \rightarrow X_1 \ldots X_k$, the production $(A \rightarrow X_1 \ldots X_k) \rightarrow X_1 \ldots X_k$. (which is a production of the form $r \rightarrow \alpha$ for some $r \in R$ and $\alpha \in (V \cup \Sigma)^*$) is a production of G'.

Note that each element of $V \cup R$ is the left-hand side of exactly one production in R', hence there exists exactly one locally consistent tree of G'. Also, since the acceptance condition is $P(V \cup R)$, this tree is a valid derivation tree of the grammar, thus $\mathcal{T}$ is well-defined and has at most $|V| + |R| + |\Sigma_\varepsilon|$ subtrees up to isomorphism. Hence it is a regular tree.

Example 3. For the MCFG of Example 1, this tree $\mathcal{T}$ is depicted in Fig. 2.

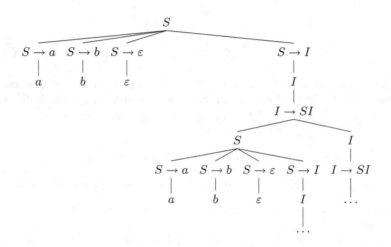

Fig. 2. Grammar tree of the MCFG of Example 1

Moreover, each locally consistent tree t of G can be *embedded* into $\mathcal{T}$ in the following sense: there exists a mapping $h_t : \mathrm{dom}(t) \rightarrow \mathrm{dom}(\mathcal{T})$ with $h_t(\varepsilon) = \varepsilon$ and $t(u) = \mathcal{T}(h_t(u))$ for each $u \in \mathrm{dom}(t)$, moreover $h_t(u \cdot i)$ is a descendant (in particular, a grandchild) of $h_t(u)$ in $\mathcal{T}$ for each $u \cdot i \in \mathrm{dom}(t)$, moreover, when u and v are siblings in t, then so are $h_t(u)$ and $h_t(v)$ in $\mathcal{T}$. Indeed, assume $u \cdot i \in \mathrm{dom}(t)$ and that $u' = h_t(u)$ is already defined. Then since u is an inner node of t, we have $t(u) = A \in V$. By $\mathcal{T}(u') = t(u) = A$, each production $r = A \rightarrow \alpha$ occurs as $\mathcal{T}(u' \cdot k_r)$ for some $k_r \in \mathbb{N}$. In particular, let $r = A \rightarrow X_1 \ldots X_n$ be the production corresponding to u, so that u has n children and $t(u \cdot j) = X_j$ for each $j = 1, \ldots, n$ (subsuming the case when $n = 0$ as $t(u \cdot 1) = \varepsilon$). Then, we define $h_t(u \cdot j)$ as $u' \cdot k_r \cdot j$.

We call a prefix-closed nonempty set $T \subseteq \mathrm{dom}(\mathcal{T})$ *derivation-like* iff it satisfies the following conditions:

1. For each $u \in T$ with $\mathcal{T}(u) \in R$, each child of u is in T.
2. For each $u \in T$ with $\mathcal{T}(u) \in V$, exactly one child of u is in T.

It is clear that derivation-like subsets of $\mathrm{dom}(\mathcal{T})$ are in one-to-one correspondence with the locally consistent trees of G: with any locally consistent tree t of G we associate the derivation-like set $T \subseteq \mathrm{dom}(\mathcal{T})$ which is the closure of $\mathrm{im}(h_t)$ with respect to the prefix relation. Given a derivation-like set T, the corresponding locally consistent tree is denoted t.

Example 4. Figure 3 shows a (part of a) derivation tree t of the grammar of Example 1 and the corresponding derivation-like subset T of $\mathrm{dom}(\mathcal{T})$ (as nodes in boldface).

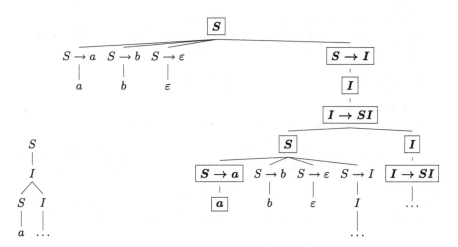

Fig. 3. A part of a derivation tree t of Example 1 and the corresponding derivation-like subset T of $\mathcal{T}$

Proposition 1. *There is an MSO formula $d(X)$ with the free variable $X \in \mathcal{X}_2$ such that $(\mathcal{T}, \Pi_1, \Pi_2) \models d(X)$ if and only if $\underline{X}$ is a derivation-like set.*

  (In short, it is MSO-definable whether some set $\underline{X}$ is derivation-like.)

Proof. We can define $d(X)$ as the conjunction of the formulas stating that $\underline{X}$ is nonempty and prefix-closed (see Example 2), the formula

$$\forall x \Big((x \in X \wedge \bigvee_{r \in R} r(x)) \;\rightarrow\; \forall y (\mathsf{child}(x, y) \rightarrow y \in X) \Big)$$

stating that all the children of the nodes labeled by productions are members of $\underline{X}$, and the formula

$$\forall x \Big((x \in X \wedge \bigvee_{A \in V} A(x)) \rightarrow \exists y \big(\mathsf{child}(x, y) \wedge \forall z (\mathsf{child}(x, z) \wedge z \in X \leftrightarrow z = y) \big) \Big)$$

stating that exactly one child of the nodes labeled by nonterminals is in $\underline{X}$. $\square$

Also, from a derivation-like set T determining a locally consistent tree t of G, one can select a subset of nodes corresponding to a path of t:

Proposition 2. *There exists an MSO-formula $p(\boldsymbol{X}, \boldsymbol{Y})$ with the free second-order variables $\boldsymbol{X}, \boldsymbol{Y}$ such that $(T, \Pi_1, \Pi_2) \models p(\boldsymbol{X}, \boldsymbol{Y})$ if and only if $\underline{\boldsymbol{X}} = T$ is a derivation-like subset of T with corresponding locally consistent tree t and $\underline{\boldsymbol{Y}} = h_t(\pi)^1$ for some infinite path π of t.*

Proof. $p(\boldsymbol{X}, \boldsymbol{Y})$ expresses the following:

  (i) $\underline{\boldsymbol{X}}$ is a derivation-like subset of T, i.e. $d(\boldsymbol{X})$ holds,
  (ii) $\underline{\boldsymbol{Y}} \subseteq \underline{\boldsymbol{X}}$,
  (iii) each of the nodes of $\underline{\boldsymbol{Y}}$ is labeled by some member of V,
  (iv) whenever v is a grandparent of some node $u \in \underline{\boldsymbol{Y}}$, then $v \in \underline{\boldsymbol{Y}}$ as well, and
  (v) each $u \in \underline{\boldsymbol{Y}}$ has exactly one grandchild in $\underline{\boldsymbol{Y}}$.

These properties can clearly be defined in MSO. □

Proposition 3. *There is an MSO formula $d'(\boldsymbol{X})$ expressing that $\underline{\boldsymbol{X}}$ is a derivation-like set corresponding to an actual derivation tree of G.*

Proof. The descendant relation $x \preceq y$ can be defined in MSO by a formula expressing that whenever $\underline{\boldsymbol{Y}}$ is a prefix-closed set containing $\underline{y}$, then $\underline{\boldsymbol{Y}}$ contains $\underline{x}$ as well.

Then, for each $F \in \mathcal{F}$ we can construct a formula $m_F(\boldsymbol{Y})$ stating that if $\underline{\boldsymbol{Y}} = h_t(\pi)$ for some path π of some locally consistent tree t of G, then infLabels$(\pi) = F$:

$$m_F(\boldsymbol{Y}) = \bigwedge_{A \in F} i_A(\boldsymbol{Y}) \wedge \bigwedge_{A \notin F} \neg i_A(\boldsymbol{Y})$$

where $i_A(\boldsymbol{Y})$ is the formula

$$\forall \boldsymbol{x}(\boldsymbol{x} \in \boldsymbol{Y} \rightarrow \exists \boldsymbol{y}(\boldsymbol{x} \preceq \boldsymbol{y} \wedge \boldsymbol{y} \in \boldsymbol{Y} \wedge A(\boldsymbol{y})))$$

stating that A occurs infinitely many times on the path given by $\underline{\boldsymbol{Y}}$.

Now, we can define $d'(\boldsymbol{X})$ as

$$d(\boldsymbol{X}) \wedge \forall \boldsymbol{Y}(p(\boldsymbol{X}, \boldsymbol{Y}) \rightarrow \bigvee_{F \in \mathcal{F}} m_F(\boldsymbol{Y}))$$

expressing that $\underline{\boldsymbol{X}}$ is a derivation-like set corresponding to some locally consistent tree t of G such that any infinite path π of t satisfies the Muller acceptance condition. □

Now a set $\underline{\boldsymbol{Y}} \subseteq \mathrm{dom}(T)$ corresponds to a frontier word of some derivation tree of G (i.e. belongs to $L(G)$) if and only if there exists some $\underline{\boldsymbol{X}} \subseteq \mathrm{dom}(T)$ satisfying

---

[1] Here, $h_t(\pi)$ denotes the set of images of the nodes of π with respect to the embedding h_t.

$d'(\boldsymbol{X})$ such that a node $v \in \mathrm{dom}(\mathcal{T})$ is in $\underline{\boldsymbol{Y}}$ if and only if $v \in \underline{\boldsymbol{X}}$ and is a Σ-labeled node of $\mathcal{T}$, which is an MSO-definable property:

$$\exists \boldsymbol{X} \big(d'(\boldsymbol{X}) \wedge \forall y (y \in \boldsymbol{Y} \leftrightarrow (y \in \boldsymbol{X} \wedge \bigvee_{a \in \Sigma} a(y)))\big).$$

Moreover, the lexicographic ordering on these leaves can also be defined in MSO as $u \prec_\ell v$ iff there exists some common ancestor w of u and v such that $w \cdot i$ and $w \cdot j$ are respectively ancestors of u and v for some $i < j$:

$$\boldsymbol{x} \prec_\ell \boldsymbol{y} \;=\; \exists z_1 \exists z_2 \big(z_1 \preceq \boldsymbol{x} \wedge z_2 \preceq \boldsymbol{y} \wedge \mathsf{sibling}^+(z_1, z_2)\big)$$

where $\mathsf{sibling}^+(\boldsymbol{x}, \boldsymbol{y})$ is the transitive closure of sibling:

$$\boldsymbol{x} \neq \boldsymbol{y} \;\wedge\; \forall \boldsymbol{X}\big(\boldsymbol{x} \in \boldsymbol{X} \wedge \forall z_1 \forall z_2 (z_1 \in \boldsymbol{X} \wedge \mathsf{sibling}(z_1, z_2) \to z_2 \in \boldsymbol{X})\big) \to \boldsymbol{y} \in \boldsymbol{X}$$

Hence we have shown:

Proposition 4. *For any MCFG G, there exists an effectively constructible MSO formula $w(\boldsymbol{Y})$ such that $(\mathcal{T}, \Pi_1, \Pi_2) \models w(\boldsymbol{Y})$ if and only $\underline{\boldsymbol{Y}}$ is the set of Σ-labeled leaves of some derivation-like subset of $\mathcal{T}$ corresponding to a derivation tree of G.*

As a corollary, we obtain the main result of this note.

Theorem 1. *It is decidable for a given MCFL L and an MSO-definable property φ of words whether every member of L satisfies φ.*

Proof. The question can be reduced to checking whether $\mathcal{T}$ satisfies the formula $\forall \boldsymbol{Y}(w(\boldsymbol{Y}) \to \varphi(\boldsymbol{Y}))$ where $\varphi(\boldsymbol{Y})$ is obtained from φ by replacing all the subformulas of the form $\exists \boldsymbol{x} \varphi'$ and $\exists \boldsymbol{X} \varphi'$ respectively to $\exists \boldsymbol{x}(\boldsymbol{x} \in \boldsymbol{Y} \wedge \varphi')$ and $\exists \boldsymbol{X}(\forall \boldsymbol{x}(\boldsymbol{x} \in \boldsymbol{X} \to \boldsymbol{x} \in \boldsymbol{Y}) \wedge \varphi')$ and substituting the formula defining the lexicographic ordering $\boldsymbol{x} \prec_\ell \boldsymbol{y}$ in place of the atomic formulas $\boldsymbol{x} < \boldsymbol{y}$. Since $\mathcal{T}$ is regular, model checking the resulting formula on $\mathcal{T}$ is decidable. $\square$

(We remark that thus it is also decidable whether *there exists* a word in L satisfying an MSO formula φ since such a word exists if and only if *not* all members of L satisfy $\neg \varphi$.)

In particular, our former decidability results (without complexity bounds) regarding whether an MCFG generates scattered (or well-ordered) words are corollaries of this general theorem. However, since model-checking MSO formulas on regular trees has a high complexity in general ($\mathrm{tower}(n)$ when n is the alternation depth of the second-order quantifiers), no polytime decision procedures follow from the present theorem. Nevertheless several interesting properties are decidable in polynomial time, including whether every word generated by an MFCG is scattered or well-ordered, cf. [11].

As another example, let $\Sigma = \{a, b\}$ be an alphabet. The following formula $\mathsf{segment}(\boldsymbol{X})$ expresses that $\underline{\boldsymbol{X}}$ is a nonempty interval, or *segment* of a given word:

$$(\exists \boldsymbol{x} \; \boldsymbol{x} \in \boldsymbol{X}) \wedge \big(\forall \boldsymbol{x} \forall \boldsymbol{y} \forall \boldsymbol{z}(\boldsymbol{x} < \boldsymbol{y} \wedge \boldsymbol{y} < \boldsymbol{z} \wedge \boldsymbol{x} \in \boldsymbol{X} \wedge \boldsymbol{z} \in \boldsymbol{X}) \to \boldsymbol{y} \in \boldsymbol{X}\big).$$

The following formula $\mathsf{dense}(X)$ expresses that $\underline{X}$ is a dense subset (containing at least two elements) of the word:

$$\exists x \exists y(x < y \wedge x \in X \wedge y \in X) \wedge$$
$$\forall x \forall y\Big(x < y \wedge x \in X \wedge y \in X \rightarrow \exists z(x < z \wedge z < y \wedge z \in X)\Big).$$

Thus, the property "there exists a dense segment X of the word such that for all $x < y$ in X there exists z, z' with $x < z, z' < y$ and z is labeled by a, z' is labeled by b" is also expressible in MSO as

$$\exists X \Big(\;\mathsf{dense}(X) \wedge \mathsf{segment}(X)\;\wedge$$
$$\forall x \forall y\big(x < y \wedge x \in X \wedge y \in X$$
$$\rightarrow \exists z \exists z'(x < z \wedge z < y \wedge x < z' \wedge z' < y \wedge a(z) \wedge b(z'))\big)\Big)$$

In other words, $w \in \{a, b\}^{\sharp}$ satisfies the above formula iff $w = u\{a, b\}^{\eta}v$ for some words $u, v \in \{a, b\}^{\sharp}$ where $\{a, b\}^{\eta}$ is the so-called *shuffle* of a and b. Thus, it is also decidable for a MCFL L whether every word in L is of the form $u\{a, b\}^{\eta}v$.

Another expressible property is that whether a word is the *shuffle product* of, say, a dense word and a scattered word consisting only of a's, that is, whether the underlying linear order can be partitioned into two subsets such that the two subwords determined by the partitions satisfy the appropriate property:

$$\exists X \Big(\mathsf{dense}(X) \wedge \forall x(x \notin X \rightarrow a(x)) \wedge \neg \exists Y \big(\forall x(x \in Y \rightarrow x \notin X) \wedge \mathsf{dense}(Y)\big)\Big).$$

Thus, it is also decidable for a given MCFL L whether every member of L is a shuffle product of a dense word and a scattered one consisting only of a's.

4   Conclusion

We have proved that there is an algorithm to decide for a Muller context-free language L generated by an MCFG and an MSO-definable property P of words whether every word in L has property P. We obtained this result by assigning a regular tree t to an MCFG such that the derivation trees of the grammar have an MSO-interpretation in t. We then used the fact that the MSO-theory of a regular tree is decidable.

There is an alternative method. First, we can prove that the MCFLs are exactly the frontier languages of the tree languages recognizable by Muller tree automata. This is similar to the well-known fact that ordinary context-free languages are the frontier languages of the languages of finite trees recognizable by finite tree automata. Also, there is an algorithm to decide, for a Muller tree automaton and an MSO-definable property of trees whether every tree in the language L recognized by the automaton has property P. This follows using the fact that every Muller tree automaton can be converted to an MSO-formula φ, and if P is definable by the formula ψ, then it holds that every tree in L satisfies P iff there is no tree satisfying $\varphi \wedge \neg\psi$, which is decidable.

References

1. Bedon, N.: Finite automata and ordinals. Theor. Comput. Sci. **156**(1–2), 119–144 (1996)
2. Bedon, N., Bès, A., Carton, O., Rispal, C.: Logic and rational languages of words indexed by linear orderings. In: Hirsch, E.A., Razborov, A.A., Semenov, A., Slissenko, A. (eds.) CSR 2008. LNCS, vol. 5010, pp. 76–85. Springer, Heidelberg (2008)
3. Bès, A., Carton, O.: A kleene theorem for languages of words indexed by linear orderings. In: De Felice, C., Restivo, A. (eds.) DLT 2005. LNCS, vol. 3572, pp. 158–167. Springer, Heidelberg (2005)
4. Boasson, L.: Context-free sets of infinite words. In: Weihrauch, K. (ed.) Theoretical Computer Science 4th GI Conference. LNCS, vol. 67, pp. 1–9. Springer, Heidelberg (1979)
5. Bruyère, V., Carton, O.: Automata on linear orderings. J. Comput. Syst. Sci. **73**(1), 1–24 (2007)
6. Richard Büchi, J.: The monadic second order theory of ω_1. In: Müller, G.H., Siefkes, D. (eds.) Decidable Theories II. Lecture Notes in Mathematics, vol. 328, pp. 1–127. Springer, Heidelberg (1973)
7. Choueka, Y.: Finite automata, definable sets, and regular expressions over ω^n-tapes. J. Comput. Syst. Sci. **17**(1), 81–97 (1978)
8. Cohen, R.S., Gold, A.Y.: Theory of ω-languages, parts one and two. J. Comput. Syst. Sci. **15**, 169–208 (1977)
9. Courcelle, B.: Frontiers of infinite trees. RAIRO - Theor. Inf. Appl. - Informatique Théorique et Applications **12**(4), 319–337 (1978)
10. Ésik, Z., Iván, S.: Büchi context-free languages. Theor. Comput. Sci. **412**(8–10), 805–821 (2011)
11. Ésik, Z., Iván, S.: On Müller context free grammars. Theor. Comput. Sci. **416**, 17–32 (2012)
12. Nivat, M.: Sur les ensembles de mots infinis engendrés par une grammaire algébrique. RAIRO - Theor. Inf. Appl. - Informatique Théorique et Applications **12**(3), 259–278 (1978)
13. Perrin, D., Pin, J.É.: Infinite Words: Automata, Semigroups, Logic and Games. Pure and Applied Mathematics. Elsevier Science (2004)
14. Rabin, M.O.: Decidability of second order theories and automata on infinite trees. Trans. AMS **141**, 1–35 (1969)
15. Rosenstein, J.G.: Linear Orderings. Pure and Applied Mathematics. Academic Press, London (1982)
16. Wojciechowski, J.: Classes of transfinite sequences accepted by finite automata. Fundamenta Informaticae **7**(2), 191–223 (1984)
17. Wojciechowski, J.: Finite automata on transfinite sequences and regular expressions. Fundamenta Informaticae **8**(3–4), 379–396 (1985)

Contextual Array Grammars with Matrix and Regular Control

Henning Fernau[1]([✉]), Rudolf Freund[2], Rani Siromoney[3],
and K.G. Subramanian[4]

[1] FB 4 – Abteilung Informatikwissenschaften, Universität Trier,
54296 Trier, Germany
fernau@uni-trier.de

[2] Institut für Computersprachen, Technische Universität Wien, 1040 Wien, Austria
rudi@emcc.at

[3] Chennai Mathematical Institute, Kelambakkam 603103, India
siromoney@cmi.ac.in

[4] Department of Mathematics and Computer Science, Faculty of Science,
Liverpool Hope University, Liverpool L16 9JD, UK
kgsmani1948@yahoo.com

Abstract. We investigate the computational power of d-dimensional contextual array grammars with matrix control and regular control languages. For $d \geq 2$, d-dimensional contextual array grammars are less powerful than matrix contextual array grammars, which themselves are less powerful than contextual array grammars with regular control languages. Yet in the 1-dimensional case, for a one-letter alphabet, the family of 1-dimensional array languages generated by contextual array grammars with regular control languages coincides with the family of regular 1-dimensional array languages, whereas for alphabets with more than one letter, we obtain the array images of the linear languages.

1 Introduction

Contextual string grammars were introduced by Solomon Marcus [14] with motivations arising from descriptive linguistics. A contextual string grammar consists of a finite set of strings (*axioms*) and a finite set of productions, which are pairs (s, c) where s is a string, the *selector,* and c is the *context,* i. e., a pair of strings, $c = (u, v)$, over the alphabet under consideration. Starting from an axiom, contexts iteratively are added as is indicated by the productions, which yields new strings. In contrast to usual sequential string grammars in the Chomsky hierarchy (e.g., see [20]), these contextual string grammars are pure grammars where new strings are not obtained by rewriting, but by adjoining strings. Several classes of contextual grammars have been introduced and investigated, e.g., see [3,17] for surveys on the area.

The idea of contextual productions then was also introduced for multi-dimensional array grammars, for instance, to carry over ideas from formal

© IFIP International Federation for Information Processing 2016
Published by Springer International Publishing Switzerland 2016. All Rights Reserved
C. Câmpeanu et al. (Eds.): DCFS 2016, LNCS 9777, pp. 98–110, 2016.
DOI: 10.1007/978-3-319-41114-9_8

languages to the processing of digital images. In the area of two-dimensional picture languages, e.g., see [12,16,18,19], different kinds of array grammars, both isometric and non-isometric ones, have been proposed, motivated by many applications such as character recognition (also confer [4]), cluster analysis of patterns, and so on. Isometric contextual array grammars were introduced in [11].

Regulated rewriting with different control mechanisms has been studied extensively especially for string grammars (e.g., see [2]), for example, grammars with control languages and matrix grammars, but then also for array grammars, e.g., see [9]. Non-isometric contextual array grammars (with regulation) were considered in [7,8,13].

In this paper we consider matrix contextual array grammars and contextual array grammars with regular control and examine their generative power. In the 1-dimensional case, we obtain special results: the family of 1-dimensional array languages generated by contextual array grammars with regular control languages coincides with the family of regular 1-dimensional array languages over unary alphabets and with array images of the linear languages over alphabets with more than one letter; already for binary alphabets, regular control is strictly more powerful than matrix control, a phenomenon rarely observed in regulated rewriting (confer [10]).

2   Definitions

For notions and notations as well as results related to formal language theory we refer to books like [2]. The families of λ-free (λ denotes the empty string) regular string languages (over a k-letter alphabet) is denoted by $\mathcal{L}\left(REG\right)$ ($\mathcal{L}\left(REG^k\right)$). For the definitions and notations for arrays and sequential array grammars we refer to [9,18,22].

Let $\mathbb{Z}$ be the set of integers and $\mathbb{N}$ be the set of positive integers. Let $d \in \mathbb{N}$. A d-*dimensional array* $\mathcal{A}$ over the alphabet V is a mapping $\mathcal{A} : \mathbb{Z}^d \to V \cup \{\#\}$ where $shape\left(\mathcal{A}\right) = \left\{v \in \mathbb{Z}^d \mid \mathcal{A}\left(v\right) \neq \#\right\}$ is finite and $\# \notin V$ is called the *blank symbol*. We usually write $\mathcal{A} = \{(v, \mathcal{A}\left(v\right)) \mid v \in shape\left(\mathcal{A}\right)\}$. The set of all d-dimensional arrays over V is denoted by V^{*d}. The *empty array* Λ_d in V^{*d} satisfies $shape(\Lambda_d) = \emptyset$. Moreover, we define $V^{+d} = V^{*d} \setminus \{\Lambda_d\}$.

Let $v \in \mathbb{Z}^d$. Then the *(linear) translation* $\tau_v : \mathbb{Z}^d \to \mathbb{Z}^d$ is defined by $\tau_v\left(w\right) = w + v$ for all $w \in \mathbb{Z}^d$, and for any array $\mathcal{A} \in V^{*d}$ we define $\tau_v\left(\mathcal{A}\right)$, the corresponding d-dimensional array translated by v, by $\left(\tau_v(\mathcal{A})\right)\left(w\right) = \mathcal{A}\left(w - v\right)$ for all $w \in \mathbb{Z}^d$. The vector $(0, ..., 0) \in \mathbb{Z}^d$ is denoted by Ω_d.

Usually (see [18]) arrays are regarded as equivalence classes of arrays with respect to linear translations. The equivalence class $[\mathcal{A}]$ of an array $\mathcal{A} \in V^{*d}$ satisfies $[\mathcal{A}] = \left\{\mathcal{B} \in V^{*d} \mid \mathcal{B} = \tau_v\left(\mathcal{A}\right) \text{ for some } v \in \mathbb{Z}^d\right\}$. The set of all equivalence classes of d-dimensional arrays over V with respect to linear translations is denoted by $\left[V^{*d}\right]$, and this bracket notation carries over to classes of array languages, as well.

As many results for d-dimensional arrays for a specific d can be taken over immediately for higher dimensions, we introduce special notions:

Let $n, m \in \mathbb{N}$ with $n \leq m$. For $n < m$, *the natural embedding* $i_{n,m} : \mathbb{Z}^n \to \mathbb{Z}^m$ is defined by $i_{n,m}(v) = (v, \Omega_{m-n})$ for all $v \in \mathbb{Z}^n$; for $n = m$ we define $i_{n,n} : \mathbb{Z}^n \to \mathbb{Z}^n$ by $i_{n,n}(v) = v$ for all $v \in \mathbb{Z}^n$. To an n-dimensional array $\mathcal{A} \in V^{+n}$ with $\mathcal{A} = \{(v, \mathcal{A}(v)) \mid v \in shape(\mathcal{A})\}$ we assign the m-dimensional array $i_{n,m}(\mathcal{A}) = \{(i_{n,m}(v), \mathcal{A}(v)) \mid v \in shape(\mathcal{A})\}$.

We can use the well-known graph-theoretic notion of a connected graph to define connected arrays. Let W be a non-empty finite subset of $\mathbb{Z}^d$. We associate a graph $g(W)$ to W with vertex set W and an edge between $v, w \in W$ if and only if $\|v - w\| = 1$, where the norm $\|u\|$ of a vector $u \in \mathbb{Z}^d$, $u = (u(1), ..., u(d))$, is defined by $\|u\| = \max\{|u(i)| \mid 1 \leq i \leq d\}$. Then W is said to be *connected* if $g(W)$ is connected. There is a natural bijection between the (equivalence classes of) 1-dimensional connected arrays and strings: for any equivalence class of 1-dimensional arrays $\mathcal{A} = [\{((i-1), a_i) \mid 1 \leq i \leq n\}]$ we define its *string image* as $str(\mathcal{A}) = a_1 \ldots a_n$; the string $w = a_1 \ldots a_n$ can be interpreted as the array $arr(w) = \{\{((i-1), a_i)\} \mid 1 \leq i \leq n\}$. In the standard way, these notions are extended from strings and arrays to sets of strings and arrays.

Example 1. Consider the language L_1 of connected 2-dimensional arrays

$$L_1 = \left\{ \left\{ ((0,i), a) \mid 0 \leq i \leq n \right\} \cup \left\{ ((j,0), a) \mid 1 \leq j \leq m \right\} \;\middle|\; n, m \in \mathbb{N} \right\}.$$

$\begin{array}{l} a \\ a \\ a \\ a\ a\ a\ a\ a \end{array}$ An example of these L-shaped arrays (for $n = 3$ and $m = 4$) from $[L_1]$ can be depicted as shown on the left. Observe that both arms of these arrays can have arbitrary lengths.   □

Definition 1. *A regular d-dimensional array grammar is specified as* $G = (d, N, T, \#, P, \{(v_S, S)\})$ *where N is the alphabet of* non-terminal symbols, *T is the alphabet of* terminal symbols, *$N \cap T = \emptyset$, $\# \notin N \cup T$; P is a finite non-empty set of regular d-dimensional array productions over $N \cup T$, as well as $v_S \in \mathbb{Z}^d$ and $S \in N$ is the start symbol. A regular d-dimensional array production either is of the form $A \to b$, $A \in N$, $b \in T$, or $Av\# \to bC$, $A, C \in N$, $b \in T$, $v \in \mathbb{Z}^d$ with $\|v\| = 1$. The application of $A \to b$ means replacing A by b in a given array. $Av\# \to bC$ can be applied if in the underlying array we find a position u occupied by A and a blank symbol at position $u + v$; A then is replaced by b, and $\#$ by C. The array language generated by G is the set of all d-dimensional arrays derivable from the initial array $\{(v_S, S)\}$. The family of Λ-free d-dimensional array languages (of equivalence classes) of arrays over a k-letter alphabet generated by regular d-dimensional array grammars is denoted by $\mathcal{L}(d\text{-}REGA^k)$ $([\mathcal{L}(d\text{-}REGA^k)])$. For arbitrary alphabets, we omit the superscript k.*

The following results for 1-dimensional array languages are folklore:

Theorem 1. *For all $k \geq 1$, $[\mathcal{L}(1\text{-}REGA^k)] = [arr(\mathcal{L}(REG^k))]$ and $str([\mathcal{L}(1\text{-}REGA^k)]) = \mathcal{L}(REG^k)$.*

Let us mention the close similarities of the work of 1-dimensional regular array grammars and Lindenmayer systems with apical growth [21]. Another similar development can be found within Watson-Crick systems [15].

3   Contextual Array Grammars

We now turn our attention to the main variants of contextual array grammars considered in this paper.

Definition 2. *A d-dimensional contextual array grammar $(d \in \mathbb{N})$ is a construct $G = (d, V, \#, P, A)$ where V is an alphabet not containing the blank symbol $\#$, A is a finite set of axioms, i. e., of d-dimensional arrays in V^{+d}, and P is a finite set of rules of the form $(U_\alpha, \alpha, U_\beta, \beta)$ where*

(i) $U_\alpha, U_\beta \subseteq \mathbb{Z}^d$, $U_\alpha \cap U_\beta = \emptyset$, *and U_α, U_β are finite and non-empty;*
(ii) $\alpha : U_\alpha \to V$ *and* $\beta : U_\beta \to V$.

(U_α, α) corresponds with the selector and (U_β, β) with the context of the production $(U_\alpha, \alpha, U_\beta, \beta)$; U_α is called the selector area, and U_β is the context area. As the sets U_α and U_β are uniquely determined by α and β, we will also represent $(U_\alpha, \alpha, U_\beta, \beta)$ by (α, β) only.

For $\mathcal{C}_1, \mathcal{C}_2 \in V^{+d}$ we say that $\mathcal{C}_2$ is directly derivable from $\mathcal{C}_1$ by the contextual array production $p \in P$, $p = (U_\alpha, \alpha, U_\beta, \beta)$ (we write $\mathcal{C}_1 \Longrightarrow_p \mathcal{C}_2$), if there exists a vector $v \in \mathbb{Z}^d$ such that

- $\mathcal{C}_1(w) = \mathcal{C}_2(w) = \alpha(\tau_{-v}(w))$ *for all $w \in \tau_v(U_\alpha)$,*
- $\mathcal{C}_1(w) = \#$ *for all $w \in \tau_v(U_\beta)$,*
- $\mathcal{C}_2(w) = \beta(\tau_{-v}(w))$ *for all $w \in \tau_v(U_\beta)$,*
- $\mathcal{C}_1(w) = \mathcal{C}_2(w)$ *for all $w \in \mathbb{Z}^d \setminus \tau_v(U_\alpha \cup U_\beta)$.*

Hence, if in $\mathcal{C}_1$ we find a subpattern that corresponds with the selector α and only blank symbols at the places corresponding with β, we can add the context β thus obtaining $\mathcal{C}_2$. For every $\mathcal{B}_1, \mathcal{B}_2 \in [V^{+d}]$ we say that $\mathcal{B}_2$ is directly derivable from $\mathcal{B}_1$ by the contextual array production $p \in P$, $p = (U_\alpha, \alpha, U_\beta, \beta)$, denoted $\mathcal{B}_1 \Longrightarrow_p \mathcal{B}_2$, if and only if $\mathcal{C}_1 \Longrightarrow_p \mathcal{C}_2$ for some $\mathcal{C}_1 \in \mathcal{B}_1$ and $\mathcal{C}_2 \in \mathcal{B}_2$. $\mathcal{C}_1 \Longrightarrow_G \mathcal{C}_2$ $(\mathcal{B}_1 \Longrightarrow_G \mathcal{B}_2)$ means that $\mathcal{C}_1 \Longrightarrow_p \mathcal{C}_2$ $(\mathcal{B}_1 \Longrightarrow_p \mathcal{B}_2)$ for some $p \in P$.

The array language generated by G is defined as

$$L(G) = \{ \mathcal{C} \in V^{+d} \mid \mathcal{A} \Longrightarrow_G^* \mathcal{C} \text{ for some } \mathcal{A} \in A \}.$$

The special type of d-dimensional contextual array grammars where axioms are connected and rule applications preserve connectedness is denoted by d-ContA, the corresponding family of d-dimensional array languages by $\mathcal{L}(d\text{-}ContA)$; by $\mathcal{L}(d\text{-}ContA^k)$ we denote the corresponding family of d -dimensional array languages over a k-letter alphabet.

Remark 1. As we mostly are interested in (families of) equivalence classes of arrays, a d-dimensional contextual array grammar $[G]$ for generating $[L]$ for $L \in \mathcal{L}(d\text{-}ContA)$ being generated by a d-dimensional contextual array grammar $G = (d, V, \#, P, A)$ with $A = \{\mathcal{A}_i \mid 1 \le i \le n\}$ will be specified by writing $[G] = (d, V, \#, P, A')$ where $A' = \{\mathcal{A}'_i \mid 1 \le i \le n\}$ such that $\mathcal{A}'_i \in [\mathcal{A}_i]$, $1 \le i \le n$, which means specifying an axiom $\mathcal{A}_i$ by one array from $[\mathcal{A}_i]$.

Example 2. Any finite d-dimensional array language of connected arrays $L \subset T^{+d}$ is in $\mathcal{L}\,(d\text{-}ContA)$ as $L = L\,(G_L)$ where $G_L = (d, T, \#, \emptyset, L)$. $\qquad\square$

Example 3. We now show how the language L_1 from Example 1 can be generated by the contextual array grammar G_1, i.e., $L_1 \in \mathcal{L}\,(2\text{-}ContA^1)$: $G_1 = (2, \{a\}, \#, P_1, \{\mathcal{A}_1\})$ where $\mathcal{A}_1 = \{((0,0), a), ((0,1), a), ((1,0), a)\}$ is the only axiom and P_1 consists of the two productions p_u and p_r:

$$p_u = (\{(0,0), (0,1)\}, \{((0,0), a), ((0,1), a)\}, \{(0,2)\}, \{((0,2), a)\}),$$
$$p_r = (\{(0,0), (1,0)\}, \{((0,0), a), ((1,0), a)\}, \{(2,0)\}, \{((2,0), a)\}).$$

As the selector area U_α and the context area U_β in a contextual array production of the form $(U_\alpha, \alpha, U_\beta, \beta)$ are disjoint, both α and β can be represented within only one pattern, i.e., p_u and p_r can be represented in a more depictive way by the patterns shown on the right (the symbols of the selector are enclosed in boxes).

$$p_u = \begin{array}{c} a \\ \boxed{a} \\ a \end{array}, \quad p_r = \boxed{a}\ \boxed{a}\ a.$$

The example of the L-shaped array for $n = 3$ and $m = 4$ then is generated by twice applying rule p_u and three times applying rule p_r, in any order. We also observe that every intermediate array obtained by applying these rules is in L_1, too. Obviously, by the definition of equivalence classes of arrays, we also have $[L\,(G_1)] = [L_1] \in [\mathcal{L}\,(2\text{-}ContA^1)]$.

$[\mathcal{A}_1]$ can be described in a more depictive way by $\begin{array}{c} a \\ a\ a \end{array}$, i.e., the contextual array grammar $[G_1]$ for $[L\,(G_1)]$ can also be written as $[G_1] = \left(2, \{a\}, \#, P_1, \left\{ \begin{array}{c} a \\ a\ a \end{array} \right\}\right)$ (see Remark 1). In the following, the axiom(s) often will just be given in such a pictorial variant. $\qquad\square$

Example 4. For the singleton language $L_\perp = \left\{ \begin{array}{c} a \\ a \\ a\ a\ a\ a\ a \end{array} \right\} \subset \left[\{a\}^{+2}\right]$, we have $L_\perp \in [\mathcal{L}\,(2\text{-}ContA)] \setminus [\mathcal{L}\,(2\text{-}REGA)]$. As we can take $L_\perp$ (as any finite language) as a set of axioms, containment in $[\mathcal{L}\,(2\text{-}ContA)]$ is clear. Conversely, any regular array grammar has to scan the non-blank symbols of the array A, which is impossible, as the underlying graph $g(shape(A))$ is not Hamiltonian. $\qquad\square$

Theorem 2. $[\mathcal{L}\,(1\text{-}REGA^1)] \subseteq [\mathcal{L}\,(1\text{-}ContA^1)].$

Proof. Due to the results from Theorem 1, it only remains to show that $[arr\,(\mathcal{L}\,(REG^1))] \subseteq [\mathcal{L}\,(1\text{-}ContA^1)].$

From [1, Theorem 4.4], we deduce that any infinite language $L \subseteq \{a\}^+$ in $\mathcal{L}\,(REG^1)$ can be written in the form $L = \{a^{s_1}, a^{s_2}, \ldots, a^{s_t}\} \cup \bigcup_{i=1}^m \{a^{k \cdot n + d_i} \mid n \geq 0\}$ for some numbers $k, k \leq d_1 < d_2 < \ldots < d_m < 2k, 0 \leq s_1 < s_2 < \ldots < s_t < k$. The 1-dimensional contextual array grammar now is constructed using a context of length k and putting the words a^{s_j}, $1 \leq j \leq t$, and a^{d_i}, $1 \leq i \leq m$, into the set of axioms, i.e., we define the 1-dimensional contextual array grammar $G\left(L\right) = \left(1, \{a\}, \#, P, A\right)$ with $A = \{arr\left(a^{s_j}\right) \mid 1 \leq j \leq t\} \cup \{arr\left(a^{d_i}\right) \mid 1 \leq i \leq m\}$ and $P = \left\{\boxed{a}^k a^k\right\}$. Obviously, $[L\left(G\left(L\right)\right)] = [arr\left(L\right)]$. The 1-dimensional contextual array grammar $[G\left(L\right)]$ for $[L\left(G\left(L\right)\right)]$ can also be written as $[G\left(L\right)] = \left(1, \{a\}, \#, P, A'\right)$ with $A' = \{a^{s_j} \mid 1 \leq j \leq t\} \cup \{a^{d_i} \mid 1 \leq i \leq m\}$ (compare with Remark 1).

For the sake of completeness we mention that every finite array language $A = \{arr\left(a^{s_j}\right) \mid 1 \leq j \leq t\}$ is generated by the 1-dimensional contextual array grammar $G\left(L\right) = \left(1, \{a\}, \#, P, A\right)$ with $P = \emptyset$. □

Remark 2. Following the definition already given in [11], our *d*-dimensional extension of (external) contextual grammars only appends at one location, while external contextual string grammars as originally defined by Solomon Marcus, see [14], append to both ends of a string at the same time. This design decision has two main reasons. First, it is not quite clear what the *d*-dimensional counterpart of external contextual grammars would really mean: for instance, for $d = 2$, should we allow appending on both ends of a row or column at the same time, as we did in [8] for the case of non-isometric contextual array grammars? Or, should we rather append on 'all ends'? Obviously, this situation becomes even more intricate for higher dimensions. Yet second and even more important, appending at both sides of a string, i.e., a 1-dimensional array, in parallel can easily be simulated sequentially by a matrix with two components. It is therefore easy to see that in the 1-dimensional case, the string images of the arrays generated by contextual array grammars with matrix control exactly correspond with the string languages generated by external contextual string grammars. This means that for the regulated variants discussed in the following, any variant that can be conceivably defined for the *d*-dimensional analogue of external contextual grammars, in the 1-dimensional case should lead to the same results as the original variant of contextual array grammars defined in [11] and taken as the basis in this paper, too.

3.1 Matrix Contextual Array Grammars

Definition 3. *A d-dimensional matrix contextual array grammar is a pair $G_M = (G, M)$ where $G = (d, V, \#, P, A)$ is a d-dimensional contextual array grammar and M is a finite set of sequences, called matrices, of rules from P, i.e., each element of M is of the form $\langle p_1, \cdots, p_n \rangle$, $n \geq 1$, where $p_i \in P$ for $1 \leq i \leq n$. Derivations in a matrix contextual array grammar are defined as in a contextual array grammar except that a single derivation step now consists of the sequential application of the rules of one of the matrices in M, in the order in which the rules are given in the matrix. The array language generated by G_M is the set of all d-dimensional arrays which can be derived from any of the axioms in A. The family of d-dimensional array languages of arrays generated by d-dimensional matrix contextual array grammars (over a k-letter alphabet) is denoted by $\mathcal{L}\left(d\text{-}MContA\right)$ $\left(\mathcal{L}\left(d\text{-}MContA^k\right)\right)$.*

Example 5. Consider the language L_2 of connected arrays given by

$$L_2 = \left\{ \{ ((0,0),a) \} \cup \{ ((0,i),a), ((i,0),a) \mid 1 \le i \le n \} \;\middle|\; n \in \mathbb{N} \right\},$$

which contains L-shaped arrays as L_1 from Example 1, but now with both arms having the same length. $L_2 \in \mathcal{L}(2\text{-}MContA^1)$, as it can be generated by the 2-dimensional matrix contextual array grammar $G_M = (G_1, M)$ where G_1 is the 2-dimensional contextual array grammar from Example 3 and $M = \{\langle p_u, p_r \rangle\}$. The only derivations possible in G'_M for $[L_2] \in \left[\mathcal{L}(2\text{-}MContA^1) \right]$ (see Remark 1) are:

$$
\begin{array}{l}
a \\
a \\
a \quad \Longrightarrow_{G'_M} \quad a \\
a\,a \qquad\qquad a\,a\,a
\end{array}
\qquad
\begin{array}{l}
a \\
a \\
a \\
\Longrightarrow_{G'_M} \quad a \\
a\,a\,a\,a
\end{array}
\qquad \Longrightarrow_{G'_M} \cdots
$$

The single matrix $\langle p_u, p_r \rangle$, $p_u = \begin{array}{c} a \\ \boxed{a} \\ a \end{array}$, $p_r = \boxed{a}\,\boxed{a}\,a$, guarantees that both arms of the array grow in a synchronized way. □

Theorem 3. *For any $d \ge 2$ and any $k \ge 1$, we have $\mathcal{L}(d\text{-}ContA^k) \subsetneq \mathcal{L}(d\text{-}MContA^k)$ and $\left[\mathcal{L}(d\text{-}ContA^k) \right] \subsetneq \left[\mathcal{L}(d\text{-}MContA^k) \right]$.*

Proof. The inclusion $\mathcal{L}(d\text{-}ContA^k) \subseteq \mathcal{L}(d\text{-}MContA^k)$ and therefore also $\left[\mathcal{L}(d\text{-}ContA^k) \right] \subseteq \left[\mathcal{L}(d\text{-}MContA^k) \right]$ is obvious from general results for grammars working on various kinds of objects and with specific regulating mechanisms, see [10].

For showing the strictness of the inclusion, we prove that the array language L_2 from Example 5 cannot be generated by a 2-dimensional contextual array grammar; for dimensions $d > 2$, we just take $[i_{2,d}(L_2)]$.

Now assume we could find a 2-dimensional contextual array grammar $[G = (2, \{a\}, \#, P, A)]$ that generates $[L_2]$. As contextual grammars are pure grammars, $[A]$ is a finite subset of $[L(G)]$. As $[L(G)]$ is infinite, we would need an infinite number of rules to get $[L_2]$ which resembles the case of *external* contextual string grammars; in fact, as soon as the arms get long enough, we have to apply a rule which only grows the arm going up or only grows the arm going to the right, resulting in an array which contradicts the definition of $[L_2]$. It is obvious that we also have $[i_{2,d}(L_2)] \in \left[\mathcal{L}(d\text{-}MContA^k) \right] \setminus \left[\mathcal{L}(d\text{-}ContA^k) \right]$; this observation completes the proof. □

In the 1-dimensional case, the situation is different: as we shall prove later, see Theorem 6, $\left[\mathcal{L}(1\text{-}ContA^1) \right] = \left[\mathcal{L}(1\text{-}MContA^1) \right]$, but for $k \ge 2$, we still have $\left[\mathcal{L}(1\text{-}ContA^k) \right] \subsetneq \left[\mathcal{L}(1\text{-}MContA^k) \right]$, as the following example shows.

Example 6. Consider the non-regular language $L_n = \{a^n b a^n \mid n \ge 1\}$. By Theorem 1, there cannot exist an array grammar G of type $1\text{-}REGA^2$ such that $[L(G)] = [arr(L_n)]$. Even more, there is no 1-dimensional contextual array grammar for L_n. Namely, if this would be the case, then first observe that there must be rules that append something to the right, as well as to the left of the array, and this should be possible infinitely often. Otherwise, the sequence of context additions would happen (finally) only on one side, which means that this behavior can again be simulated by some regular array grammar, contradicting our previous reasoning. Hence, there must be a rule that contains a sequence of a's as its selector, say, $arr(a^{r_s})$, and also a sequence of a's, say, $arr(a^{r_c})$ as its context in order to append a^{r_c} to the right of the current array, and likewise, there must be a rule that contains a sequence of a's as its selector, say, $arr(a^{\ell_s})$, and also a sequence of a's, say, $arr(a^{\ell_c})$ as its context in order to append a^{l_c} to the left of the current array. For sufficiently long arrays $arr(a^n b a^n)$, both rules can be applied, and arrays like $arr(a^n b a^{n+r_c})$ can generated that do not belong to L_n. Hence, $L_n \notin \mathcal{L}(1\text{-}ContA)$.

Yet for the 1-dimensional matrix contextual array grammar $[G_M] = (G_n, M_n)$ with $[G_n] = (1, \{a, b\}, \#, P, \{aba\})$ where $p_l = a\,\boxed{a}$, $p_r = \boxed{a}\,a$, and $M_n = \{\langle p_l, p_r \rangle\}$, we have $[L(G_n)] = [arr(L_n)]$. The single matrix $\langle p_l, p_r \rangle$ guarantees that the number of symbols a grows to the left and to the right in a synchronized way. $\qquad\square$

In addition, the following example even yields that for any $k \geq 2$, $[\mathcal{L}(1\text{-}MContA^k)]$ is incomparable with $[\mathcal{L}(1\text{-}REGA^k)]$.

Example 7. Consider the regular string language $L_r = \{ba^n b \mid n \geq 1\}$. Due to Theorem 1, there exists an array grammar of type $1\text{-}REGA^2$ G_r such that $[L(G_r)] = [arr(L_r)]$. Yet on the other hand, there cannot exist an array grammar of type $1\text{-}MContA^2$ $[G]$ such that $L([G]) = [arr(L_r)]$, which can be proved by a simple pumping argument: The number of symbols a between the two symbols b can become arbitrarily large, but we only have a finite set of axioms A; as $[G]$ is a pure grammar, $[A] \subset [L]$; yet $[G]$ can only grow these arrays in an external way, i.e., by adding symbols on the left or on the right, but in this way we are not able to grow the number of symbols a in the middle. $\qquad\square$

3.2   Contextual Array Grammars with Regular Control

Definition 4. *A d-dimensional contextual array grammar with regular control is a pair $G_C = (G, L)$ where $G = (d, V, \#, P, A)$ is a d-dimensional contextual array grammar and L is a regular string language over P. Derivations in a d-dimensional contextual array grammar with regular control are defined as in the contextual array grammar G except that in a successful derivation the sequence of applied rules has to be a word from L. The array language generated by G_C is the set of all d-dimensio nal arrays which can be derived from any of the axioms in A following a control word from L. The family of d-dimensional array languages of arrays generated by d-dimensional contextual array grammars over a k-letter alphabet with regular control is denoted by $\mathcal{L}((d\text{-}ContA, REG))$. The corresponding family of array languages of equivalence classes of arrays is denoted by using brackets in the notations.*

As a general result (following [10]) we can state:

Theorem 4. *For any $d \geq 1$ and any $k \geq 1$,*

$$[\mathcal{L}\,(d\text{-}ContA^k)] \subseteq [\mathcal{L}\,(d\text{-}MContA^k)] \subseteq [\mathcal{L}\,((d\text{-}ContA^k, REG))]\,.$$

Example 8. Consider the regular string language $L_r = \{ba^n b \mid n \geq 1\}$ from Example 7. We have shown that $[arr\,(L_r)] \in [\mathcal{L}\,(1\text{-}REGA^2)] \setminus [\mathcal{L}\,(1\text{-}MContA^2)]$. Moreover, $[arr\,(L_r)] \in [\mathcal{L}\,((1\text{-}ContA^1, REG))] \setminus [\mathcal{L}\,(1\text{-}MContA^2)]$: Consider $G'_r = (G_r, C_r)$ with $G_r = (1, \{a, b\}, \#, P, \{arr\,(ba)\})$ and $P = \{p_{aa}, p_{ab}\}$ with $p_{aa} = \boxed{a}\,a$, and $p_{ab} = \boxed{a}\,b$, as well as $C_r = \{p_{aa}\}^* \{p_{ab}\}$. It is easy to see that $[L\,(G'_r)] = [arr\,(L_r)]$. □

Theorem 5. *For any $d \geq 1$ and any $k \geq 2$, we have:*

$$[\mathcal{L}\,(d\text{-}ContA^k)] \subsetneq [\mathcal{L}\,(d\text{-}MContA^k)] \subsetneq [\mathcal{L}\,((d\text{-}ContA^k, REG))]\,.$$

Proof. The inclusions directly follow from Theorem 4. The strictness of the first inclusion follows from Example 6 by taking the non-regular string language $L_n = \{a^n b a^n \mid n \geq 1\}$. Then $[i_{1,d}\,(arr\,(L_n))] \in [\mathcal{L}\,(d\text{-}MContA^2)] \setminus [\mathcal{L}\,(d\text{-}ContA^k)]$. The strictness of the second inclusion follows from Example 8 by taking $[i_{1,d}\,(arr\,(L_r))]$. □

On the other hand, in the 1-dimensional case, the following theorem says that even with the regulating mechanisms of matrix control or regular control languages, with 1-dimensional contextual array grammars over a one-letter alphabet we cannot go beyond regularity, i.e., beyond $[\mathcal{L}\,(1\text{-}REGA^1)]$.

Theorem 6. $[\mathcal{L}\,(1\text{-}REGA^1)] = $
$[\mathcal{L}\,(1\text{-}ContA^1, REG)] = [\mathcal{L}\,(1\text{-}MContA^1)] = [\mathcal{L}\,(1\text{-}ContA^1)]\,.$

Proof. (Sketch) According to Theorems 4 and 2, we only have to show that $[\mathcal{L}\,(1\text{-}REGA^1)] \supseteq [\mathcal{L}\,((1\text{-}ContA^1, REG))]$. The main ideas of the corresponding technically non-trivial proof can be described as follows:

- Without loss of generality, right-hand sides of rules have the form $\boxed{a}^{\,m} a^n$.
- Context information is irrelevant for the unary 1-dimensional case, assuming that the set of axioms collects all arrays of sufficient size.
- The state information of the regular control is then encoded in the nonterminals of the regular array grammar. □

Allowing for more than one symbol, 1-dimensional contextual array grammars can generate exactly the array images of linear languages. The proof is based on the following normal form:

Lemma 1. *For any 1-dimensional contextual array grammar with regular control $G_C = (G, L)$, where $G = (1, V, \#, P, A)$, $L \subseteq P^*$, we can construct an equivalent 1-dimensional contextual array grammar with regular control $G'_C = (G', L')$ with $G' = (1, V, \#, P', A')$, $L' \subseteq P'^*$, such that for P' we have:*

- *All rules in P' are of the form $\boxed{a}\,b$ or $b\,\boxed{a}$ for some $a, b \in V$, i.e., we only have the minimal non-empty size of selectors and minimal contexts of size 1.*
- *If there is a rule of the form $\boxed{a}\,b$ / $b\,\boxed{a}$ in P', then also all rules of the form $\boxed{c}\,b$ or $b\,\boxed{c}$ are in P', for any $c \in V$, i.e., the selector contents is irrelevant, only direction of growth of the array is important.*

The rules in this normal form nicely correspond with the operations of left and right insertions for strings, which operations together with regular control languages also characterize the family of linear languages.

Theorem 7. $[\mathcal{L}\,(1\text{-}ContA, REG)] = arr\,(\mathcal{L}\,(LIN))$.

Proof. (Sketch) The main ideas of the proof can be described as follows:

- Adding strings in a controlled way "on both ends" corresponds to applying linear rules, but in reverse order.
- The information about the finitely many selectors possible can be stored in the nonterminal; on the other hand, the nonterminal can be stored in the state of the finite automaton of the control language.   □

For $d \geq 2$, i.e., in the case of at least two symbols, we can prove the incomparability of the families of array languages generated by contextual array grammars and those equipped with control mechanisms:

Theorem 8. *For any $d \geq 2$ and any $k \geq 1$, all the three families*
$$[\mathcal{L}\,(d\text{-}ContA^k)],\ [\mathcal{L}\,(d\text{-}MContA^k)],\ and\ [\mathcal{L}\,((d\text{-}ContA^k, REG))]$$
are incomparable with $[\mathcal{L}\,(d\text{-}REGA^k)]$.

Proof. For the singleton language $L_\perp$ from Example 4, we have $i_{2,d}\,(L_\perp) \in ([\mathcal{L}\,(d\text{-}ContA^1)] \cap [\mathcal{L}\,(d\text{-}MContA^1)] \cap [\mathcal{L}\,((d\text{-}ContA^1, REG))]) \setminus [\mathcal{L}\,(d\text{-}REGA^1)]$. On the other hand, for L_r from Example 7 we have $i_{2,d}\,([arr\,(L_r)]) \in [\mathcal{L}\,(1\text{-}REGA^2)] \setminus ([\mathcal{L}\,(d\text{-}ContA^1)] \cup [\mathcal{L}\,(d\text{-}MContA^1)] \cup [\mathcal{L}\,((d\text{-}ContA^1, REG))])$. Yet even for the case of one-letter alphabets we can find an array language of 2-dimensional arrays in $[\mathcal{L}\,(2\text{-}REGA^1)] \setminus [\mathcal{L}\,((2\text{-}ContA^1, REG))]$: we consider ⊔-shaped arrays with the left vertical line having a length being a multiple of 3 and the right vertical line having a length being a multiple of 5. These arrays can easily be generated by a regular array grammar by first generating the left vertical line from up to down, followed by the horizontal line, finally generating the right vertical line upwards. On the other hand, this set of 2-dimensional arrays cannot be generated by a contextual array grammar even when using regular control: as soon as the vertical lines have become long enough, we cannot distinguish any more between the left and the right one, so either the lengths will not necessarily fulfill the constraints of being a multiple of 3 and 5, respectively, any more, or even worse, the lines might even be prolonged below the horizontal line yielding arrays of the shape of an H.   □

4   Decidability Questions

As the size of the arrays generated by contextual array grammars (even with any control mechanism) increases with every derivation step, the generated array languages are computable (i.e., recursive).

As an immediate consequence of Theorem 7, we obtain:

Corollary 1. *Emptiness is decidable for* $\mathcal{L}$ (1-*ContA, REG*).

Yet for higher dimensions, we obtain a completely different situation:

Theorem 9. *Emptiness is not decidable for* $\mathcal{L}$ $(d\text{-}ContA^k, REG)$ *for* $d \geq 2$, *even for* $k = 1$.

Proof. (Sketch) As, for example, described in [5], the derivation carpet of a Turing machine can be described using 2-dimensional contextual array productions in the t-mode of derivation, i.e., a derivation only stops if no rule can be applied any more. The goal of only halting with specific conditions being fulfilled can also be obtained using suitable regular control languages, as we can require specific final rules to be applied. Hence, we will obtain a non-empty array language if and only if there is a derivation simulating the acceptance of a string by the given Turing machine. The proof given in [5] does not bound the number of symbols used. Yet m symbols can be encoded by $2 \times m$ rectangles with the k-th of these m symbols being encoded by leaving the k-th position in the second vertical line free, which then can be checked by the selector in the contextual array productions. Hence, simulating successful computations of the given Turing machine will result in the generation of $2k \times mn$ rectangles for accepting computations. □

5   Picture Generation

Another interesting topic is to consider the generation of geometric objects such as solid rectangles and squares, which has been used to exhibit the generative power of various array grammar variants. Both of them, i.e., the 2-dimensional array language L_{rect} of all solid rectangles of size $m \times n$, $m, n \geq 2$, made of a single symbol a and the 2-dimensional array language L_{square} of all solid squares of side length n, $n \geq 2$, made of a single symbol a are well-known to be in $\left[\mathcal{L}\left(2\text{-}REGA^1\right)\right]$, see [23], but as we are able to show they can also be generated by 2-dimensional contextual array grammars with regular control, i.e., $\{L_{rect}, L_{square}\} \subset \left[\mathcal{L}\left((2\text{-}ContA^1, REG)\right)\right]$. We now only exhibit the contextual array grammar with regular control for the squares.

Example 9. L_{square} is generated by the 2-dimensional contextual array grammar with regular control $G_{squareRC} = (G_{square}, C_{square})$ with $G_{square} = (\{a\}, P_{square}, A_{square})$, where A_{square} collects the 2×2 and 3×3 squares,

$$P_{square} = \{s_{ul}, s_{dr}, s_{ur}, s_{dl}, r_{ul}, r_{uu}, r_{dr}, r_{dd}\}, \text{ and}$$
$$C_{square} = (\{s_{ul}s_{dr}\}\{r_{ul}r_{dr}\}^* \{r_{uu}r_{dd}\}^* \{s_{ur}s_{dl}\})^+.$$

The rules are listed in the following:

$$s_{ul} = \begin{matrix} a & a & \\ a & \boxed{a} & \boxed{a} \\ & \boxed{a} & \end{matrix} \;,\; s_{dr} = \begin{matrix} & \boxed{a} & \\ \boxed{a} & \boxed{a} & a \\ a & a & \end{matrix} \;,\; s_{ur} = \begin{matrix} \boxed{a} & a & a \\ \boxed{a} & \boxed{a} & a \\ \boxed{a} & \boxed{a} & \end{matrix} \;,\; s_{dl} = \begin{matrix} \boxed{a} & \boxed{a} & \\ a & \boxed{a} & \boxed{a} \\ a & a & \boxed{a} \end{matrix} \;,$$

$$r_{ul} = \begin{matrix} & \boxed{a} & a \\ \boxed{a} & \boxed{a} & \boxed{a} \end{matrix} \;,\; r_{uu} = \begin{matrix} & \boxed{a} & \boxed{a} \\ a & \boxed{a} \\ \boxed{a} \end{matrix} \;,\; r_{dr} = \begin{matrix} \boxed{a} & \boxed{a} & \boxed{a} \\ a & \boxed{a} \end{matrix} \;,\; r_{dd} = \begin{matrix} \boxed{a} & a \\ \boxed{a} & \boxed{a} \end{matrix} \;.$$

How to derive a 4×4 square is shown below:

$$\begin{matrix} a\,a \\ a\,a \\ a\,a \end{matrix} \Rightarrow_{s_{ul}} \begin{matrix} a\,a \\ a\,a\,a \\ a\,a\,a \\ a\,a \end{matrix} \Rightarrow_{s_{dr}} \begin{matrix} a\,a \\ a\,a\,a \\ a\,a\,a \\ a\,a \end{matrix} \Rightarrow_{s_{ur}} \begin{matrix} a\,a\,a\,a \\ a\,a\,a\,a \\ a\,a\,a \\ a\,a \end{matrix} \Rightarrow_{s_{dl}} \begin{matrix} a\,a\,a\,a \\ a\,a\,a\,a \\ a\,a\,a\,a \\ a\,a\,a\,a \end{matrix}$$

Notice that the rules s_{ur} and s_{dl} check if a complete new border layer was actually generated, so they provide "keystones" as used in architecture, and it somehow replaces the t-mode of derivation, e.g., see [6].   □

As already with the t-mode of derivation, e.g., see [6], only eight contextual array rules were needed in Example 9 to generate the squares. This shows that the ability of contextual array grammars to insert new parts on different positions in the current array allows for a significantly smaller number of rules when using specific control mechanisms as the t-mode of derivation or regular control languages, in comparison with the construction of an extended regular array grammar as described in [23], where the construction has to be carried out along a Hamiltonian path. The inserted pieces used in [23] in fact could also be used as arrays inserted by a contextual array grammar with regular control, yet even for the subset of squares of side lengths $5k + 16$, $k \geq 0$, as exhibited in [23], 27 rules (arrays) were used. As these are in fact a kind of macro-rules, a complete list of regular array rules based on [23] would correspond to about one thousand rules. This is an example showing that contextual array grammars may allow for a succinct description of specific picture languages with rather small descriptional complexity.

References

1. Chrobak, M.: Finite automata and unary languages. Theor. Comput. Sci. **47**, 149–158 (1986)
2. Dassow, J., Păun, Gh.: Regulated Rewriting in Formal Language Theory. EATCS Monographs in Theoretical Computer Science, vol. 18. Springer, Heidelberg (1989)
3. Ehrenfeucht, A., Păun, Gh., Rozenberg, G.: Contextual grammars and formal languages. In: [20], vol. 2, pp. 237–293 (1997)
4. Fernau, H., Freund, R.: Bounded parallelism in array grammars used for character recognition. In: Perner, P., Rosenfeld, A., Wang, P. (eds.) SSPR 1996. LNCS, vol. 1121, pp. 40–49. Springer, Heidelberg (1996)
5. Fernau, H., Freund, R., Holzer, M.: Representations of recursively enumerable array languages by contextual array grammars. Fundam. Inf. **64**, 159–170 (2005)

6. Fernau, H., Freund, R., Schmid, M.L., Subramanian, K.G., Wiederhold, P.: Contextual array grammars and array P systems. Ann. Math. Artif. Intell. **75**(1–2), 5–26 (2015)
7. Fernau, H., Freund, R., Siromoney, R., Subramanian, K.G.: Contextual array grammars with matrix and regular control. Ann. Univ. Bucharest (Seria Informatica) **LXII**(3), 63–78 (2015)
8. Fernau, H., Freund, R., Siromoney, R., Subramanian, K.G.: Non-isometric contextual array grammars with regular control and local selectors. In: Durand-Lose, J., Nagy, B. (eds.) MCU 2015. LNCS, vol. 9288, pp. 61–78. Springer, Switzerland (2015)
9. Freund, R., Kogler, M., Oswald, M.: Control mechanisms on #–context-free array grammars. In: Păun, Gh. (ed.) Mathematical Aspects of Natural and Formal Languages, pp. 97–136. World Scientific Publ., Singapore (1994)
10. Freund, R., Kogler, M., Oswald, M.: A general framework for regulated rewriting based on the applicability of rules. In: Kelemen, J., Kelemenová, A. (eds.) Computation, Cooperation, and Life. LNCS, vol. 6610, pp. 35–53. Springer, Heidelberg (2011)
11. Freund, R., Păun, Gh., Rozenberg, G.: Contextual array grammars. In: Subramanian, K., Rangarajan, K., Mukund, M. (eds.) Formal Models, Languages and Applications. Series in Machine Perception and Articial Intelligence, vol. 66, chap. 8, pp. 112–136. World Scientific (2007)
12. Giammarresi, D., Restivo, A.: Two-dimensional languages. In: Rozenberg, G., Salomaa, A. (eds.) Handbook of Formal Languages, vol. 3, pp. 215–267. Springer, Heidelberg (1997)
13. Krithivasan, K., Balan, M., Rama, R.: Array contextual grammars. In: Martín-Vide, C., Păun, Gh. (eds.) Recent Topics in Mathematical and Computational Linguistics, pp. 154–168. Editura Academiei Române, Bucureşti (2000)
14. Marcus, S.: Contextual grammars. Revue Roumaine de Mathématiques Pures et Appliquées **14**, 1525–1534 (1969)
15. Nagy, B.: On a hierarchy of $5' \rightarrow 3'$ sensing Watson-Crick finite automata languages. J. Logic Comput. **23**(4), 855–872 (2013)
16. Pradella, M., Cherubini, A., Crespi-Reghizzi, S.: A unifying approach to picture grammars. Inform. Comput. **209**, 1246–1267 (2011)
17. Păun, Gh.: Marcus Contextual Grammars, Studies in Linguistics and Philosophy, vol. 67. Springer, Dordrecht (1997)
18. Rosenfeld, A.: Picture Languages. Academic Press, Reading (1979)
19. Rosenfeld, A., Siromoney, R.: Picture languages - a survey. Lang. Des. **1**(3), 229–245 (1993)
20. Rozenberg, G., Salomaa, A. (eds.): Handbook of Formal Languages, vol. 1–3. Springer, Heidelberg (1997)
21. Subramanian, K.G.: A note on regular controlled apical growth filamentous systems. Int. J. Comput. Inform. Sci. **14**, 235–242 (1985)
22. Wang, P.S.-P.: Some new results on isotonic array grammars. Inf. Process. Lett. **10**, 129–131 (1980)
23. Yamamoto, Y., Morita, K., Sugata, K.: Context-sensitivity of two-dimensional regular array grammars. Int. J. Pattern Recognit. Artif. Intell. **3**, 295–319 (1989)

Descriptional Complexity of Graph-Controlled Insertion-Deletion Systems

Henning Fernau[1], Lakshmanan Kuppusamy[2]([✉]), and Indhumathi Raman[3]

[1] Fachbereich 4 – Abteilung Informatikwissenschaften,
Universität Trier, 54286 Trier, Germany
fernau@uni-trier.de
[2] School of Computing Science and Engineering,
VIT University, Vellore 632 014, India
klakshma@vit.ac.in
[3] School of Information Technology and Engineering,
VIT University, Vellore 632 014, India
indhumathi.r@vit.ac.in

Abstract. We consider graph-controlled insertion-deletion systems and prove that the systems with sizes (i) $(3; 1, 1, 1; 1, 0, 1)$, (ii) $(3; 1, 1, 1; 1, 1, 0)$ and (iii) $(2; 2, 0, 0; 1, 1, 1)$ are computationally complete. Moreover, graph-controlled insertion-deletion systems simulate linear languages with sizes $(2; 2, 0, 1, 1, 0, 0)$, $(2; 2, 1, 0; 1, 0, 0)$, $(3; 1, 0, 1; 1, 0, 0)$, or $(3; 1, 1, 0; 1, 0, 0)$. Simulations of metalinear languages are also studied. The parameters in the size $(k; n, i', i''; m, j', j'')$ of a graph-controlled insertion-deletion system denote (from left to right) the maximum number of components, the maximal length of the insertion string, the maximal length of the left context for insertion, the maximal length of the right context for insertion; a similar list of three parameters concerning deletion follows.

Keywords: Insertion-deletion systems · Graph-controlled systems · Descriptional complexity measures · Computational completeness

1 Introduction

Insertion and deletion operations frequently occur in DNA processing and RNA editing. In the theoretical process of mismatched annealing of DNA sequences, certain segments of the strands are either inserted or deleted [18]. During RNA editing, some fragments of messenger RNA are inserted or deleted [2,3]. The motivation for insertion operations can be found in [7], where this operation and its iterated variant were introduced as a generalization of concatenation and Kleene's closure. The deletion operation was introduced in [10]. Insertion and deletion operations together were introduced into formal language theory in [11]. The corresponding grammatical mechanism is called *insertion-deletion system* (abbreviated as ins-del system). Informally, if a string η is inserted between two

Published by Springer International Publishing Switzerland 2016. All Rights Reserved
C. Câmpeanu et al. (Eds.): DCFS 2016, LNCS 9777, pp. 111–125, 2016.
DOI: 10.1007/978-3-319-41114-9_9

parts w_1 and w_2 of a string $w_1 w_2$ to get $w_1 \eta w_2$, we call the operation *insertion*, whereas if a substring δ is deleted from a string $w_1 \delta w_2$ to get $w_1 w_2$, we call the operation *deletion*. Suffixes of w_1 and prefixes of w_2 are called *contexts*.

Several variants of ins-del systems have been considered in literature, like ins-del P systems [1], tissue P systems with ins-del rules [14], context-free ins-del systems [16], matrix ins-del systems [13,17], etc. All the mentioned papers (as well as [19]) attempted to characterize the recursively enumerable languages (*i.e.*, they show computational completeness) using ins-del systems. We refer to the survey article [20] for details of variants thereof.

One of the important variants of ins-del systems is *graph-controlled ins-del systems* introduced in [5] and further studied in [9]. In such a system, the concept of a component is introduced and is associated with every insertion or deletion rule. The transition is performed by choosing any applicable rule from the set of rules of the current component and by moving the resultant string to the target component specified in the rule. If the transition of strings from component to component establishes a tree structure for a given system, then this system can also be seen as an ins-del P system. The objective is to obtain computationally completeness results with few components and small descriptional complexity measures of the ins-del rules.

For an ins-del system, the descriptional complexity measures are based on the size comprising of (i) the maximal length of the insertion string, denoted by n, (ii) the maximal length of the left context and right context used in insertion rules, denoted by i' and i'', respectively, (iii) the maximal length of the deletion string, denoted by m, (iv) the maximal length of the left context and right context used in deletion rules denoted by j' and j'', respectively. The size of an ins-del system is denoted by $(n, i', i''; m, j', j'')$.

Initially, computationally completeness results for graph-controlled ins-del systems were obtained with 5 components [12], then reduced to 4 components with sizes $(1, 1, 0; 2, 0, 0)$, $(2, 0, 0; 1, 1, 0)$, $(1, 1, 0; 1, 1, 0)$, $(1, 1, 0; 1, 0, 1)$ [5] and then later reduced to 3 components with sizes $(1, 2, 0; 1, 1, 0)$, $(1, 1, 0; 1, 2, 0)$ [8]. In [9], even graph-controlled ins-del systems with only 2 components and sizes $(1, 1, 0; 1, 2, 0)$, $(1, 2, 0; 1, 1, 0)$ were shown to be computationally complete. As an ins-del system without graph-control can be seen as a graph-controlled ins-del system with just one component, it is remarkable in this context to note that such system with size $(1, 1, 1; 1, 1, 1)$ are computationally complete; see [19].

In this paper, we prove the computational completeness of the following graph-controlled ins-del systems: (i) 3 components with size $(1, 1, 1; 1, 1, 0)$ or $(1, 1, 1; 1, 0, 1)$; (ii) 2 components with size $(2, 0, 0; 1, 1, 1)$. We also simulate linear grammars by graph-controlled ins-del systems having (i) 3 components with size $(1, 0, 1; 1, 0, 0)$ or $(1, 1, 0; 1, 0, 0)$; (ii) 2 components with size $(2, 0, 1; 1, 0, 0)$ or $(2, 1, 0; 1, 0, 0)$. We also extend the simulation technique to metalinear languages.

2   Preliminaries

We assume that the readers are familiar with the standard notations used in formal language theory. However, we now recall a few notations here.

Let $\mathbb{N}$ denote the set of positive integers, and $[1 \ldots k] = \{i \in \mathbb{N}: 1 \leq i \leq k\}$. Given an *alphabet* (finite set) Σ, Σ^* denotes the free monoid generated by Σ. The elements of Σ^* are called *strings* or *words*; λ denotes the empty string. For a string $w \in \Sigma^*$, $|w|$ denotes the length of a string w and w^R denotes the reversal (mirror image) of w. Likewise, L^R and $\mathcal{L}^R$ are understood for languages L and language families $\mathcal{L}$. RE denotes the family of the recursively enumerable languages, The family of linear and metalinear languages is denoted by LIN, $MLIN$, respectively, where $MLIN$ is the smallest language class containing LIN and is closed under concatenation. It is known from [15] that LIN is neither closed under concatenation nor under Kleene closure whereas $MLIN$ is not closed under Kleene closure but closed under concatenation. Also, both LIN and $MLIN$ are closed under reversal.

For the computational completeness results, we are using the fact that type-0 grammars in the special Geffert normal form are known to characterize the recursively enumerable languages. According to [5], a type-0 grammar $G = (N, T, P, S)$ is said to be in *special Geffert normal form*, SGNF for short, if

- N decomposes as $N = N' \cup N''$, where $N'' = \{A, B, C, D\}$ and N' contains at least the two nonterminals S and S',
- the only non-context-free rules in P are the two erasing rules $AB \rightarrow \lambda$ and $CD \rightarrow \lambda$,
- the context-free rules are of the following forms:
  $X \rightarrow Yb$ or $X \rightarrow bY$ where $X, Y \in N'$, $X \neq Y$, $b \in T \cup N''$, or $S' \rightarrow \lambda$.

How to construct this normal form is described in [5] and is based on [6]. Also, the derivation of a string is done in two phases. First, the context-free rules are applied repeatedly and the phase I is completed by applying the rule $S' \rightarrow \lambda$ in the derivation. In phase II, only the non-context-free erasing rules are applied repeatedly and the derivation ends. It is to be noted that as these context-free rules are more of a linear type, it is easy to see that there can be at most only one nonterminal from N' present in the derivation of G. We exploit this observation in the proofs of Theorems 2 and 4. Also, note that $X \neq Y, X, Y \in N'$ in the context-free rules.

2.1 Insertion-Deletion Systems

We now give the basic definition of insertion-deletion systems, following [11, 18].

Definition 1. *An* insertion-deletion system *is a construct* $\gamma = (V, T, A, R)$, *where V is an alphabet, $T \subseteq V$ is the terminal alphabet, A is a finite language over V, R is a finite set of triplets of the form $(u, \eta, v)_{ins}$ or $(u, \delta, v)_{del}$, where $(u, v) \in V^* \times V^*$, $\eta, \delta \in V^+$.*

The pair (u, v) is called the *context*, η is called the *insertion string*, δ is called the *deletion string* and $x \in A$ is called an *axiom*. For all contexts of t where $t \in \{ins, del\}$, if $u = \lambda$ $(v = \lambda)$, then we call the operation t to be right context (left context). If both $u, v = \lambda$ for a rule, then it means, the corresponding insertion/deletion can be done freely anywhere in the string and is called context-free insertion/deletion. An insertion rule will be of the form $(u, \eta, v)_{ins}$, which means that the string η is inserted between u and v. A deletion rule will be of the form $(u, \delta, v)_{del}$, which means that the string δ is deleted between u and v. In other words, $(u, \eta, v)_{ins}$ corresponds to the rewriting rule $uv \to u\eta v$, and $(u, \delta, v)_{del}$ corresponds to the rewriting rule $u\delta v \to uv$.

Consequently, for $x, y \in V^*$ we can write $x \Rightarrow y$ if y can be obtained from x by using either an insertion rule or a deletion rule which is given as follows:

1. $x = x_1 u v x_2$, $y = x_1 u \eta v x_2$, for some $x_1, x_2 \in V^*$ and $(u, \eta, v)_{ins} \in R$.
2. $x = x_1 u \delta v x_2$, $y = x_1 u v x_2$, for some $x_1, x_2 \in V^*$ and $(u, \delta, v)_{del} \in R$.

The language generated by γ is defined by

$$L(\gamma) = \{w \in T^* \mid x \Rightarrow^* w, \text{ for some } x \in A\},$$

where $\Rightarrow^*$ is the reflexive and transitive closure of the relation $\Rightarrow$.

2.2　Graph-Controlled Insertion-Deletion Systems

A *graph-controlled insertion-deletion system* with k components, or $(k\text{-})$GCID for short, is a construct $\Pi = (k, V, T, A, H, i_0, i_f, R)$ where

- k is the number of components,
- V is an alphabet,
- $T \subseteq V$ is the terminal alphabet,
- $A \subseteq V$ is a finite set of axioms,
- H is a set of labels associated (in a one-to-one manner) to the rules in R,
- $i_0 \in [1 \ldots k]$ is the initial component,
- $i_f \in [1 \ldots k]$ is the final or target component, and
- R is a finite set of rules of the form (i, r, j) where r is an insertion rule of the form $(u, \eta, v)_{ins}$ or deletion rule of the form $(u, \delta, v)_{del}$ and $i, j \in [1 \ldots k]$.

A rule of the form $l : (i, r, j)$, where $l \in H$ is the label associated to the rule, denotes that the string is sent from component i (for short denoted as Ci) to Cj after the application of the insertion or deletion rule r on the string.

A configuration of Π is represented by $(w)_i$ where i is the number of the current component (initially i_0) and w is the current string. A transition $(w)_i \Rightarrow (w')_j$ is performed if there exists a rule $l : (i, r, j)$ in R such that $w \Rightarrow w'$ on applying the insertion or deletion rule r; in this case, we also write $(w)_i \Rightarrow_l (w')_j$ or $(w')_j \Leftarrow_l (w)_i$. By $(w)_i \overset{\Rightarrow l}{\underset{\Leftarrow l'}{}} (w')_j$, we mean that $(w')_j$ is derivable from $(w)_i$ using rule l and $(w)_i$ is derivable from $(w')_j$ using rule l'. The language of a graph-controlled insertion-deletion system is the set of all terminal strings in the target component i_f reachable from an axiom and the initial component i_0. Formally,

$$L(\Pi) = \{w \in T^* \mid (x)_{i_0} \Rightarrow^* (w)_{i_f} \text{ for some } x \in A\}.$$

Next, we discuss about the size of a graph-controlled ins-del system. A graph-controlled ins-del system Π is of size $(k; n, i', i''; m, j', j'')$ (with the corresponding language classes denoted by $GCID(k; n, i', i''; m, j', j'')$) if

k = the number of components
$n = \max\{|\eta|: (i, (u, \eta, v)_{ins}, j) \in R\}$ (max. length of the inserted string)
$i' = \max\{|u|: (i, (u, \eta, v)_{ins}, j) \in R\}$ (max. length of the left context)
$i'' = \max\{|v|: (i, (u, \eta, v)_{ins}, j) \in R\}$ (max. length of the right context)
$m = \max\{|\delta|: (i, (u, \delta, v)_{del}, j) \in R\}$ (max. length of the deleted string)
$j' = \max\{|u|: (i, (u, \delta, v)_{del}, j) \in R\}$ (max. length of the left context)
$j'' = max\{|v|: (i, (u, \delta, v)_{del}, j) \in R\}$ (max. length of the right context)

Let us give some examples for GCID systems.

Example 1. The following GCID system Π_1 of size $(2; 1, 0, 0; 0, 0, 0)$ generates the language $L_1 = \{w \in \{a, b\}^*: |w|_a = |w|_b\}$.

$$\Pi_1 = (2, \{a, b\}, \{a, b\}, \{\lambda\}, \{r1, r2\}, 1, 1, R),$$

where the rules of R are: $r1: (1, (\lambda, a, \lambda)_{ins}, 2), r2: (2, (\lambda, b, \lambda)_{ins}, 1)$.   □

Example 2. With axiom $A = \{ab, \lambda\}$, two rules grouped in singleton components $C1 = \{(1, (a, a, b)_{ins}, 2)\}$, $C2 = \{(2, (a, b, b)_{ins}, 1)\}$, initial and target component $C1$, the GCID system Π_2 can describe $L_2 = \{a^n b^n: n \geq 0\}$, *i.e.*, $L_2 \in GCID(2; 1, 1, 1; 0, 0, 0)$.   □

Example 3. Consider the GCID system Π_3 of size $(3; 1, 0, 1; 1, 0, 0)$ as follows:

$$\Pi_3 = (3, \{S, S', a, b\}, \{a, b\}, \{SS'\}, H, 1, 1, R),$$

where the rules of R are the following ones:

$$r1.1: (1, (\lambda, a, S)_{ins}, 2) \quad r1.2: (1, (\lambda, S, \lambda)_{del}, 3)$$
$$r2.1: (2, (\lambda, b, S')_{ins}, 1)$$
$$r3.1: (3, (\lambda, S, S')_{ins}, 1) \quad r3.2: (3, (\lambda, S', \lambda)_{del}, 1)$$

We claim that Π_3 generates $L_3 = \{a^n b^n: n \geq 1\}^*$. We prove our claim by discussing the working of the rules of Π_3 here. Starting with the axiom SS' in $C1$, a is inserted before S and then b is inserted before S' in order, repeatedly, and this leads to $(a^n S b^n S')$ in $C1$. After $n(\geq 0)$ cycles of repetitions, rule $r1.2$ is applied and this deletes S and we move to $C3$ with the string $a^n b^n S'$. We now have a choice of applying rule $r3.1$ or $r3.2$. In the latter case, S' is deleted and the process terminates at the target component $C1$. In the former case, we are back to the starting point in order to generate $a^n b^n a^m b^m SS'$. On repeating this process several times as desired, the process can be terminated by applying the rule $r3.2$. With these arguments, one can see that this system generates L_3.   □

Observe the similarities between the examples: L_1 is the iterated shuffle closure of $(L_2 \cup L_2^R)$, while L_3 is the Kleene closure of L_2. Notice that $L_1 \notin LIN$ and $L_3 \notin MLIN$, and the latter can be proved in the same way as argued in [4, p. 137] for the Łukasiewicz language.

3   Auxiliary Results

In order to simplify the proofs of some of our main results, the following observations are helpful.

Theorem 1. *For all non-negative integers k, n, i', i'', m, j, j'', we have that*

$$GCID(k; n, i', i''; m, j', j'') = [GCID(k; n, i'', i'; m, j'', j')]^R .$$

Proof. To an ins-del rule $(x, y, z)_\mu$ with $\mu \in \{ins, del\}$, we associate the reversed rule $\rho(r) = (z^R, y^R, x^R)_\mu$. Let $\Pi = (k, V, T, A, H, i_0, i_f, R)$ be a graph-controlled insertion-deletion system with k components. Map a rule $l: (i, r, j) \in \Pi$ to $l: (i, \rho(r), j)$ in $\rho(R)$. Define $\Pi^R = (k, V, T, A^R, H, i_0, i_f, \rho(R))$. Then, an easy inductive argument shows that $L(\Pi^R) = (L(\Pi))^R$. Observing the sizes of the system now shows the claim.   □

Corollary 1. *Let $\mathcal{L}$ be a language class that is closed under reversal. Then, for all non-negative integers $k, n, i', i'', m, j', j''$, we conclude that*

1. *$\mathcal{L} = GCID(k; n, i', i''; m, j', j'')$ if and only if*
   $\mathcal{L} = GCID(k; n, i'', i'; m, j'', j')$;
2. *$\mathcal{L} \subseteq GCID(k; n, i', i''; m, j', j'')$ if and only if*
   $\mathcal{L} \subseteq GCID(k; n, i'', i'; m, j'', j')$.

4   Computational Completeness Results

In this section, we prove the computational completeness results for GCID systems of sizes (i) $(3; 1, 1, 1; 1, 1, 0)$ (ii) $(3; 1, 1, 1; 1, 0, 1)$ and (iii) $(2; 2, 0, 0; 1, 1, 1)$. One may note that, in the first (second) system, the deletion is left context (right context) and in the third system, the insertions are performed in a context-free manner.

Theorem 2. $GCID(3; 1, 1, 1; 1, 1, 0) = RE.$

Proof. Consider a type-0 grammar $G = (N, T, P, S)$ in SGNF. We build a GCID system Π such that $L(\Pi) = L(G)$. Let $\Pi = (3, V, T, \{S\}, H, 1, 1, R)$. The rules in P are labelled injectively with labels from $[1 \ldots |P|]$. Let $V = N \cup T \cup \{p: p \in [1 \ldots |P|]\}$. R is defined as follows. The rules are classified into components $C1$, $C2$ and $C3$ as indicated by the first character following the rule label. We simulate the rule $p: X \to bY$ by the following ins-del rules:

$$
\begin{aligned}
&p1.1: \ (1, (\lambda, p, X)_{ins}, 2) \\
&p2.1: \ (2, (\lambda, b, p)_{ins}, 3), \quad p2.2: \ (2, (Y, X, \lambda)_{del}, 1) \\
&p3.1: \ (3, (b, Y, p)_{ins}, 3), \quad p3.2: \ (3, (Y, p, \lambda)_{del}, 2)
\end{aligned}
$$

We simulate the rule $q: X \to Yb$ by the following ins-del rules:

$$
\begin{aligned}
&q1.1: \ (1, (\lambda, q, X)_{ins}, 2) \\
&q2.1: \ (2, (\lambda, Y, q)_{ins}, 3), \quad q2.2: \ (2, (\lambda, q, \lambda)_{del}, 1) \\
&q3.1: \ (3, (q, b, X)_{ins}, 3), \quad q3.2: \ (3, (b, X, \lambda)_{del}, 2)
\end{aligned}
$$

We simulate the rule $f: AB \to \lambda$ by the following ins-del rules:

$$f1.1: \ (1,(\lambda, f, A)_{ins}, 2)$$
$$f2.1: \ (2,(\lambda, f, \lambda)_{del}, 1), \ \ f2.2: \ (2,(f, A, \lambda)_{del}, 3)$$
$$f3.1: \ (3,(f, B, \lambda)_{del}, 2)$$

We simulate the rule $g: CD \to \lambda$ by the following ins-del rules:

$$g1.1: \ (1,(\lambda, g, C)_{ins}, 2)$$
$$g2.1: \ (2,(\lambda, g, \lambda)_{del}, 1), \ \ g2.2: \ (2,(g, C, \lambda)_{del}, 3)$$
$$g3.1: \ (3,(g, D, \lambda)_{del}, 2)$$

We simulate the rule $h: S' \to \lambda$ by the ins-del rule $h1.1: (1,(\lambda, S', \lambda)_{del}, 1)$.

We now proceed to prove that $L(\Pi) = L(G)$. We do this by explaining how the simulation of the rules of G should work and why no other malicious derivations are possible in Π.

Working of $p: X \to bY$: Consider the string $\alpha X \beta$ in $C1$. Then there is a unique sequence of rule applications in Π as follows.

$$(\alpha X \beta)_1 \Rightarrow_{p1.1} (\alpha p X \beta)_2 \Rightarrow_{p2.1} (\alpha b p X \beta)_3 \Rightarrow_{p3.1} (\alpha b Y p X \beta)_3$$
$$\Rightarrow_{p3.2} (\alpha b Y X \beta)_2 \Rightarrow_{p2.2} (\alpha b Y \beta)_1.$$

Note that though applying the rule $p3.1$ leaves the string in $C3$ itself, rule $p3.1$ cannot be applied again (the benefit of using double-sided context). Also, only one X of N' is present in the derivation until a $Y \in N'$ is introduced, thus, $p2.2$ cannot be used before the rule $p2.1$ is applied.

Working of $q: X \to Yb$: Consider the string $\alpha X \beta$ in $C1$. On applying rule $q1.1$, we insert q before X and we get $\alpha q X \beta$ in $C2$. Now, we can apply either $q2.1$ or $q2.2$. In the latter case, we delete the just inserted marker q and end up with $\alpha X \beta$ in $C1$ (back to the starting point). Hence, we choose rule $q2.1$ eventually to move on. In this case, consider the following sequence of rule applications in Π.

$$(\alpha X \beta)_1 \underset{\Leftarrow q2.2}{\overset{\Rightarrow q1.1}{}} (\alpha q X \beta)_2 \Rightarrow_{q2.1} (\alpha Y q X \beta)_3 \Rightarrow_{q3.1} (\alpha Y q b X \beta)_3 \Rightarrow_{q3.2} (\alpha Y q b \beta)_2$$

At this point, we again have a choice of applying rule $q2.1$ or $q2.2$. In the former case, we will again insert a Y before q yielding $\alpha Y Y q b \beta$ in $C3$. As $Y \in N'$ is the only nonterminal in the string, the first symbol of β cannot be X. Thus, we cannot apply any rule in $C3$ and the derivation stops with nonterminals in a non-target component. In the latter case, by applying $q2.2$ we delete q and get $\alpha Y b \beta$ in $C1$, which is the target component.

We next proceed to discuss the simulation of the non context-free erasing rules $AB \to \lambda$ and $CD \to \lambda$.

Working of $f: AB \to \lambda$: The working of the rule is shown by the following sequence of rule applications.

$$(\alpha A B \beta)_1 \underset{\Leftarrow f2.1}{\overset{\Rightarrow f1.1}{}} (\alpha f A B \beta)_2 \Rightarrow_{f2.2} (\alpha f B \beta)_3 \Rightarrow_{f3.1} (\alpha f \beta)_2 \Rightarrow_{f2.1} (\alpha \lambda \beta)_1$$

Working of $g : CD \rightarrow \lambda$: Similar to the working of the rule $f : AB \rightarrow \lambda$.

The rule $(1, (\lambda, S', \lambda)_{del}, 1)$ directly erases S'. We start at S in $C1$ and by repeatedly applying the rules p, q, f, g, h, we eventually get $(S)_1 \Rightarrow_* (w)_1$. This proves that $L(G) \subseteq L(\Pi)$.

To prove the reverse relation $(L(\Pi) \subseteq L(G))$, we observe that the rules of Π are applied in groups and each group of rules corresponds to one of p, q, f, g, h. Also, it is not possible to switch between the simulation of some p, say, to that of f, as we always use unique marker symbols to prevent this from happening. This observation completes the proof.   □

As RE is known to be closed under reversal, we conclude with Corollary 1:

Theorem 3. $GCID(3; 1, 1, 1; 1, 0, 1) = RE$.

Theorem 4. $GCID(2; 2, 0, 0; 1, 1, 1) = RE$.

Proof. Consider a type-0 grammar $G = (N, T, P, S)$ in SGNF. We construct a GCID system Π such that $L(\Pi) = L(G)$. Let $\Pi = (2, V, T, \{S\}, H, 1, 1, R)$. The rules from P in G are labelled injectively with labels from $[1 \ldots |P|]$. The alphabet of Π is $V = N \cup T \cup \{p, p' : p \in [1 \ldots |P|]\}$. R is defined as follows. We simulate the rule $p : X \rightarrow bY$, with $X, Y \in N'$, by the following ins-del rules:

$$p1.1 : \ (1, (\lambda, bY, \lambda)_{ins}, 2)$$
$$p2.1 : \ (2, (Y, X, \lambda)_{del}, 1)$$

We simulate the rule $q : X \rightarrow Yb$, with $X, Y \in N'$, by the following ins-del rules:

$$q1.1 : \ (1, (\lambda, Yb, \lambda)_{ins}, 2)$$
$$q2.1 : \ (2, (\lambda, X, Y)_{del}, 1)$$

We simulate the rule $f : AB \rightarrow \lambda$, with $A, B \in N''$, by the following ins-del rules:

$$f1.1 : \ (1, (\lambda, ff', \lambda)_{ins}, 2), \quad f1.2 : \ (1, (A, f, f')_{del}, 1), \quad f1.3 : \ (1, (\lambda, A, f')_{del}, 2)$$
$$f2.1 : \ (2, (f', B, \lambda)_{del}, 1), \quad f2.2 : \ (2, (\lambda, f', \lambda)_{del}, 1)$$

We simulate the rule $g : CD \rightarrow \lambda$ by the following ins-del rules:

$$g1.1 : \ (1, (\lambda, gg', \lambda)_{ins}, 2), \quad g1.2 : \ (1, (C, g, g')_{del}, 1), \quad g1.3 : \ (1, (\lambda, C, g')_{del}, 2)$$
$$g2.1 : \ (2, (g', D, \lambda)_{del}, 1), \quad g2.2 : \ (2, (\lambda, g', \lambda)_{del}, 1)$$

We simulate the rule $h : S' \rightarrow \lambda$ by the ins-del rule $h1.1 : (1, (\lambda, S', \lambda)_{del}, 1)$. We now proceed to reason why $L(\Pi) = L(G)$.

Working of $p : X \rightarrow bY$: Consider a string $\alpha X \beta$ in $C1$. The string bY is free to be inserted anywhere in the string using rule $p1.1$ and the derivation moves to $C2$. Rule $p2.1$ can be applied only if bY is inserted before X. Recall that $X, Y \in N'$ and these types of nonterminals only occur once in valid sentential forms of G (SGNF property). In this case, the X is deleted yielding bY and the derivation ends at the target component $C1$. If bY has been inserted elsewhere, then no rule of $C2$ can be applied and we are trapped in a non-target component with nonterminals in the string.

Working of $q : X \to Yb$: Similar to the working of the rule p as explained above.

Working of $f : AB \to \lambda$: Consider the string $\alpha AB\beta$ in $C1$. We introduce two markers f, f' together anywhere in the string using the rule $f1.1$ and move to $C2$. Suppose that ff' has been inserted between A and B. Now, there is a choice of applying rule $f2.1$ or $f2.2$. In the latter case, we will delete the marker f' and come to the target component $C1$ with $\alpha A f B\beta$. If we introduce ff' again, this will eventually lead to a string having the nonterminals f and A in it, thus not deriving any terminal string. This observation forces one to choose rule $f2.1$ before applying $f2.2$. In this case, there is a unique sequence of rule applications:

$$(\alpha A f f' B\beta)_2 \Rightarrow_{f2.1} (\alpha A f f' \beta)_1 \Rightarrow_{f1.2} (\alpha A f' \beta)_1 \Rightarrow_{f1.3} (\alpha f' \beta)_2 \Rightarrow_{f2.2} (\alpha \lambda \beta)_1$$

Suppose that ff' has not been inserted between A and B, then it is not difficult to see that the derived string will always contain some nonterminals.

Working of $g : CD \to \lambda$: Similar to the working of the rule $f : AB \to \lambda$.

The rule $(1, (\lambda, S', \lambda)_{del}, 1)$ directly erases S'. We start at S in $C1$ and by repeatedly applying the rules p, q, f, g, h, we eventually get $(S)_1 \Rightarrow_* (w)_1$. As argued above, no malicious derivations can lead to terminal strings in $C1$.   □

5   (Meta)linear Languages

We next prove that GCID systems of sizes $(2; 2, 1, 0; 1, 0, 0)$, $(2; 2, 0, 1; 1, 0, 0)$, $(3; 1, 1, 0; 1, 0, 0)$, or $(3; 1, 0, 1; 1, 0, 0)$ can simulate all linear languages. In these systems, deletions are performed in a context-free manner. While comparing the last two sizes with the first two sizes, one may note that the length of the inserted string is reduced at the cost of increasing the number of components. We also show how to extend the simulations beyond linear languages.

Theorem 5. $LIN \subsetneq GCID(2; 2, 1, 0; 1, 0, 0)$.

Proof. Consider a linear grammar $G = (N, T, P, S)$, where every rule of P is of the form $X \to Ya$ or $X \to aY$ or $X \to a$ or $X \to \lambda$. We construct a GCID system $\Pi = (2, V, T, \{S\}, H, 1, 1, R)$ for G. The rules from P in G are labelled injectively with labels from $[1 \ldots |P|]$. The alphabet of Π is $V = N \cup T \cup \{p : p \in [1 \ldots |P|]\}$. The set of rules R of Π is defined as follows.
We simulate the rule $p : X \to Ya$ by the following ins-del rules:

$$p1.1 : \ (1, (X, p, \lambda)_{ins}, 2), \ \ p1.2 : \ (1, (p, Ya, \lambda)_{ins}, 2)$$
$$p2.1 : \ (2, (\lambda, X, \lambda)_{del}, 1), \ \ p2.2 : \ (2, (\lambda, p, \lambda)_{del}, 1)$$

We simulate the rule $q : X \to aY$ by the following ins-del rules:

$$q1.1 : \ (1, (X, q, \lambda)_{ins}, 2), \ \ q1.2 : \ (1, (q, aY, \lambda)_{ins}, 2)$$
$$q2.1 : \ (2, (\lambda, X, \lambda)_{del}, 1), \ \ q2.2 : \ (2, (\lambda, q, \lambda)_{del}, 1)$$

We next simulate the rule $f : X \to a$ by the following ins-del rules:

$$f1.1 : \ (1, (X, a, \lambda)_{ins}, 2)$$
$$f2.1 : \ (2, (\lambda, X, \lambda)_{del}, 1)$$

We now prove the theorem by discussing the working of the above rules.

<u>Working of $p : X \to Ya$</u>: Consider the string $\alpha X \beta$ in $C1$. On applying rule $p1.1$, we insert p after X and get $\alpha X p \beta$ in $C2$. At this point, we have a choice of applying rule $p2.1$ or $p2.2$. In the latter case, the marker p is deleted and we move to $C1$ with $\alpha X \beta$ in the string and this is our starting point. Hence we have to use rule $p2.1$ eventually to proceed. In this case, X is deleted and move to $C1$ with $\alpha p \beta$. At this point, we note that the rule $p1.1$ cannot be applied since in linear grammar there is at most one nonterminal (in this case, X) in the string; this was already deleted in the previous step. With these arguments, we simulate the rule $X \to Ya$ as follows:

$$(\alpha X \beta)_1 \overset{\Rightarrow p1.1}{\underset{\Leftarrow p2.2}{}} (\alpha X p \beta)_2 \Rightarrow_{p2.1} (\alpha p \beta)_1 \Rightarrow_{p1.2} (\alpha p Y a \beta)_2 \Rightarrow_{p2.2} (\alpha Y a \beta)_1.$$

In the above sequence, we note that before the derivation $(\alpha p \beta)_1 \Rightarrow_{p1.2} (\alpha p Y a \beta)_2$, the rule $p1.1$ cannot be applied since in a linear grammar there is at most one nonterminal (in this case, X) in the string and it is already deleted in the previous step.

<u>Working of $q : X \to aY$</u>: Similar to the working of the above rule $p : X \to Ya$. The sequence of rule applications in Π is given below for a better understanding.

$$(\alpha X \beta)_1 \overset{\Rightarrow q1.1}{\underset{\Leftarrow q2.2}{}} (\alpha X q \beta)_2 \Rightarrow_{q2.1} (\alpha q \beta)_1 \Rightarrow_{q1.2} (\alpha p a Y \beta)_2 \Rightarrow_{q2.2} (\alpha a Y \beta)_1.$$

The working of rule $f : X \to a$ is simple and straightforward. Since we start at S in $C1$ and if we repeatedly apply the rules p, q, f, we eventually get $(S)_1 \Rightarrow_* (w)_1$. This proves that $L(G) \subseteq L(\Pi)$.

For the converse direction $L(G) \supseteq L(\Pi)$, observe the remarks that we gave above when explaining the working of the simulations; apart from unnecessary additional loops in the simulation, no successful derivations are possible in Π other than those intended for the simulation of G.

The strictness of the inclusion follows from Examples 1 and 3.   □

Remark 1. By allowing for a few more components, we can extend the previous simulation result to cover Kleene closures of linear languages or also *MLIN*. For instance, starting with axiom $S'S$ and a third component containing rules $r3.1 : \ (3, (S', S, \lambda)_{ins}, 1)$ and $r3.2 : \ (3, (\lambda, S', \lambda)_{del}, 1)$ and changing $f2.1$ to transit to $C3$, the modified system Π' would describe $(L(G))^+$, or, by having S' as the axiom and starting in $C3$, we can get $(L(G))^*$.   □

Likewise, we can describe metalinear languages with three or four components.

Theorem 6. *MLIN* $\subsetneq$ *GCID*$(4; 2, 1, 0; 1, 0, 0) \cap$ *GCID*$(3; 2, 1, 0; 1, 0, 1)$.

Proof. If $L \in MLIN$ happens to be a linear language, we can proceed as in Theorem 5. So, we assume that $L \in MLIN - LIN$ is given. We can think of the work of a metalinear grammar G with $L(G) = L \subseteq T^*$ (generating the concatenation of k linear languages $L(G_1), \ldots, L(G_k)$ with start symbols $S_1, \ldots, S_k$, respectively, and k pairwise disjoint nonterminal alphabets $N_1, \ldots, N_k$) as follows: starting with $S_1 S_2'$ as the axiom, first, G_1 generates a terminal word. Then, $S_2' \to S_2 S_3'$ is executed, and G_2 generates a terminal word, starting from S_2. This strategy continues, until $S_{k-1}' \to S_{k-1} S_k'$ is executed, followed by the generation of a terminal word by G_{k-1} and finally $S_k' \to S_k$ initiates the last grammar G_k to append a terminal word.

Let us first focus on $GCID(4; 2, 1, 0; 1, 0, 0)$. More formally, we construct a GCID system $\Pi = (4, V, T, \{S_1 S_2'\}, H, 1, 1, R)$ for G. Let $V_1, \ldots, V_k$ be the alphabets resulting from the construction of GCID systems Π_i for $G_1, \ldots, G_k$ according to Theorem 5. Let $N_i = V_i - T$ and assume (w.l.o.g.) that $N_1, \ldots, N_k$ are pairwise disjoint. Let $V = \bigcup_{i=1}^k V_i \cup \{S_i' : i \in [1 \ldots k]\}$. Let R_i be the rule set of Π_i. R_i' coincides with R_i except for (possibly) terminating rules of the type $f2.1$ that target at $C3$ for $i \in [1 \ldots (k-1)]$. Let $R = \bigcup_{i=1}^k R_i' \cup R_T$, where R_T collects transition rules that are described in details in the following.

The work of grammar G_i, say, of G_1, is simulated (as described in the proof of Theorem 5). Then, (in general) we transit to the third component. We perform the following transition rules:

$r_{1 \to 2} 2.1 : (2, (\lambda, r_{1 \to 2}, \lambda)_{del}, 1)$
$r_{1 \to 2} 3.1 : (3, (S_2', r_{1 \to 2}, \lambda)_{ins}, 4), \quad r_{1 \to 2} 3.2 : (3, (r_{1 \to 2}, S_2 S_3', \lambda)_{ins}, 2)$
$r_{1 \to 2} 4.1 : (4, (\lambda, S_2', \lambda)_{del}, 3)$

Similar transition rules are added to start simulations of $G_3, \ldots, G_{k-1}$. Finally, we have the rules:

$r_{k-1 \to k} 2.1 : (2, (\lambda, r_{k-1 \to k}, \lambda)_{del}, 1)$
$r_{k-1 \to k} 3.1 : (3, (S_k', r_{k-1 \to k}, \lambda)_{ins}, 4), \quad r_{k-1 \to k} 3.2 : (3, (r_{k-1 \to k}, S_k, \lambda)_{ins}, 2)$
$r_{k-1 \to k} 4.1 : (4, (\lambda, S_k', \lambda)_{del}, 3)$

Observe that the applications of the new rules (in comparison to what is inherited from Theorem 5) is deterministic, and due to the new components, no interference with previously introduced rules is possible. Furthermore, the context-free deletion rules in $C2$ of Theorem 5 will delete only nonterminals of N_i, $i \in [1 \ldots k]$, in the present simulation; hence, they do not interfere with the new nonterminals like S_i'.

We now turn to $GCID(3; 2, 1, 0; 1, 0, 1)$. The only real problem merging $C2$ and $C4$ was that during the simulation of G_i, possibly the symbol S_{i+1}' gets deleted. This can be prevented by requiring the right context of $r_{i \to i+1}$ in the rule that deletes S_{i+1}'. More precisely, the modified rules for P_i will be $r_{i \to i+1} 3.1 : (3, (S_{i+1}', r_{i \to i+1}, \lambda)_{ins}, 2)$ and $r_{i \to i+1} 2.2 : (2, (\lambda, S_{i+1}', r_{i \to i+1})_{del}, 3)$. The remaining technical details are left to the reader.

Remark 1 and more concretely Example 3 shows the claimed strictness of the inclusion. $\square$

Since *LIN* and *MLIN* are known to be closed under reversal [15], by using Corollary 1, we can immediately conclude the next two Theorems (7 and 8):

Theorem 7. $LIN \subsetneq GCID(2; 2, 0, 1; 1, 0, 0)$.

Theorem 8. $MLIN \subsetneq GCID(4; 2, 0, 1; 1, 0, 0) \cap GCID(3; 2, 0, 1; 1, 1, 0)$.

Theorem 9. $LIN \subsetneq GCID(3; 1, 1, 0; 1, 0, 0)$.

Proof. Consider a linear grammar $G = (N, T, P, S)$. We construct a GCID system $\Pi = (3, V, T, \{S\}, H, 1, 1, R)$. The rules from P in G are assumed to be labelled injectively with labels from the set $[1 \ldots |P|]$. The alphabet of Π is $V = N \cup T \cup \{p, p' : p \in [1 \ldots |P|]\}$. The set of rules R of Π is defined as follows.

We simulate the rule $p : X \to Ya$ by the following ins-del rules:

$p1.1 : \ (1, (X, p, \lambda)_{ins}, 3), \quad p1.2 : \ (1, (p, a, \lambda)_{ins}, 2), \quad p1.3 : \ (1, (p', Y, \lambda)_{ins}, 2)$
$p2.1 : \ (2, (p, p', \lambda)_{ins}, 3), \quad p2.2 : \ (2, (\lambda, p', \lambda)_{del}, 1)$
$p3.1 : \ (3, (\lambda, X, \lambda)_{del}, 1), \quad p3.2 : \ (3, (\lambda, p, \lambda)_{del}, 1)$

We simulate the rule $q : X \to aY$ by the following ins-del rules:

$q1.1 : \ (1, (X, q, \lambda)_{ins}, 3), \quad q1.2 : \ (1, (q, q', \lambda)_{ins}, 2), \quad q1.3 : \ (1, (q', Y, \lambda)_{ins}, 2)$
$q2.1 : \ (2, (q, a, \lambda)_{ins}, 3), \quad q2.2 : \ (2, (\lambda, q', \lambda)_{del}, 1)$
$q3.1 : \ (3, (\lambda, X, \lambda)_{del}, 1), \quad q3.2 : \ (3, (\lambda, q, \lambda)_{del}, 1)$

We simulate the rule $f : X \to a$ by the following ins-del rules:

$$f1.1 : \ (1, (X, a, \lambda)_{ins}, 3)$$
$$f3.1 : \ (3, (\lambda, X, \lambda)_{del}, 1)$$

Working of $p : X \to Ya$: Consider the string $\alpha X \beta$ in $C1$. On applying rule $p1.1$, we insert p after X and get $\alpha Xp\beta$ in $C3$. At this point, we have a choice of applying rule $p3.1$ or $p3.2$. In the latter case, the marker p is deleted and we move to $C1$ with $\alpha X \beta$ as the string and this is our starting point. Hence, we use rule $p3.1$ eventually to proceed. Then, X is deleted and we move to $C1$ with $\alpha p \beta$. Now, the rule $p1.1$ cannot be applied since in linear grammars there is at most one nonterminal (in this case, X) in the string that was already deleted in the previous step. Hence, we simulate the rule $X \to Ya$ as follows:

$$(\alpha X \beta)_1 \underset{\Leftarrow_{p3.2}}{\overset{\Rightarrow_{p1.1}}{}} (\alpha Xp\beta)_3 \Rightarrow_{p3.1} (\alpha p \beta)_1 \Rightarrow_{p1.2} (\alpha pa\beta)_2 \Rightarrow_{p2.1} (\alpha pp'a\beta)_3$$
$$\Rightarrow_{p3.2} (\alpha p'a\beta)_1 \Rightarrow_{p1.3} (\alpha p'Ya\beta)_2 \Rightarrow_{p2.2} (\alpha Ya\beta)_1.$$

Working of $q : X \to aY$: Consider the string $\alpha X \beta$ in $C1$. On applying rule $q1.1$, we insert q after X and get $\alpha Xq\beta$ in $C3$. At this point, we have a choice of applying rule $q3.1$ or $q3.2$. In the latter case, the marker q will be deleted and we move back to the starting point. Hence we have to use rule $q3.1$ eventually to proceed. In this case, X is deleted and we move to $C1$ with $\alpha q\beta$ where q' is inserted after q and the string moves to $C2$ with $\alpha qq'\beta$. In $C2$, we can apply the rule $q2.1$ or $q2.2$. On applying $q2.2$, q' is deleted and the string $\alpha q\beta$ will be in $C1$ and we are back to the previous step. This is also depicted in the following derivation. This forces us to apply the rule $q2.1$ and the sequence of rule applications is shown in the derivation. With these arguments, we simulate the rule $X \to aY$ as follows:

$$(\alpha X\beta)_1 \overset{\Rightarrow_{q1.1}}{\underset{\Leftarrow_{q3.2}}{}} (\alpha Xq\beta)_3 \Rightarrow_{q3.1} (\alpha q\beta)_1 \overset{\Rightarrow_{q1.2}}{\underset{\Leftarrow_{q2.2}}{}} (\alpha qq'\beta)_2 \Rightarrow_{q2.1} (\alpha qaq'\beta)_3$$

$$\Rightarrow_{q3.1} (\alpha aq'\beta)_1 \Rightarrow_{q1.3} (\alpha aq'Y\beta)_2 \Rightarrow_{q2.2} (\alpha aY\beta)_1.$$

The working of rule $f : X \to a$ is simple and straightforward. By repeatedly applying p, q, f, we eventually get $(S)_1 \Rightarrow_* (w)_1$. Thus $L(G) \subseteq L(\Pi)$. Moreover, as argued above, no other derivations are possible for Π, entering $C1$ with a string $\alpha X\beta$. So, by induction, $L(G) \supseteq L(\Pi)$ also follows. $\qquad\square$

As LIN is known to be closed under reversal, by using Corollary 1, we have:

Theorem 10. $LIN \subsetneq GCID(3; 1, 0, 1; 1, 0, 0)$.

In the literature, $GCID(4; 1, 1, 0; 1, 0, 1)$, $GCID(4; 1, 0, 1; 1, 1, 0)$ (see [5]) and i) $GCID(5; 1, 1, 0; 1, 1, 0)$, ii) $GCID(5; 1, 1, 0; 1, 0, 1)$, iii) $GCID(5; 1, 1, 0; 2, 0, 0)$, iv) $GCID(5; 1, 0, 1; 2, 0, 0)$, v) $GCID(5; 2, 0, 0; 1, 1, 0)$, vi) $GCID(5; 2, 0, 0; 1, 0, 1)$ (see [12]) describe RE. Thus, the generative power of $GCID(4; 1, 1, 0; 1, 0, 0)$, $GCID(4; 1, 0, 1; 1, 0, 0)$, $GCID(5; 1, 1, 0; 1, 0, 0)$, $GCID(5; 1, 0, 1; 1, 0, 0)$ is open. In the following, we discuss the power of these systems.

Remark 2. As in Remark 1, one can see that the Kleene star of each of the linear languages lies in $GCID(4; 1, 1, 0; 1, 0, 0) \cap GCID(4; 1, 0, 1; 1, 0, 0)$. Inheriting the proof idea of Theorem 6, we deduce the following from Theorems 9 and 10:

Theorem 11. $MLIN \in GCID(5; 1, 1, 0; 1, 0, 0) \cap GCID(5; 1, 0, 1; 1, 0, 0)$.

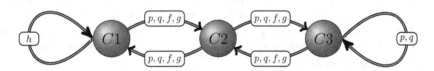

Fig. 1. Control graph structure of Theorem 2; the corresponding simple undirected graph is a path on three vertices, which corresponds to three nested membranes.

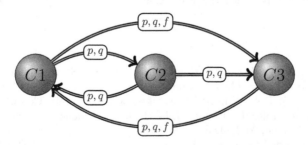

Fig. 2. Control graph structure of Theorem 9; the corresponding simple undirected graph is a cycle on three vertices, which cannot correspond to any nested membrane structure.

6 Conclusions

We have studied GCID systems of various small sizes, proving them to be either computationally complete or able to simulate at least all (meta-)linear languages. Example 2 shows (together with [20]) that two components are more powerful than one for systems of size $(1,1,1;x,y,z)$ with $y+z \leq 1$, $x \in \{0,1\}$. Proving a non-trivial simulation result for the family of context-free languages (say, by GCID systems with size $(3;1,1,0;1,1,0)$) is left open. Also, we have indicated how to simulate Kleene closures of meta-linear languages; it would be therefore interesting to see if the regular closure of the linear languages can be also simulated; refer to [15] for details of this language class.

The *underlying control graph* of a k-GCID system Π is defined to be a graph with k nodes labelled $C1$ through Ck. There exists a directed edge from Ci to Cj if and only if there exists a rule of the form (i,r,j) in R of Π. If the undirected simple graph corresponding to this underlying directed graph is a tree, then Π can be viewed as an insertion-deletion P system (see [5]). In this paper, the underlying graphs of the GCID systems that simulate the families RE and LIN (in Theorems 2, 4 and 5) are trees. Hence, the corresponding results can be immediately also read as results on insertion-deletion P systems. However, one may note that the control graph of the construction of Theorem 9 contains a triangle ($q1.3$ leads from $C1$ to $C2$, $q2.1$ from $C2$ to $C3$ and $q3.1$ from $C3$ to $C1$ in the proof of Theorem 9) and is hence not a tree. Whether or not similar results hold for insertion-deletion P systems remains open. The control graphs of the graph-controlled ins-del systems discussed in this paper are visualized in Figs. 1 and 2 for the case of Theorems 2 and 9, respectively. The annotations given on the edges tells what part of the simulation is responsible for this edge. The according pictures of the simulations in the metalinear cases are even a bit more involved (as we have four components in the first part of Theorem 6) and is hence omitted. However, as there are only connections between $C1$ and $C2$, between $C2$ and $C3$, and between $C3$ and $C4$, this corresponds again to an insertion-deletion P system.

Acknowledgement. The second author would like to acknowledge the project SR/S3/EECE/054/2010, Department of Science and Technology, New Delhi, India.

References

1. Alhazov, A., Krassovitskiy, A., Rogozhin, Y., Verlan, S.: P systems with minimal insertion and deletion. Theoret. Comput. Sci. **412**(1–2), 136–144 (2011)
2. Benne, R. (ed.): RNA Editing: The Alteration of Protein Coding Sequences of RNA. Series in Molecular Biology. Ellis Horwood, Chichester (1993)
3. Biegler, F., Burrell, M.J., Daley, M.: Regulated RNA rewriting: modelling RNA editing with guided insertion. Theoret. Comput. Sci. **387**(2), 103–112 (2007)
4. Chomsky, N., Schützenberger, M.P.: The Algebraic Theory of Context-Free Languages. Studies in Logic and the Foundations of Mathematics, pp. 118–161. North-Holland, Amsterdam (1970)

5. Freund, R., Kogler, M., Rogozhin, Y., Verlan, S.: Graph-controlled insertion-deletion systems. In: McQuillan, I., Pighizzini, G. (eds.) Proceedings Twelfth Annual Workshop on Descriptional Complexity of Formal Systems, DCFS. EPTCS, vol. 31, pp. 88–98 (2010)
6. Geffert, V.: How to generate languages using only two pairs of parentheses. J. Inf. Process. Cybern. EIK **27**(5/6), 303–315 (1991)
7. Haussler, D.: Insertion languages. Inf. Sci. **31**(1), 77–89 (1983)
8. Ivanov, S., Verlan, S.: About one-sided one-symbol insertion-deletion P systems. In: Alhazov, A., Cojocaru, S., Gheorghe, M., Rogozhin, Y., Rozenberg, G., Salomaa, A. (eds.) CMC 2013. LNCS, vol. 8340, pp. 225–237. Springer, Heidelberg (2014)
9. Ivanov, S., Verlan, S.: Universality of graph-controlled leftist insertion-deletion systems with two states. In: Durand-Lose, J., Nagy, B. (eds.) Machines, Computations, and Universality - 7th International Conference, MCU. LNCS, vol. 9288, pp. 79–93. Springer, Switzerland (2015)
10. Kari, L.: On insertion and deletion in formal languages. Ph.D. thesis, University of Turku, Finland (1991)
11. Kari, L., Thierrin, G.: Contextual insertions/deletions and computability. Inf. Comput. **131**(1), 47–61 (1996)
12. Krassovitskiy, A., Rogozhin, Y., Verlan, S.: Computational power of insertion-deletion (P) systems with rules of size two. Nat. Comput. **10**, 835–852 (2011)
13. Kuppusamy, L., Mahendran, A., Krishna, S.N.: Matrix insertion-deletion systems for bio-molecular structures. In: Natarajan, R., Ojo, A. (eds.) ICDCIT 2011. LNCS, vol. 6536, pp. 301–312. Springer, Heidelberg (2011)
14. Kuppusamy, L., Rama, R.: On the power of tissue P systems with insertion and deletion rules. Pre-Proceedings of Workshop on Membrane Computing. Report RGML, vol. 28, pp. 304–318. Univ. Tarragona, Spain (2003)
15. Kutrib, M., Malcher, A.: Finite turns and the regular closure of linear context-free languages. Discrete Appl. Math. **155**(16), 2152–2164 (2007)
16. Margenstern, M., Păun, G., Rogozhin, Y., Verlan, S.: Context-free insertion-deletion systems. Theoret. Comput. Sci. **330**(2), 339–348 (2005)
17. Petre, I., Verlan, S.: Matrix insertion-deletion systems. Theoret. Comput. Sci. **456**, 80–88 (2012)
18. Păun, G., Rozenberg, G., Salomaa, A.: DNA Computing: New Computing Paradigms. Springer, Heidelberg (1998)
19. Takahara, A., Yokomori, T.: On the computational power of insertion-deletion systems. Nat. Comput. **2**(4), 321–336 (2003)
20. Verlan, S.: Recent developments on insertion-deletion systems. Comput. Sci. J. Moldova **18**(2), 210–245 (2010)

Operations on Weakly Recognizing Morphisms

Lukas Fleischer$^{(\boxtimes)}$ and Manfred Kufleitner

FMI, University of Stuttgart, Universitätsstraße 38, 70569 Stuttgart, Germany
{fleischer,kufleitner}@fmi.uni-stuttgart.de

Abstract. Weakly recognizing morphisms from free semigroups onto finite semigroups are a classical way for defining the class of ω-regular languages, i.e., a set of infinite words is weakly recognizable by such a morphism if and only if it is accepted by some Büchi automaton. We consider the descriptional complexity of various constructions for weakly recognizing morphisms. This includes the conversion from and to Büchi automata, the conversion into strongly recognizing morphisms, and complementation. For some problems, we are able to give more precise bounds in the case of binary alphabets or simple semigroups.

1   Introduction

Büchi automata define the class of ω-regular languages. They were introduced by Büchi for deciding the monadic second-order theory of $(\mathbb{N}, <)$ [2]. Since then, ω-regular languages have become an important tool in formal verification, and many other automata models for this language class have been considered; see e.g. [10,13]. Each automaton model has its merits and its disadvantages. Recently, the authors have shown that recognizing morphisms have many nice algorithmic properties [5]. Such morphisms come in two different flavors. Strongly recognizing morphisms admit efficient minimization and complementation, whereas weakly recognizing morphisms can be exponentially more succinct (but there is no minimal weak recognizer and there is no efficient complementation). The situation is similar to the behavior of deterministic and nondeterministic finite automata. The major difference to both nondeterministic finite automata and Büchi automata is that there is an efficient inclusion test for weakly recognizing morphisms [5]. Every strongly recognizing morphism is also weakly recognizing, but the converse is false.

In this paper, we consider the descriptional complexity of various operations on weakly recognizing morphisms and conversions involving nondeterministic Büchi automata (BA) and strongly recognizing morphisms. In each case, we give asymptotically tight bounds. For the conversion of a BA into a weakly recognizing morphism, we give a lower bound which matches the naive upper bound. Our results are summarized in Table 1.

There are some similarities between recognizing morphisms over finite and over infinite words. Strong recognition is the natural counterpart to recognition

---

This work was supported by the DFG grants DI 435/5-2 and KU 2716/1-1.

Published by Springer International Publishing Switzerland 2016. All Rights Reserved
C. Câmpeanu et al. (Eds.): DCFS 2016, LNCS 9777, pp. 126–137, 2016.
DOI: 10.1007/978-3-319-41114-9_10

Table 1. Bounds for the descriptional complexity of various operations.

Operation	Lower bound	Upper bound
BA to weak recognition	2^{n^2} [new]	2^{n^2} [9]
BA to weak recognition, binary alphabet	$2^{(n-1)^2/4}$ [new]	2^{n^2} [9]
Weak recognition to BA	$(n-3)(n+1)/32$ [new]	$n(n+1)$ [9]
Weak recognition to strong recognition	$n2^{n-1}$ [new]	2^{n^2} [10]
Complementation of weak recognition	$n2^{n-1}$ [new]	2^{n^2} [10]
Complementation for simple semigroups	$n2^{n-1}$ [new]	$n2^n$ [new]

for finite words. Nevertheless, in order to prove lower bounds for the conversion of Büchi automata to weakly recognizing morphisms, we first show that bounds for converting nondeterministic finite automata to recognizing morphisms over finite words (with some limitations) also hold for the conversion of Büchi automata to weakly recognizing morphisms. We then use techniques of Sakoda and Sipser [12] and of Yan [14] to obtain tight bounds for the conversion of nondeterministic finite automata to recognizing morphisms. This step is similar to the work of Holzer and König [6]. To the best of our knowledge, our lower bound over finite words for the conversion of an NFA into a recognizing morphism is also a new result.

2   Preliminaries

This section gives a brief overview of some basic definitions from the fields of formal languages, finite automata and semigroup theory. We refer to [10,11] for more detailed introductions.

Words. Let A be a finite *alphabet*. The elements of A are called *letters*. A *finite word* is a sequence $a_1a_2 \cdots a_n$ of letters of A and an *infinite word* is an infinite sequence $a_1a_2 \cdots$. The empty word is denoted by ε. Given an infinite word $\alpha = a_1a_2 \cdots$, we let $\inf(\alpha) \subseteq A$ denote the set of letters in α which occur infinitely often.

Let K be a set of finite words and let L be a set of infinite words. We set $KL = \{u\alpha \mid u \in K, \alpha \in L\}$, $K^n = \{u_1u_2 \cdots u_n \mid u_i \in K\}$, $K^+ = \bigcup_{n \geqslant 1} K^n$ and $K^* = K^+ \cup \{\varepsilon\}$. Moreover, if $\varepsilon \notin K$ we define the *infinite iteration* $K^\omega = \{u_1u_2 \cdots \mid u_i \in K\}$. A natural extension to $K \subseteq A^*$ is $K^\omega = (K \setminus \{\varepsilon\})^\omega \cup \{\varepsilon\}$.

Automata. A *finite automaton* is a 5-tuple $\mathcal{A} = (Q, A, \delta, I, F)$ where Q is a finite set of *states* and A is a finite alphabet. The *transition relation* δ is a subset of $Q \times A \times Q$ and its elements are called *transitions*. The sets I and F are subsets of Q and are called *initial states* and *final states*, respectively.

A *finite run* of a word $a_1a_2 \cdots a_n$ on $\mathcal{A}$ is a sequence $q_0a_1q_1a_1 \cdots q_{n-1}a_nq_n$ such that $q_0 \in I$ and $(q_i, a_{i+1}, q_{i+1}) \in \delta$ for all $i \in \{0, \ldots, n-1\}$. The run is said to *start* in q_0 and *end* in q_n. The word $a_1 a_2 \cdots a_n$ is the *label* of the run. A finite run is called *accepting* if it ends in a final state. A finite word u is said to be *accepted by* $\mathcal{A}$ if there exists an accepting finite run of u on $\mathcal{A}$ and the language *accepted by* $\mathcal{A}$ is the set of all finite words over A^* accepted by $\mathcal{A}$. It is denoted by $L_{\mathrm{NFA}}(\mathcal{A})$.

Analogously, an *infinite run* of a word $a_1 a_2 \cdots$ on $\mathcal{A}$ is an infinite sequence $q_0 a_1 q_1 a_1 \cdots$ such that $q_0 \in I$ and $(q_i, a_{i+1}, q_{i+1}) \in \delta$ for all $i \geqslant 0$. It is called *accepting* if $\inf(q_0 q_1 q_2 \cdots) \cap F \neq \emptyset$. An infinite word α is said to be *Büchi-accepted by* $\mathcal{A}$ if there exists an accepting infinite run of α on $\mathcal{A}$. The language *Büchi-accepted by* $\mathcal{A}$ is the set of all infinite words Büchi-accepted by $\mathcal{A}$ and it is denoted by $L_{\mathrm{BA}}(\mathcal{A})$.

We use the term *run* for both finite and infinite runs if the reference is clear from the context. A language $L \subseteq A^*$ (resp. $L \subseteq A^\omega$) is *regular* (resp. *ω-regular*) if it is accepted (resp. Büchi-accepted) by some finite automaton.

Finite semigroups. A *semigroup morphism* is a mapping $h \colon S \to T$ between two (not necessarily finite) semigroups S and T such that $h(s)h(t) = h(st)$ for all $s, t \in S$. Since we do not consider morphisms of other objects, we use the term *morphism* synonymously. A *subsemigroup* of a semigroup S is a subset that is closed under multiplication. We say that a semigroup T *divides* a semigroup S if there exists a surjective morphism from a subsemigroup of S onto T.

Green's relations are an important tool in the study of semigroups. For the remainder of this subsection, let S be a finite semigroup. We let S^1 denote the monoid that is obtained by adding a new neutral element 1 to S. For $s, t \in S$ let

$s \mathrel{\mathcal{R}} t$ if there exist $q, q' \in S^1$ such that $sq = t$ and $tq' = s$,

$s \mathrel{\mathcal{L}} t$ if there exist $p, p' \in S^1$ such that $ps = t$ and $p't = s$,

$s \mathrel{\mathcal{J}} t$ if there exist $p, q, p', q' \in S^1$ such that $psq = t$ and $p'tq' = s$,

$s \mathrel{\mathcal{H}} t$ if $s \mathrel{\mathcal{R}} t$ and $s \mathrel{\mathcal{L}} t$.

These relations are equivalence relations. The equivalence classes of $\mathcal{R}$ (resp. $\mathcal{L}$, $\mathcal{J}$, $\mathcal{H}$) are called *$\mathcal{R}$-classes* (resp. *$\mathcal{L}$-classes*, *$\mathcal{J}$-classes*, *$\mathcal{H}$-classes*). For $s \in S$, we denote the $\mathcal{R}$-class (resp. $\mathcal{L}$-class) of s by R_s (resp. L_s) and we let $S/\mathcal{R} = \{R_s \mid s \in S\}$ as well as $S/\mathcal{L} = \{L_s \mid s \in S\}$.

A semigroup is called *$\mathcal{J}$-trivial* if each of its $\mathcal{J}$-classes contains exactly one element. A semigroup is called *simple* if it consists of a single $\mathcal{J}$-class. In a finite simple semigroup, the relations $s \mathrel{\mathcal{R}} st \mathrel{\mathcal{L}} t$ hold for all $s, t \in S$. Moreover, each $\mathcal{H}$-class forms a group and all such groups are isomorphic [11]. We will also utilize the following lemma:

Lemma 1. *Let S be a finite simple semigroup and let $x, y, z \in S$ such that $y \mathrel{\mathcal{R}} z$. Then $xy = xz$ implies $y = z$.*

Proof. Suppose that $xy = xz$. Since S is simple, we have $y \mathrel{\mathcal{L}} xy$ and thus, there exists an element $p \in S^1$ such that $pxy = y$. Since $y \mathrel{\mathcal{R}} z$, there exists an element $q \in S^1$ with $yq = z$. It follows that $y = pxy = pxz = pxyq = yq = z$. $\qquad\square$

Recognition by morphisms. Let $h: A^+ \to S$ be a morphism to a finite semi-group S. A pair (s, e) of elements of S is a *linked pair* if $se = s$ and $e^2 = e$. For $s \in S$, we set $[s]_h = h^{-1}(s)$ and if h is understood from the context, we may skip the reference to the morphism in the subscript. A language $L \subseteq A^+$ is *recognized* by a morphism $h : A^+ \to S$ if L is a union of sets $[s_i]$ with $s_i \in S$. A language $L \subseteq A^\omega$ is *weakly recognized* by a morphism $h : A^+ \to S$ if it is a union of sets $[s_i][e_i]^\omega$ where (s_i, e_i) are linked pairs of S. A language $L \subseteq A^\omega$ is *strongly recognized* by a morphism $h : A^+ \to S$ if $[s][t]^\omega \cap L \neq \emptyset$ implies $[s][t]^\omega \subseteq L$ for all $s, t \in S$. It is easy to see that strong recognition implies weak recognition, see e.g. [10, Theorem 2.2]. Moreover, if a morphism strongly recognizes L, it also strongly recognizes its complement $A^\omega \setminus L$. By extension, we also say that a semigroup S recognizes (resp. weakly recognizes, strongly recognizes) a language L if there exists a morphism $h: A^+ \to L$ that recognizes (resp. weakly recognizes, strongly recognizes) L.

For a language $L \subseteq A^+ \cup A^\omega$, we have $u \equiv_L v$ if and only if

$$(xuy)z^\omega \in L \Leftrightarrow (xvy)z^\omega \in L \text{ and}$$
$$z(xuy)^\omega \in L \Leftrightarrow z(xvy)^\omega \in L$$

for all finite words $x, y, z \in A^*$. Keep in mind that $\varepsilon^\omega = \varepsilon$. The relation $\equiv_L$ was introduced by Arnold [1]; it is called the *syntactic congruence* of L. The congruence classes of $\equiv_L$ form the so-called *syntactic semigroup* $A^+/\equiv_L$ and the *syntactic morphism* $h_L: A^+ \to A^+/\equiv_L$ is the natural quotient map. If $L \subseteq A^*$ (resp. $L \subseteq A^\omega$) is regular (resp. ω-regular), the syntactic semigroup of L is finite and h_L recognizes (resp. strongly recognizes) the language L; see [1,10].

3 Lower Bound Techniques

3.1 Proving Lower Bounds for Weakly Recognizing Morphisms

We first consider the general problem of proving lower bounds for the size of weakly recognizing semigroups for a given language L. In the case of recognizing morphisms over finite words and in the case of strongly recognizing morphisms, this is easy since one only needs to compute the syntactic semigroup, which immediately yields a tight lower bound. On the contrary, weakly recognizing morphisms do not admit minimal objects. However, it turns out that one can still use a relaxed version of Arnold's syntactic congruence.

We first prove a combinatorial lemma and then give the main result of this section.

Lemma 2. *Let $u, v \in A^+$ and let (s, e) be a linked pair. Then uv^ω is contained in $[s][e]^\omega$ if and only if there exists a factorization $v = v_1 v_2$ and powers $k, \ell \geqslant 0$ such that ℓ is odd, $h(uv^k v_1) = s$ and $h(v_2 v^\ell v_1) = e$.*

Proof. Let $v = a_1 a_2 \cdots a_n$ with $n \geqslant 1$ and $a_i \in A$. If uv^ω is contained in $[s][e]^\omega$, there exists a factorization $uv^\omega = u'v_1'v_2' \cdots$ such that $h(u') = s$ and $h(v_i') = e$

for all $i \geqslant 1$. Since u and v are finite words, there exist indices $j > i \geqslant 1$, powers $k, \ell \geqslant 1$ and a position $m \in \{1, \ldots, n\}$ such that $u'v_1'v_2' \cdots v_{i-1}' = uv^k a_1 a_2 \cdots a_m$ and $v_i'v_{i+1}' \cdots v_j' = a_{m+1}a_{m+2} \cdots a_n v^\ell a_1 a_2 \cdots a_m$. We set $v_1 = a_1 a_2 \cdots a_m$ and $v_2 = a_{m+1}a_{m+2} \cdots a_n$. Then $v_1 v_2 = v$,

$$h(uv^k v_1) = h(uv^k a_1 a_2 \cdots a_m) \qquad\qquad = h(u'v_1'v_2' \cdots v_{i-1}') = se^{i-1} = s,$$
$$h(v_2 v^\ell v_1) = h(a_{m+1}a_{m+2} \cdots a_n v^\ell a_1 a_2 \cdots a_m) = h(v_i'v_{i+1}' \cdots v_j') = e^{j-i+1} = e.$$

If ℓ is even, we can replace ℓ by $2\ell + 1$ since $h(v_2 v^{2\ell+1} v_1) = h(v_2 v^\ell v_1 v_2 v^\ell v_1) = e^2 = e$. The converse implication is trivial. □

Theorem 3. *Let $L \subseteq A^\omega$ be a language weakly recognized by some morphism $h \colon A^+ \to S$ and let $u, v, z \in A^+$ and $x, y \in A^*$ be words such that one of the following two properties holds:*

1. *$xuyz^\omega \in L$ and $xvyz^\omega \notin L$*
2. *$x(uy)^\omega \in L$ and $x(uyvy)^\omega \notin L$ and $x(vyuy)^\omega \notin L$.*

Then $h(u) \neq h(v)$.

Proof. We consider finite words $u, v \in A^+$ such that $h(u) = h(v)$ and show that in this case, neither of the properties can hold.

If the first property holds, there exists a linked pair (s, e) such that $xuyz^\omega \in [s][e]^\omega \subseteq L$. Thus, by Lemma 2, we have $h(xuyz^k z_1) = s$ and $h(z_2 z^\ell z_1) = e$ for some factorization $z = z_1 z_2$ and powers $k, \ell \geqslant 0$. Now, since $h(xvyz^k z_1) = h(xuyz^k z_1) = s$, we obtain $xvyz^\omega \in [s][e]^\omega \subseteq L$, a contradiction.

If the second property holds, there exists a linked pair (s, e) of S such that $xw^\omega \in [s][e]^\omega \subseteq L$ where $w = uy$. Thus, by Lemma 2, we have $h(xw^k w_1) = s$ and $h(w_2 w^\ell w_1) = e$ for some factorization $w = w_1 w_2$, some power $k \geqslant 0$ and some odd power $\ell \geqslant 0$. Since ℓ is odd $(\ell - 1)/2$ is an integer and we have $h(w_2 (vyuy)^{(\ell-1)/2} vyw_1) = h(w_2 (uy)^\ell w_1) = e$. Now, if k is odd as well, we obtain $h(x(vyuy)^{(k-1)/2} vyw_1) = h(x(uy)^k w_1) = s$ and therefore, $x(vyuy)^\omega \in L$. Equivalently, if k is even, we have $h(x(uyvy)^{k/2} w_1) = h(x(uy)^k w_1) = s$ and hence, $x(uyvy)^\omega \in L$. Both cases contradict Property 2 above. □

The next proposition is another simple, yet useful, tool for proving lower bounds. It allows to transfer bounds from the setting of finite words to infinite words.

Proposition 4. *Let $\mathcal{A} = (Q, A, \delta, I, F)$ and let $a \in A$ be a letter such that for all $q \in Q$ and $q_f \in F$, we have $(q, a, q_f) \in \delta$ if and only if $q = q_f$. Let $K = L_{BA}(\mathcal{A})$ and let $L = L_{NFA}(\mathcal{A})$. Then each semigroup weakly recognizing K has at least $|A^+/{\equiv_L}|$ elements.*

Proof. Let $h \colon A^+ \to S$ be a morphism weakly recognizing K and consider two words $u, v \in A^+$ such that $u \not\equiv_L v$. Then, without loss of generality, there exist $x, y \in A^*$ such that $xuy \in L$ and $xvy \notin L$. This implies $xuya^\omega \in K$ since $(q_f, a, q_f) \in \delta$ for all $q_f \in F$. Equivalently, because of $(q, a, q_f) \notin \delta$ for all $q \in Q \setminus F$ and $q_f \in F$, we have $xvya^\omega \notin K$. By Theorem 3, this yields $h(u) \neq h(v)$. □

3.2   The Full Automata Technique

The *full automata technique* is a useful tool for proving lower bounds for the conversion of automata to other objects. It was introduced by Yan [14] who attributes it to Sakoda and Sipser [12]. The technique works for both accepted and Büchi-accepted languages. However, we will prove the main result of this section only for the setting of finite words and use Proposition 4 to obtain analogous results for infinite words.

Let Q be a finite set and let I, F be subsets of Q. The *full automaton* $\mathcal{F}(Q, I, F)$ is the finite automaton (Q, B, Δ, I, F) defined by $B = 2^{Q^2}$ and by the transition relation $\Delta = \{(p, T, q) \in Q \times B \times Q \mid (p, q) \in T\}$.

Theorem 5. *Let $\mathcal{A} = (Q, A, \delta, I, F)$ be a finite automaton and let $\mathcal{F}(Q, I, F) = (Q, B, \Delta, I, F)$ be the corresponding full automaton. Then the syntactic semigroup of $L_{NFA}(\mathcal{A})$ divides the syntactic semigroup of $L_{NFA}(\mathcal{F}(Q, I, F))$.*

Proof. We first define a morphism $\pi \colon A^+ \to B^+$ by $\pi(a) = \{(p, q) \mid (p, a, q) \in \delta\}$. Let $K = L_{NFA}(\mathcal{F}(Q, I, F))$ and let $L = L_{NFA}(\mathcal{A})$. It suffices to show that $\pi(u) \equiv_K \pi(v)$ implies $u \equiv_L v$. Thus, consider $u, v \in A^+$ such that $\pi(u) \equiv_K \pi(v)$. In particular, for all $x, y \in A^*$, we have $\pi(xuy) \in K$ if and only if $\pi(xvy) \in K$. By the definition of π, we have $\pi(w) \in K$ if and only if $w \in L$ for all $w \in A^+$. Using the equivalence from above, this yields $xuy \in L$ if and only if $xvy \in L$ for all $x, y \in A^*$, thereby proving that $u \equiv_L v$. $\square$

4   From Automata to Weakly Recognizing Morphisms

The standard construction for converting a finite automaton $\mathcal{A}$ to a recognizing morphism is the so-called *transition semigroup* of $\mathcal{A}$. For a given word $u \in A^+$, it encodes for each pair (p, q) of states whether there is a run of u on $\mathcal{A}$ starting in p and ending in q. Thus, for a finite automaton with n states the transition semigroup has 2^{n^2} elements. For details on the construction, we refer to [10,11]. We show that this construction is optimal.

Theorem 6. *Let $\mathcal{A}$ be a finite automaton with n states. Then there exists a semigroup recognizing $L_{NFA}(\mathcal{A})$ (resp. weakly recognizing $L_{BA}(\mathcal{A})$) which has at most 2^{n^2} elements and this bound is tight.*

Proof. Each language that is accepted (resp. Büchi-accepted) by $\mathcal{A}$ is recognized (resp. weakly recognized) by the transition semigroup of $\mathcal{A}$ which has size 2^{n^2}.

To show that this is optimal, we consider the full automaton $\mathcal{F}(N, N, N) = (N, B, \Delta, N, N)$ where $N = \{1, \ldots, n\}$ and let $L = L_{NFA}(\mathcal{F}(N, N, N))$. For two different letters $X, Y \in B$ we may assume, without loss of generality, that there exist $p, q \in N$ such that $(p, q) \in X \setminus Y$. With $P = \{(p, p)\}$ and $Q = \{(q, q)\}$, we then have $PXQ \in L$ and $PYQ \notin L$. Thus, $X \not\equiv_L Y$. This shows that $B^+/{\equiv_L}$ has at least $|B| = 2^{n^2}$ elements.

Noting that the transitions labeled by the letter $\{(q, q) \mid q \in N\}$ form self-loops at each state, the Büchi case immediately follows by Proposition 4. $\square$

The proof of the optimality result requires a large alphabet that grows super-exponentially in the number of states of the automaton. A natural restriction is considering automata over fixed-size alphabets.

By a result of Chrobak [3], the size of the syntactic semigroup of an unary language accepted by a finite automaton of size n is in $2^{O(\sqrt{n \log n})}$ (note that since unary languages are commutative, the syntactic monoid is isomorphic to the minimal deterministic automaton). Over infinite words, the unary case is uninteresting since the only language over the alphabet $A = \{a\}$ is $\{a^\omega\}$.

For binary alphabets, a lower bound can be obtained by combining the full automata technique with a result from the study of semigroups of binary relations [7, Proposition 6]. In order to keep the paper self-contained, we present a proof that is adapted to finite automata and does not require any knowledge of binary relations.

Theorem 7. *Let $A = \{a, b\}$ and let n be an odd natural number. There exists a language $L \subseteq A^+$ (resp. $L \subseteq A^\omega$) and a finite automaton with n states accepting (resp. Büchi-accepting) L, such that each semigroup recognizing (resp. weakly recognizing) L has at least $2^{(n-1)^2/4}$ elements.*

Proof. We first analyze the case of finite words. Let $m = (n-1)/2$ and let $M = \{1, \ldots, m\}$. We consider the automaton $\mathcal{A}$ depicted below and let $L = L_{\mathrm{NFA}}(\mathcal{A})$.

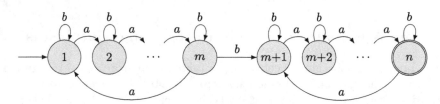

For $1 \leqslant i, j \leqslant m$ we first define $p_{i,j} = (m+j-i)m-i$ and $q_{i,j} = (m+i-j+2)m+i$. Furthermore, we set $u_{i,j} = a^{p_{i,j}} b a^{q_{i,j}}$. We claim that for each i, j there exists a path from state k to ℓ labeled by $u_{i,j}$ if and only if $(k, \ell) = (i, j+m)$ or $k = \ell$.

The two a-cycles have length m and $m+1$, respectively. Since for each pair (i, j) we have $p_{i,j} + q_{i,j} = 2m(m+1)$ and since one can always stay in the same state when reading the letter b, there clearly exists a path from each state to itself labeled by $u_{i,j}$. Now, fix some (i, j) and let $(k, \ell) = (i, j+m)$. We have $i + p_{i,j} = (m+j-i)m$ which means that, when starting in state i, one can reach state m by reading $a^{p_{i,j}}$. Being in state m, one of the b-transitions leads to state $m+1$. From there on, we make a single step backwards whenever reading the factor a^m. Thus, by reading the word $a^{q_{i,j}}$, we perform $(m+i-j+2)-i = m-j+2$ backward steps in total, finally reaching state $n+1-(m-j+2) = 2m+2-(m-j+2) = m+j = \ell$. The converse direction of our claim follows immediately since the automaton is deterministic when restricted to a-transitions and since one can only reach states $\ell > m$ by using the transition $(m, b, m+1)$.

For $X \subseteq M \times M$, we now define u_X as the concatenation of all $u_{i,j}$ with $(i, j) \in X$, where the factors are ordered according to their indices (i, j).

By the above argument, it is easy to see that there is a path from state i to $j + m$ labeled by u_X if and only if $(i, j) \in X$. Since there are $2^{m^2} = 2^{(n-1)^2/4}$ subsets of the Cartesian product $M \times M$, it remains to show that for different subsets $X, Y \subseteq M \times M$, we have $u_X \not\equiv_L v_Y$. To this end, assume without loss of generality that $(i, j) \in X \setminus Y$. Then $a^{i-1} u_X a^{n-j} \in L$ but $a^{i-1} u_Y a^{n-j} \notin L$, as desired.

For the Büchi case note that for all $i \in Q$, we have $(i, b, n) \in \delta$ if and only if $i = n$. Therefore, by Proposition 4 and the arguments above, the smallest semigroup weakly recognizing $L_{\mathrm{BA}}(\mathcal{A})$ has at least $2^{(n-1)^2/4}$ elements.   □

The construction above does not reach the 2^{n^2} bound obtained when using a larger alphabet. However, this is not surprising, given the following result.

Proposition 8. *Let $m \in \mathbb{N}$ be a fixed integer and let A be an alphabet of size m. Then there exists an integer $n_m \geqslant 1$ such that for each finite automaton $\mathcal{A}$ over A with $n \geqslant n_m$ states, the language $L_{NFA}(\mathcal{A}) \subseteq A^*$ (resp. $L_{BA}(\mathcal{A}) \subseteq A^\omega$) is recognized (resp. weakly recognized) by a morphism onto a semigroup with less than 2^{n^2} elements.*

We do not give a full proof of the proposition here, but the claim essentially follows from a careful analysis of the subsemigroup of the transition semigroup generated by the transitions corresponding to the letters in A. Applying Devadze's Theorem [4,8] to the matrix representation of this subsemigroup shows that it is proper, i.e., smaller than the full transition semigroup itself.

5   From Weakly Recognizing Morphisms to Automata

The well-known construction to convert weakly recognizing morphisms to finite automata with a Büchi-acceptance condition has quadratic blow-up [10]. We show that this is optimal up to a constant factor.

Theorem 9. *Let $A = \{a, b\}$, let $n \geqslant 3$, and let $L = \bigcup_{i=1}^{n} (ba^i b A^*)^\omega$. Then there exists a semigroup with $4n + 3$ elements that weakly recognizes L and every finite automaton Büchi-accepting L has at least $n(n+1)/2$ states.*

Proof. We first define a semigroup $S = \{a^i, a^i b, ba^i, ba^i b \mid 1 \leqslant i \leqslant n\} \cup \{b, bb, 0\}$ by the multiplication $0 \cdot s = s \cdot 0 = 0$ for all $s \in S$ and

$$b^\ell a^i b^r \cdot b^m a^j b^s = \begin{cases} bb & \text{if } i = j = 0 \\ b^\ell a^{i+j} b^s & \text{if } r = m = 0 \text{ and } 1 \leqslant i + j \leqslant n \\ 0 & \text{if } r = m = 0 \text{ and } i + j > n \\ b^\ell a^i b & \text{otherwise} \end{cases}$$

where $\ell, m, r, s \in \{0, 1\}$ and $i, j \in \{0, \ldots, n\}$. The morphism $h \colon A^+ \to S$ defined by $h(a) = a$ and $h(b) = b$ now weakly recognizes L since L is the union of all sets $[ba^i b][ba^i b]^\omega$ with $1 \leqslant i \leqslant n$.

Now assume that we are given a finite automaton $\mathcal{A} = (Q, A, \delta, I, F)$ such that $L_{BA}(\mathcal{A}) = L$. For each $i \in \{1, \ldots, n\}$, we consider the word $\alpha_i = (ba^i b)^\omega$ and let r_i be an accepting run of α_i. We first show that for $i \neq j$, we have $\inf(r_i) \cap Q \cap \inf(r_j) = \emptyset$, and then prove that $|\inf(r_i) \cap Q| \geq i$ for $1 \leq i \leq n$. Together, this yields

$$|Q| \geq \sum_{i=1}^{n} |\inf(r_i) \cap Q| \geq \sum_{i=1}^{n} i = n(n+1)/2.$$

Let $i, j \in \{1, \ldots, n\}$ such that $i \neq j$. We assume for the sake of contradiction that there exists a state $q \in Q$ with $q \in \inf(r_i)$ and $q \in \inf(r_j)$. Let $u \in ba^i bA^*$ be a prefix of α_i such that r_i visits q after reading u. Let $v \in A^*$ be a factor of α_j such that there exists a finite run labeled by v, which starts and ends in q, visits at least one final state and such that $v^\omega = (ba^j b)^\omega$ or $v^\omega = a^k b(ba^j b)^\omega$ for some $k \in \{0, \ldots, j\}$. Obviously, we then have $uv^\omega \in L_{BA}(\mathcal{A})$ but $uv^\omega \notin L$, a contradiction.

For the second part of the proof, assume again for the sake of contradiction that $|\inf(r_i) \cap Q| < i$ for some accepting run r_i of α_i. Then inside each $ba^i b$-factor, a state is visited twice and we can apply the standard pumping argument to show that a word in $A^\omega \setminus L_{BA}(\mathcal{A})$ has an accepting run as well.   □

6   Complementation

To date, the best construction for complementing weakly recognizing morphisms is the so-called *strong expansion* [10]. Given a morphism $h \colon A^+ \to S$, the strong expansion of h is a morphism $g \colon A^+ \to T$ which strongly recognizes all languages weakly recognized by h. If S has n elements, the size of T is 2^{n^2}. The purpose of this section is to give a lower bound for complementation. At the same time, the established bound also serves as a lower bound for the conversion of weak recognition to strong recognition since each morphism strongly recognizing a language also strongly recognizes its complement.

Complementing weakly recognizing morphisms is easy in the case of $\mathcal{J}$-trivial semigroups since each language weakly recognized by a $\mathcal{J}$-trivial semigroup S is already strongly recognized by S, i.e., there is no need the compute the strong expansion if the $\mathcal{J}$-classes of the input are trivial already. In order to establish a lower bound, we thus consider the class of simple semigroups, which is dual to $\mathcal{J}$-trivial semigroups in the sense that simple semigroups consist of a single $\mathcal{J}$-class only.

Proposition 10. *Let $n \geq 1$ be an arbitrary integer and let $A = \{a_1, a_2, \ldots, a_n\}$. The language $L = \bigcup_{i=1}^{n} (a_i A^*)^\omega$ is weakly recognized by a simple semigroup with n elements and every semigroup weakly recognizing $A^\omega \setminus L$ has at least $n2^{n-1}$ elements.*

Proof. The alphabet A can be extended to a semigroup by defining an associative operation $a \circ b = a$ for all $a, b \in A$. Now, the morphism $h \colon A^+ \to (A, \circ)$ given by $h(a) = a$ for all $a \in A$ weakly recognizes L. The semigroup $(A, \circ)$ contains $|A| = n$ elements and it is simple because we have $a \mathcal{L} b$ for all $a, b \in A$.

Now, let $h \colon A^+ \to S$ be a morphism weakly recognizing $A^\omega \setminus L$. For a letter $b \in A$ and a subset $B \subseteq A \setminus \{b\}$, let $u_{b,B}$ be the uniquely defined word $ba_{i_1} a_{i_2} \cdots a_{i_\ell}$ such that $i_1 < i_2 < \cdots < i_\ell$ and $\{a_{i_1}, a_{i_2}, \ldots, a_{i_\ell}\} = B$. Consider two letters $b, c \in A$ and subsets $B \subseteq A \setminus \{b\}$, $C \subseteq A \setminus \{c\}$. If $b \neq c$, we have $u_{b,B} c^\omega \notin L$ and $u_{c,C} c^\omega \in L$. If $B \neq C$ we may assume, without loss of generality, that there exists a letter $a \in B \setminus C$. In this case, we have $a u_{c,C} c^\omega \notin L$ but $a(u_{b,B} u_{c,C})^\omega \in L$ and $a(u_{c,C} u_{b,B})^\omega \in L$. By Theorem 3, this suffices to conclude that $h(u_{b,B}) \neq h(u_{c,C})$ whenever $b \neq c$ or $B \neq C$ and therefore, S contains at least $|A| \, 2^{|A|-1} = n 2^{n-1}$ elements. □

Rather surprisingly, the established lower bound turns out to be asymptotically tight in the case of simple semigroups. More generally, for simple semigroups, the construction of the strong expansion can be improved such that only $n 2^n$ elements are needed. This will be proved in the remainder of this section.

We start with a morphism $h \colon A^+ \to S$ onto a simple semigroup with $n = |S|$ elements. Since S is simple, there exists a surjective mapping $\gamma \colon S \to G$ onto a finite group G that becomes a bijection when restricted to a single $\mathcal{H}$-class. Therefore, the mapping $\pi \colon (S/\mathcal{R}) \times G \times (S/\mathcal{L}) \to S$ with $\pi^{-1}(s) = (R_s, \gamma(s), L_s)$ for all $s \in S$ is well-defined and bijective. Moreover, for $s, t \in S$, we write $R_t \cdot s$ to denote the element $\pi(R_t, \gamma(s), L_s)$.

Let $T = \{(s, X) \mid s \in S, X \subseteq S\}$ and let $g \colon A^+ \to T$ be defined by

$$g(u) = (h(u), \{R_{h(q)} \cdot h(p) \mid p, q \in A^+, pq = u\})$$

for all $u \in A^+$. The set T can be extended to a semigroup by defining an associative multiplication

$$(s, X) \cdot (t, Y) = (st, X \cup \{R_t \cdot s\} \cup \hat{Y})$$

where $\hat{Y}$ denotes the set $\{\pi(R_y, \gamma(s(R_t \cdot y)), L_y) \mid y \in Y\}$. Under this extension, the mapping g becomes a morphism.

The following three technical lemmas capture important properties of the construction and are needed for the main proof.

Lemma 11. *Let $s, t \in S$. Then $R_t \cdot s$ is the unique element x such that $x \mathcal{R} t$, $x \mathcal{L} s$ and $\gamma(x) = \gamma(s)$ or, equivalently, the unique element x such that $x \mathcal{H} ts$ and $\gamma(x) = \gamma(s)$.*

Proof. Let $x = R_t \cdot s$. We have $(R_x, \gamma(x), L_x) = \pi^{-1}(x) = \pi^{-1}(R_t \cdot s) = (R_t, \gamma(s), L_s)$. Together with the fact that π is bijective, this establishes the first claim. For the second claim, note that since S is simple, $x \mathcal{R} t$ is equivalent to $x \mathcal{R} ts$ and $x \mathcal{L} s$ is equivalent to $x \mathcal{L} ts$. □

Lemma 12. *Let $u \in A^+$ with $g(u) = (s, X)$ and let $x \in S$. Then $x \in X \cup \{s\}$ if and only if there exists a factorization $u = pq$ with $p \in A^+$ and $q \in A^*$ such that $x \mathcal{H} h(qp)$ and $\gamma(x) = \gamma(h(p))$.*

Proof. Obviously, we have $x = s$ if and only if there exists a factorization $u = pq$ with $p = u$ and $q = \varepsilon$ satisfying the properties described above. Thus, it suffices to consider factorizations where $p, q \in A^+$. By Lemma 11, such a factorization exists if and only if $x = R_{h(q)} \cdot h(p)$ which is, in turn, equivalent to $x \in X$ by the definition of g. □

Lemma 13. *Let (t, f) be a linked pair of S, let $((s, X), (e, Y))$ be a linked pair of T and let $\alpha \in [(s, X)]_g [(e, Y)]_g^\omega$. Then $\alpha \in [t]_h [f]_h^\omega$ if and only if $tq = s$, $pq = e$, $qp = f$, $R_q \cdot t \in X$ and $R_q \cdot p \in Y$ for some $p, q \in S$.*

Proof. For the direction from left to right, let $\alpha = uv_1v_1'v_2v_2' \cdots$ such that $g(u) = (s, X)$, $g(v_iv_i') = (e, Y)$, $h(uv_1) = t$ and $h(v_i'v_{i+1}) = f$ for all $i \geqslant 1$. Furthermore, we assume without loss of generality that $v_i, v_i' \neq \varepsilon$ for all $i \geqslant 1$ and that $h(v_1) = h(v_2)$. We set $p = h(v_1) = h(v_2)$ and $q = h(v_1')$. Now, $tq = h(uv_1v_1') = se = s$, $pq = h(v_1v_1') = e$ and $qp = h(v_1'v_2) = f$. Moreover, by the definition of g, we have $R_q \cdot t = R_{h(v_1')} \cdot h(uv_1) \in X$ and $R_q \cdot p = R_{h(v_1')} \cdot h(v_1) \in Y$.

For the converse implication, note that by Lemma 12, there exists a factorization $\alpha = uv_1v_1'v_2v_2' \cdots$ such that $h(u) = s$, $h(v_iv_i') = e$, $R_{h(v_i')} \cdot h(uv_1) = R_q \cdot t$ and $R_{h(v_i')} \cdot h(v_i) = R_q \cdot p$ for all $i \geqslant 1$. Since S is simple, $h(v_i) \mathcal{R} h(v_iv_i') = e \mathcal{R} p$ and $h(v_i) \mathcal{L} (R_{h(v_i')} \cdot h(v_i)) = (R_q \cdot p) \mathcal{L} p$ for all $i \geqslant 1$. Furthermore, $\gamma(h(v_i)) = \gamma(R_{h(v_i')} \cdot h(v_i)) = \gamma(R_q \cdot p) = \gamma(p)$. Together, this yields $h(v_i) = p$ by Lemma 11. Similarly, we have $h(v_i') \mathcal{R} (R_{h(v_i')} \cdot h(v_i)) = (R_q \cdot p) \mathcal{R} q$ and thus, $ph(v_i') = h(v_iv_i') = pq$ implies $h(v_i') = q$ for all $i \geqslant 1$ by Lemma 1. This shows that $h(uv_1) = sp = tqp = tf = t$ and $h(v_i'v_{i+1}) = qp = f$. We conclude that $\alpha \in [t][f]^\omega$. □

Theorem 14. *Let $h \colon A^+ \to S$ be a morphism onto a simple semigroup of size $n = |S|$ that weakly recognizes a language $L \subseteq A^\omega$. Then there exists a morphism $g \colon A^+ \to T$ to a semigroup of size $|T| = n2^n$ that strongly recognizes L.*

Proof. The construction we use is the one described in the introduction of this section. Consider a linked pair $((s, X), (e, Y))$ of T as well as two infinite words $\alpha, \beta \in [(s, X)][(e, Y)]^\omega$. If $\alpha \in L$, there exists a linked pair (t, f) of S such that $\alpha \in [t][f]^\omega \subseteq L$. Lemma 13 immediately yields $\beta \in [t][f]^\omega \subseteq L$, thereby showing that g strongly recognizes L. □

7   Discussion and Open Problems

We presented lower bound techniques and gave tight bounds for the conversion between finite automata and weakly recognizing morphisms. One can use techniques similar to those described in Sect. 4 to obtain a 3^{n^2} lower bound for the conversion of finite automata with transition-based Büchi acceptance to strongly recognizing morphisms. However, with the usual state-based Büchi acceptance criterion, the analysis becomes much more involved and it is not clear whether the 3^{n^2} upper bound can be reached. Analogously, there is no straightforward adaptation of the conversion of weakly recognizing morphisms into Büchi automata in Sect. 5 to strongly recognizing morphisms. It would be interesting to see whether the quadratic lower bound also holds in this setting.

Another open problem is to close the remaining gaps between the upper and the lower bounds. This is particularly true for the complexity of complementation and the conversion of weakly recognizing morphisms to strong recognition. We showed that there is an exponential lower bound and gave an asymptotically optimal construction for simple semigroups which was a first candidate for semigroups that are hard to complement. It is easy to adapt this construction to families of semigroups where the size of each $\mathcal{J}$-class is bounded by a constant. However, for the general case, the gap between $n2^{n-1}$ and 2^{n^2} remains.

Beyond that, another direction for future research is to investigate whether any of the bounds can be improved by considering the size of the accepting set, i.e., the number of linked pairs used to describe a language.

Acknowledgments. We thank the anonymous referees for several useful suggestions which helped to improve the presentation of this paper.

References

1. Arnold, A.: A syntactic congruence for rational ω-languages. Theoret. Comput. Sci. **39**, 333–335 (1985)
2. Büchi, J.R.: Weak second-order arithmetic and finite automata. Zeitschrift für mathematische Logik und Grundlagen der Mathematik **6**, 66–92 (1960)
3. Chrobak, M.: Finite automata and unary languages. Theoret. Comput. Sci. **47**(2), 149–158 (1986)
4. Devadze, H.M.: Generating sets of the semigroup of all binary relations in a finite set. Doklady Akademii Nauk BSSR **12**, 765–768 (1968)
5. Fleischer, L., Kufleitner, M.: Efficient algorithms for morphisms over omega-regular languages. In: Proceedings of the FSTTCS 2015. LIPIcs, vol. 45, pp. 112–124. Dagstuhl Publishing (2015)
6. Holzer, M., König, B.: On deterministic finite automata and syntactic monoid size. Theoret. Comput. Sci. **327**(3), 319–347 (2004)
7. Kim, K.H., Roush, F.W.: Two-generator semigroups of binary relations. J. Math. Psychol. **17**(3), 236–246 (1978)
8. Konieczny, J.: A proof of Devadze's theorem on generators of the semigroup of boolean matrices. Semigroup Forum **83**(2), 281–288 (2011)
9. Pécuchet, J.: Variétés de semis groupes et mots infinis. In: Proceedings of the STACS 1986, pp. 180–191 (1986)
10. Perrin, D., Pin, J.-É.: Infinite Words. Pure and Applied Mathematics, vol. 141. Elsevier, Amsterdam (2004)
11. Pin, J.-É.: Varieties of Formal Languages. North Oxford Academic, London (1986)
12. Sakoda, W.J., Sipser, M.: Nondeterminism and the size of two way finite automata. In: Proceedings of the STOC 1978, pp. 275–286. ACM Press (1978)
13. Thomas, W.: Automata on infinite objects. In: Handbook of Theoretical Computer Science, chap. 4, pp. 133–191. Elsevier (1990)
14. Yan, Q.: Lower bounds for complementation of omega-automata via the full automata technique. Logical Methods Comput. Sci. **4**(1), 1–20 (2008)

Descriptional Complexity of Bounded Regular Languages

Andrea Herrmann, Martin Kutrib, Andreas Malcher[(⊠)],
and Matthias Wendlandt

Institut für Informatik, Universität Giessen, Arndtstr. 2, 35392 Giessen, Germany
{kutrib,malcher,matthias.wendlandt}@informatik.uni-giessen.de

Abstract. We investigate the descriptional complexity of the subregular language classes of (strongly) bounded regular languages. In the first part, we study the costs for the determinization of nondeterministic finite automata accepting strongly bounded regular languages. The upper bound for the costs is larger than the costs for determinizing unary regular languages, but lower than the costs for determinizing arbitrary regular languages. In the second part, we study for (strongly) bounded languages the deterministic operational state complexity of the Boolean operations as well as the operations reversal, concatenation, and iteration. In detail, we present upper and lower bounds and we develop for the proof of the lower bounds a tool that exploits the number of different colorings of cycles occurring in deterministic finite automata accepting bounded languages.

1 Introduction

Descriptional complexity is an area of theoretical computer science in which one of the main questions is how succinctly a formal language can be described by a formalism in comparison with other formalisms. A fundamental result is the exponential trade-off between nondeterministic (NFA) and deterministic finite automata (DFA) [16]. A further exponential trade-off is known to exist between unambiguous and deterministic finite automata, whereas the trade-offs between alternating and deterministic finite automata [14] as well as between deterministic pushdown automata and deterministic finite automata [19] are bounded by doubly-exponential functions.

The question of whether the costs for determinization remain exponential even for subclasses of the regular languages, called *subregular* language classes, has been studied in [3,4] for unary languages and in [18] for finite languages. A systematic study of the problem for subregular language classes is provided in [2]. In this paper, we study with *bounded regular languages* another subregular language class which has not gained much attention yet apart from the fundamental paper [7] in which bounded regular languages are introduced and, for example, characterization theorems are established. In general, a language is called (strongly) bounded if it is a subset of $a_1^* a_2^* \cdots a_k^*$, where $a_1, a_2, \ldots, a_k$ are

© IFIP International Federation for Information Processing 2016
Published by Springer International Publishing Switzerland 2016. All Rights Reserved
C. Câmpeanu et al. (Eds.): DCFS 2016, LNCS 9777, pp. 138–152, 2016.
DOI: 10.1007/978-3-319-41114-9_11

(pairwise distinct) symbols. Bounded languages have been investigated to a large extent in the literature. We would like to mention that basic results are summarized in [8] and that there exist strong connections to counter machines which are shown, for example, in [11,13]. The descriptional complexity of bounded context-free languages has first been studied in [15] and recently in [12].

In this paper, we start to investigate the descriptional complexity of bounded regular languages. We provide the necessary definitions and notions in Sect. 2. Additionally, we summarize the closure properties for (strongly) bounded regular languages. In Sect. 3 we compute the costs for determinizing NFAs accepting strongly bounded regular languages. As bounded languages are both an extension of unary languages and a restriction of arbitrary languages, we obtain a 'similar' result for the upper bound of the determinization costs that turns out to be larger than the costs for determinizing unary NFAs, but lower than the costs for determinizing arbitrary NFAs. Finally, we study in Sect. 4 the deterministic operation problem for bounded regular languages which quantifies the costs (in terms of states of a DFA) of operations on (strongly) bounded regular languages such as union, intersection, concatenation, iteration, and reversal. The deterministic operation problem for regular languages has initially been studied in [20,21]. Nowadays, there exists a vast literature on the deterministic and nondeterministic operational state complexity of subregular languages, and we refer to the recent survey [6]. Here, we complement these findings with the results for (strongly) bounded regular languages. It should be noted that we devise a new tool to obtain lower bounds for bounded regular languages which may be of interest on its own.

2   Preliminaries and Closure Properties

Let Σ^* denote the set of all words over the finite alphabet Σ. The *empty word* is denoted by λ, and $\Sigma^+ = \Sigma^* \setminus \{\lambda\}$. The *reversal* of a word w is denoted by w^R. For the *length* of w we write $|w|$. For the number of occurrences of a symbol a in w we use the notation $|w|_a$. We denote the powerset of a set S by 2^S. By $\gcd(x_1, x_2, \ldots, x_k)$ we denote the *greatest common divisor* of the integers $x_1, x_2, \ldots, x_k$, and by $lcm(x_1, x_2, \ldots, x_n)$ their *least common multiple*. If two numbers x and y are relatively prime, that is $\gcd(x, y) = 1$, we write $x \perp y$.

A *nondeterministic finite automaton* (NFA) is a system $M = \langle S, \Sigma, \delta, s_0, F \rangle$, where S is the finite set of *internal states*, Σ is the finite set of *input symbols*, $s_0 \in S$ is the *initial state*, $F \subseteq S$ is the set of *accepting states*, and $\delta : S \times \Sigma \to 2^S$ is the partial *transition function*. The *language accepted* by M is $L(M) = \{ w \in \Sigma^* \mid \delta(s_0, w) \cap F \neq \emptyset \}$, where the transition function is recursively extended to $\delta : S \times \Sigma^* \to 2^S$.

A finite automaton is *deterministic* (DFA) if and only if $|\delta(s, a)| = 1$, for all $s \in S$ and $a \in \Sigma$. In this case we simply write $\delta(s, a) = p$ for $\delta(s, a) = \{p\}$ assuming that the transition function is a mapping $\delta : S \times \Sigma \to S$. So, any DFA is complete, that is, the transition function is total, whereas for NFAs it is possible that δ maps to the empty set.

A language $L \subseteq \Sigma^*$ is said to be *bounded* if and only if $L \subseteq a_1^* a_2^* \cdots a_k^*$, for $k \geq 1$ and $a_i \in \Sigma$, $1 \leq i \leq k$. It is *strongly bounded* if all letters $a_1, a_2, \ldots, a_k$ are pairwise different. It should be noted that in the literature bounded languages which are defined as above are often called letter-bounded languages. Moreover, if symbols $a_1, a_2, \ldots, a_k$ are replaced by fixed words $w_1, w_2, \ldots, w_k$, a language $L \subseteq w_1^* w_2^* \cdots w_k^*$ is called word-bounded. However, in this paper we confine ourselves to investigating only bounded and strongly bounded languages over symbols.

The closure properties of bounded and strongly bounded languages are summarized in Table 1. Although both language classes are not closed under all operations, it is well known that the regular languages are closed under all operations. This allows to study the deterministic state complexity of all operations for (strongly) bounded regular languages.

Table 1. Summary of closure properties of the language families discussed.

	$-$	$\cup$	$\cap$	R	$\cdot$	$*$
Bounded regular	no	yes	yes	yes	yes	no
Strongly bounded regular	no	no	yes	yes	no	no

3 Determinization

It is well known that the costs for the simulation of a nondeterministic finite automaton with n states by a deterministic finite automaton can be limited by 2^n many states using the power set construction. On the other hand, several different NFAs are known that reach this bound exactly. In the unary case the upper bound as well as the lower bound collapses to $e^{\Theta(\sqrt{n \cdot \log n})}$. Considering the costs for determinization in the strongly bounded regular case, one may expect that the bounds for the conversion might be strictly in between the bounds for the general and the unary case. In the following, we present an upper bound which is slightly more costly than in the unary case.

Theorem 1. *Let A be an NFA with n states accepting a strongly bounded regular language $L(A)$ over the alphabet Σ and $m = n \cdot |\Sigma|^2 + |\Sigma|$. Then an equivalent DFA A' with at most $|\Sigma|^2 \cdot e^{|\Sigma| \cdot \Theta(\sqrt{m \cdot \log(m)})}$ many states can be constructed.*

Proof. Given an NFA $A = \langle S, \Sigma, s_0, \delta, F \rangle$ with n states accepting a strongly bounded regular language $L(A)$ over the alphabet Σ, we will construct an equivalent DFA. We may assume that $\Sigma = \{a_1, a_2, \ldots, a_k\}$ and $L(A) \subseteq a_1^* a_2^* \cdots a_k^*$ with $k \geq 2$ and pairwise distinct $a_i \in \Sigma$ with $1 \leq i \leq k$.

The principal idea of the construction is to divide automaton A into 'unary' sections $S_{a_1}, S_{a_2}, \ldots, S_{a_k}$ according to the read input symbols, to determinize these unary subautomata, and finally to reassemble the different deterministic subautomata to an equivalent DFA. In the first step, we construct an equivalent

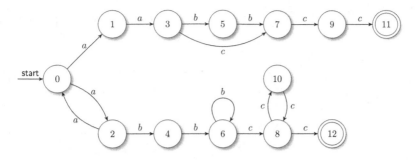

Fig. 1. An NFA A accepting a strongly bounded regular language.

NFA A' with the property that each state of A' has incoming edges with at most one type of symbol. We define $A' = \langle S', \Sigma, s_0', \delta', F' \rangle$, where s_0' is a new state, $S' = \{ s_x \mid s \in S, x \in \Sigma \} \cup \{s_0'\}$ and $F' = \{ s_x \mid s \in F, x \in \Sigma \} \cup \{ s_0' \mid s_0 \in F \}$. If $s' \in \delta(s,a)$ for some $s, s' \in S$ and $a \in \Sigma$, then define $s_a' \in \delta'(s_x, a)$ for all $x \in \Sigma$. If $s \in \delta(s_0, a)$ for some $s \in S$ and $a \in \Sigma$, then define $s_a \in \delta'(s_0', a)$. The number of states of A' is at most $n \cdot |\Sigma| + 1$. Now, each state of A' has incoming edges with at most one type of symbol and a state s_a is defined to be in section S_a of A', for $a \in \Sigma$. Furthermore, the initial state s_0' of A' is only visited in the first computation step and then never again. It is the single state in the special section S_{init}.

The next step is to modify A' in such a way that it has no states having more than one edge to another section labeled with the same symbol. Assume that there is some state $s \in S'$ having $\ell \geq 2$ edges labeled with a leading to section S_a. Then we add a new state s' to section S_a, add for every edge from s labeled with an a to some state s'' in S_a an edge from s' labeled with λ to the state s'', and replace the ℓ old edges by one edge from s to s' labeled with a. These modifications introduce at most $|\Sigma|$ new states for every state as well as λ-moves to the NFA, but preserve the given language. Moreover, for every input symbol $a \in \Sigma$, all nondeterministic moves on a take place inside section S_a. The number of states of A' is now at most $n \cdot |\Sigma|^2 + |\Sigma|$.

The first two steps of the construction based on the example NFA shown in Fig. 1 are depicted in Fig. 2.

In the following, the sections are successively determinized. We start with the determinization of the first section S_{a_1} having n_1 many states and define the set I_{a_1} of incoming states as the set of all states with incoming edges from other sections. Here, $I_{a_1} = \delta'(s_0', a_1)$ consists of one state only, since there are no states having more than one edge to another section labeled with the same symbol. Additionally, we define the set O_{a_1} of states with outgoing edges to other sections as $O_{a_1} = \{ s \in S_{a_1} \mid r \in \delta'(s, a_j) \text{ for some } r \in S' \text{ and } k \geq j > 1 \}$. For the state $s \in I_{a_1}$ we construct an NFA A_s as subautomaton of A' with state set S_{a_1} and s as initial state. Furthermore, all edges labeled with a_i such that $i \neq 1$ are removed. Finally, we eliminate λ-moves applying the construction given in [10] which does not increase the number of states. Additionally, we set all outgoing

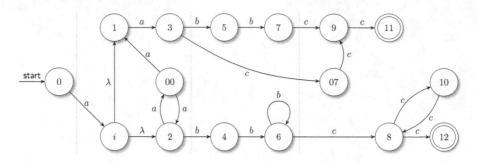

Fig. 2. The first two steps of the construction of A'. States 00 and 07 are added so that the initial state has no incoming edges and every state has only incoming edges with the same label. State i is added to ensure that there are no states having more than one edge to another section labeled with the same symbol.

states from O_{a_1} as accepting. This is done to avoid that states with outgoing edges possibly disappear in the determinization process. Such states will later be rechanged to non-accepting states and completed with the outgoing edges to other sections. Thus, the NFA A_s accepts the unary language $\{ w \in a_1^* \mid \delta'(s, w) \cap (F' \cup O_{a_1}) \neq \emptyset \}$ and has at most n_1 states. Next, we apply the construction given in [3] to obtain a DFA A'_s such that $L(A'_s) = L(A_s)$. According to [3] the costs for determinizing A_s are bounded by $e^{\Theta\left(\sqrt{n_1 \cdot \log(n_1)}\right)}$. Since n_1 is bounded by $m = n \cdot |\Sigma|^2 + |\Sigma|$, we get an upper bound of $\ell_1 = e^{\Theta\left(\sqrt{m \cdot \log(m)}\right)}$ many states. Next, we construct an NFA A'' based on A' by replacing automaton A_s in section S_{a_1} of A' by its deterministic version. Additionally, let s' be the initial state of the DFA A'_s, then all transitions in A'' that link to state s are redirected to the initial state s' of A'_s. As a result of the construction, the DFA A'_s implemented in A'' has two different types of accepting states. The first type are the original accepting states where some input is accepted in A'. The second type are the states where the outgoing edges have to be placed. These have to be changed into non-accepting states and the outgoing edges have to be added to A''.

To find out which states in section S_{a_1} have to be accepting and which states have to be connected with other sections, we do the following considerations. Let B be an NFA accepting a unary language having a single accepting state. Then the lengths of the words of the accepted language $L(B)$ can be described by a finite set $\mathcal{E}$ of equations of the form $g(x) = z_g \cdot x + y_g$, where $x, y_g, z_g \geq 0$ are integers (see, for example, [3]). Looking at the equivalent DFA B', we obtain that for each equation $g \in \mathcal{E}$ there are one or more accepting states in B' indicating the divisibility of the input words according to g.

Now, we consider again for state $s \in I_{a_1}$ the NFA A_s and its equivalent DFA A'_s. First, we set all states in A'_s non-accepting. Second, for every accepting state f in A_s such that $f \in F'$ (thus, being an original accepting state in A'), we consider an NFA $A_{s,f}$ based on A_s where f is the only accepting state and we determine the set of equations $\mathcal{E}$ for $A_{s,f}$. Based on the obtained divisibilities we set the corresponding states in A'_s as accepting. Third, for every outgoing state $o \in O_{a_1}$ linking with input symbol a_j to some state $s_j \in S_j$ for some $k \geq j > 1$, we consider an NFA $A_{s,o}$ based on A_s where o is the only accepting state and we determine the set of equations $\mathcal{E}$ for $A_{s,o}$. Based on the obtained divisibilities we add to the corresponding states in A'_s outgoing edges labeled with a_j to state s_j. We notice that this adding of edges may introduce nondeterminism in A'_s. To remove such possible nondeterministic moves, we do the following: for every state $q \in A'_s$ having more than one outgoing edges labeled by some a_j with $k \geq j > 1$, we introduce a new state p to section S_{a_j}, replace all outgoing a_j-edges from q by outgoing λ-edges from p, and add one a_j-edge from q to p. Note that the removing of nondeterministic moves adds at most ℓ_1 states to each section S_{a_j}. Finally, we rename the NFA A'' with a determinized section S_{a_1} to A' and start the determinization of the next section S_{a_2}.

Again, we define the set I_{a_2} of states with incoming edges from other sections and the set O_{a_2} of states with outgoing edges to other sections. Formally,

$$I_{a_2} = \{\, s \in S_{a_2} \mid s \in \delta'(r, a_2) \text{ for some } r \in S' \text{ and } r \notin \delta'(q, a_2) \text{ for all } q \in S' \,\},$$
$$O_{a_2} = \{\, s \in S_{a_2} \mid r \in \delta'(s, a_j) \text{ for some } r \in S' \text{ and } k \geq j > 2 \,\}.$$

For each state s in I_{a_2} we construct an automaton A_s in a similar way as above. Thus, A_s accepts the unary language $\{\, w \in a_2^* \mid \delta'(s, w) \cap (F' \cup O_{a_2}) \neq \emptyset \,\}$ and has at most $n_2 + 1$ states, if s has been added by removing nondeterministic moves in the previous step, and at most n_2 states otherwise. Next, we determinize A_s and obtain an equivalent DFA A'_s with at most

$$e^{\Theta\left(\sqrt{(m+1)\cdot\log(m+1)}\right)} = e^{\Theta\left(\sqrt{m\cdot\log(m)}\right)} = \ell_1$$

many states. Then we construct an NFA A'' based on A' by replacing automaton A_s in section S_{a_2} of A' by its deterministic version and all transitions in A' that link to state s are redirected in A'' to the initial state of A'_s. Finally, we determine the accepting states of A'_s as well as the connections from outgoing states to other sections, and we remove possibly introduced nondeterminism. Having done this for all $s \in I_{a_2}$ we rename the NFA A'' with determinized sections S_{a_1} and S_{a_2} again to A'. The size of the determinized section S_{a_2} can be calculated as follows: we have at most $m + \ell_1 = e^{\Theta\left(\sqrt{m\cdot\log(m)}\right)} = \ell_1$ states in I_{a_2}. Each determinization costs at most ℓ_1 states. Thus, we obtain $\ell_2 = \ell_1^2$ as an upper bound for the determinization costs of section S_{a_2}. Again, note that the removing of nondeterministic moves adds at most ℓ_2 states to each section S_{a_j} with $k \geq j > 2$.

We continue the construction by determinizing successively the following sections in a similar way as described above. The costs for determinizing section S_{a_i} with $3 \leq i \leq k$ can be calculated as follows. There are at most $m + \ell_1 + \ell_2 + \cdots + \ell_{i-1}$ states in I_{a_i} and each determinization costs at most ℓ_1 states. By setting $\ell_i = \ell_1^i$, we obtain $i \cdot \ell_i$ as total upper bound.

After determinizing all sections $S_{a_1}, S_{a_2}, \ldots, S_{a_k}$ we obtain a DFA A' being equivalent to A and the number of states of A' is bounded by the function

$$1 + \ell_1 + 2\ell_2 + \cdots + k \cdot \ell_k \leq k^2 \ell_1^k = |\Sigma|^2 \cdot e^{|\Sigma| \cdot \Theta\left(\sqrt{m \cdot \log(m)}\right)}. \qquad \square$$

4 Deterministic Operational State Complexity

This section is devoted to studying the deterministic operational state complexity of the family of strongly bounded regular languages, that is, the languages are given by DFAs. Clearly, the known upper bounds for general regular languages apply also here. Moreover, every unary language is also (strongly) bounded. So, the known lower bounds for unary languages apply here as well. In [21] it has been shown that the tight bounds for Boolean operations coincide for general regular and unary regular languages. In the unary case the lower bound requires the numbers of states to be relatively prime. In [17] unary regular languages are studied whose deterministic state complexities are not relatively prime. Here we can derive the following corollary for strongly bounded regular languages.

Corollary 2. *For any integers $m, n \geq 1$ let A be an m-state and B be an n-state DFA that accept strongly bounded languages.*

1. *Then m states are <u>sufficient</u> and necessary in the worst case for a DFA to accept the language $\overline{L(A)}$.*
2. *Then $m \cdot n$ states are sufficient for a DFA to accept the language $L(A) \cap L(B)$ (respectively $L(A) \cup L(B)$).*
3. *If $m \perp n$, then there exist a unary m-state DFA A and a unary n-state DFA B (with the same input symbol) such that any DFA accepting $L(A) \cap L(B)$ (respectively $L(A) \cup L(B)$) needs at least $m \cdot n$ states.*

Notice that the languages $L(A) \cup L(B)$ and $\overline{L(A)}$ are not necessarily strongly bounded. However, since they are regular they are accepted by DFAs in any case.

In the following, we turn to the operations reversal, iteration, and concatenation for which the deterministic state complexities of general and unary languages are different (see, for example, the summary in Table 2). So, an immediate question is to what extent the state complexity of strongly bounded languages is strictly in between both cases. Since the deterministic state complexities for unary languages are well known, we suppose that the strongly bounded languages that are investigated in the remainder of this section are defined over an alphabet of size at least two. In other words, we consider strongly bounded languages $L \subseteq a_1^* a_2^* \cdots a_k^*$ such that $k \geq 2$.

4.1 A Tool for Constructing Lower Bound Witnesses

A widely used method to show lower bounds is to define an infinite family of witness languages so that the sizes of the minimal automata accepting them establish the bound. In order to allow the construction of witnesses as well as to determine the necessary sizes of the automata, lower bound techniques are very helpful. For example, in proofs dealing with the nondeterministic state complexity on regular languages specified by NFAs, the so-called fooling set technique can be used [1,9].

Here we first present a tool for the definition of lower bound witnesses, that is, for the construction of DFAs accepting (strongly) bounded languages. The idea is based on the number of possibilities to color a cycle of a DFA whose edges are labeled with the same input letter. All cycles in a DFA accepting a (strongly) bounded language have this unary form. We use the two colors *gray* (g) and *white* (w).

Let S_c be a (sub)set of states of a given DFA that build a cycle on some fixed input letter. The *set of all colorings* of S_c with colors from $\{g, w\}$ is $X = \{ f \mid f : S_c \to \{g, w\} \}$. So, there are $|X| = 2^{|S_c|}$ different such colorings.

Example 3. The DFA A depicted in Fig. 3 has a cycle on input letter b, where the states of the cycle are $S_c = \{1, 2, 3, 4\}$. The coloring shown at the top of the figure is $f_1 \in X$ with $f_1(1) = g$, $f_1(2) = g$, $f_1(3) = w$, and $f_1(4) = w$. ∎

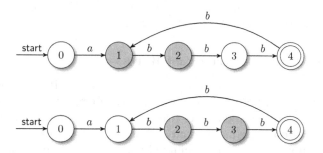

Fig. 3. Coloring of the cycle of DFA A from Example 3. Edges to the rejecting sink state are omitted.

For the clarity of presentation, colorings are written as words. For example, coloring f_1 can be written as $ggww$, and the set of all such colorings is $X = \{wwww, gwww, wgww, \ldots, gggg\}$. Moreover, a coloring can be uniquely identified by the set of states that are colored by w. For example, f_1 is given by $\{3, 4\}$.

Two colorings f_1 and f_2 are said to be equivalent if f_1 can be obtained from f_2 by applying the transition function. More precisely, two colorings f_1 and f_2 are equivalent if and only if there is some $\ell \geq 0$ so that $f_1(i) = f_2(\delta(i, x^\ell))$, for all $i \in S_c$. Here, S_c is the set of cycle states and the cycle is on input symbol x.

Example 4. Let $f_2 \in X$ with $f_2(1) = w$, $f_2(2) = g$, $f_2(3) = g$, and $f_2(4) = w$ be the coloring shown at the bottom of Fig. 3. Then f_1 and f_2 are equivalent, since $f_1(i) = f_2(\delta(i, b))$, for all $i \in \{1, 2, 3, 4\}$. ∎

Now, we turn to determine the number of possibilities to color a cycle with inequivalent colorings. The number depends only on the number of states in the cycle. For example, for four states we obtain the following equivalence classes $\{wwww\}$, $\{gwww, wgww, wwgw, wwwg\}$, $\{ggww, wggw, wwgg, gwwg\}$, $\{gggg\}$, $\{gggw, wggg, gwgg, ggwg\}$, and $\{gwgw, wgwg\}$, and thus six possibilities.

Now we consider the cyclic group G generated by the cyclic permutation $\langle(12\cdots|S_c|)\rangle$. The group naturally operates on S_c. Moreover, for $\sigma \in G$ and $f \in X$, let $\sigma f \in X$ be defined as $\sigma f(s) = f(\sigma^{-1}(s))$, for all $s \in S_c$. With this operation, G acts on the set X of colorings as well. So, two colorings are equivalent if and only if they are in the same orbit of G. Therefore, the number of possibilities to color a cycle with inequivalent colorings coincides with the number of orbits of G acting on X. This number can be determined by Polya's enumeration lemma that is a generalization of the well-known Burnside lemma on the number of orbits of a group action on a set (see, for example, [5, Chap. 8]). In the particular case of a cyclic group generated by a cyclic permutation $\langle(12\cdots n)\rangle$ and two colors, the number or orbits is $\frac{1}{n}\sum_{d|n}\varphi(d)\cdot 2^{\frac{n}{d}}$, where $d|n$ denotes the positive divisors of n and $\varphi(d) = |\{1 \leq k \leq d \mid \gcd(k,d) = 1\}|$ is Euler's function. For example, for $n = 4$ we have $d|n = \{1,2,4\}$. Since $\varphi(1) = 1$, $\varphi(2) = 1$, and $\varphi(4) = 2$, the number of orbits and, identically, the number of inequivalent colorings is $\frac{1}{4}(1\cdot 2^4 + 1\cdot 2^{\frac{4}{2}} + 2\cdot 2^{\frac{4}{4}}) = 6$.

4.2   Reversal, Concatenation, and Iteration

The first operation we consider in detail is the reversal. It turns out that the upper bound and lower bound can be described by an exponential function which is slightly smaller than in the case of arbitrary regular languages. On the other hand, in comparison with unary regular languages we obtain an exponential increase. The upper bound in the bounded case is derived from the observation that any DFA accepting some bounded language over an alphabet with at least two elements must have a rejecting sink state.

Theorem 5. *Let $k,n \geq 2$ be two integers and A be an n-state DFA that accepts a (strongly) bounded language $L(A) \subseteq a_1^* a_2^* \cdots a_k^*$. Then 2^{n-1} states are sufficient for a DFA to accept the language $L(A)^R$.*

Proof. Every non-unary DFA $A = \langle S, \Sigma, \delta, s_0, F\rangle$ accepting a (strongly) bounded language necessarily has a rejecting sink state, say $e \in S$. Now an NFA for the reversal of $L(A)$ is constructed by interchanging the initial state with the accepting states and reversing the direction of the transitions. The NFA is determinized which yields a DFA $A' = \langle 2^S, \Sigma, \delta', s_0', F'\rangle$ accepting $L(A)^R$. Since for all states $p,q \in 2^S$ so that $p = q \cup \{e\}$ we have $\delta'(p,v) \in F'$ if and only if $\delta'(q\cup\{e\},v) \in F'$ if and only if $\delta'(q,v) \in F'$, for all $v \in \Sigma^*$, the states p and q are equivalent. We conclude that A' has at most 2^{n-1} states. $\square$

In order to show the lower bound $2^{n-2}+1$ the coloring of cycles is exploited. Next, the construction of the witness DFAs is given, then we analyze a witness for six states. Finally, the general case is proven.

Let $n > 3$ be an integer. The DFA $A_n = \langle S_n, \Sigma, \delta_n, 0, \{n-2\}\rangle$ is constructed as follows (see Fig. 4): $S_n = \{0, 1, \ldots, n-2, e\}$ where e denotes the rejecting sink state, $\Sigma = \{a, a_1, a_2, \ldots, a_k\}$ with $k = \frac{1}{n-2}\left(\sum_{d|n-2} \varphi(d) \cdot 2^{\frac{n-2}{d}}\right) - 1$, and

$$\delta_n(i, a) = \begin{cases} (i+1) \bmod n-2 & \text{for } 0 \le i \le n-3 \\ e & \text{otherwise} \end{cases}.$$

The transition function is still incomplete. Now we consider the colorings of the cycle, that is, of the states $\{0, 1, \ldots, n-3\}$, whereby we disregard $gg \cdots g$. From above it is known that there remain $k = \frac{1}{n-2}\left(\sum_{d|n-2} \varphi(d) \cdot 2^{\frac{n-2}{d}}\right) - 1$ inequivalent colorings. From each equivalence class M_j one element m_j, $1 \le j \le k$, is chosen and identified by the states that are colored white. For example, $wwgw$ is identified by $\{0, 1, 3\}$. Now, the definition of the transition function is completed by setting

$$\delta_n(i, a_j) = \begin{cases} n-2 & \text{if } i \in m_j \\ e & \text{otherwise} \end{cases}$$

for $1 \le j \le k$. The DFA A_n accepts the language

$$L(A_n) = \bigcup_{j=1}^{k} \bigcup_{i \in m_j} (a^{n-2})^* a^i a_j \subseteq a^* a_1^* a_2^* \cdots a_k^*.$$

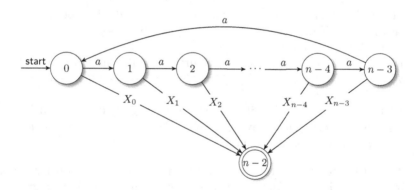

Fig. 4. The witness DFA A_n for reversal. The set of all a_j with $i \in m_j$ are denoted by X_i. Edges to the rejecting sink state are omitted.

Example 6. There are six inequivalent possibilities to color the 4-state cycle of A_6. Disregarding the coloring where all states are gray, the five equivalence classes in question are

$$M_1 = \{\{0\}, \{1\}, \{2\}, \{3\}\}, \quad M_2 = \{\{0, 1\}, \{1, 2\}, \{2, 3\}, \{3, 0\}\},$$
$$M_3 = \{\{0, 2\}, \{1, 3\}\}, \qquad M_4 = \{\{0, 1, 2\}, \{1, 2, 3\}, \{2, 3, 0\}, \{3, 0, 1\}\},$$
$$M_5 = \{\{0, 1, 2, 3\}\}.$$

Choosing $m_1 = \{3\}$, $m_2 = \{2,3\}$, $m_3 = \{1,3\}$, $m_4 = \{1,2,3\}$, $m_5 = \{0,1,2,3\}$ yields $X_0 = \{a_5\}$, $X_1 = \{a_3, a_4, a_5\}$, $X_2 = \{a_2, a_4, a_5\}$, $X_3 = \{a_1, a_2, a_3, a_4, a_5\}$ in Fig. 4.   ∎

Theorem 7. *For any integer $n > 3$, there exists an n-state DFA A that accepts a (strongly) bounded language such that any DFA accepting $L(A)^R$ needs at least $2^{n-2} + 1$ states.*

Proof. We use the DFA $A_n = \langle S_n, \Sigma, \delta_n, 0, \{n-2\} \rangle$ from above as witness. To show that A_n is minimal, consider two states $p, q \in \{0, 1, \ldots, n-3\}$. Let M_r denote the equivalence class with the colorings that color only one state white, and let $m_r = \{s\}$. Then $\delta_n(p, a^x a_r) = n - 2$ and $\delta_n(q, a^x a_r) \neq n - 2$, for $x = n - 2 - |p - s|$, since a_r sends only state s to the sole accepting state $n - 2$. Clearly, the states $n - 2$ and $p \in \{0, 1, \ldots, n-3\}$ are inequivalent.

The NFA $B_n = \langle P_n, \Sigma, \nu_n, p_{n-2}, \{p_0\} \rangle$ with $P_n = \{p_0, p_1, \ldots, p_{n-2}\}$ and

$$\nu_n(p_{n-2}, a_i) = \bigcup_{j \in m_i} p_j, \text{ for } 1 \leq i \leq k, \text{ and}$$
$$\nu_n(p_i, a) = p_j \text{ with } i = (j+1) \bmod (n-2), \text{ for } 0 \leq i \leq n-3,$$

accepts the language $L(A_n)^R$. Notice that $n - 2 \notin m_j$ for all $1 \leq j \leq k$.

By applying the powerset construction, the NFA B_n is determinized which yields the DFA $A'_n = \langle S'_n, \Sigma, \delta'_n, p_{n-2}, F'_n \rangle$, where $S'_n = 2^{P_n \setminus \{p_{n-2}\}} \cup \{p_{n-2}\}$, $F'_n = \{ T \in 2^{P_n \setminus \{p_{n-2}\}} \mid T \cap \{p_0\} \neq \emptyset \}$, $\delta'_n(\{p_{n-2}\}, a_i) = \bigcup_{j \in m_i} p_j$, for $1 \leq i \leq k$, $\delta'_n(\{p_{n-2}\}, a) = \emptyset$, and $\delta'_n(T, a) = \bigcup_{t \in T} \nu(t, a)$, for $T \in 2^{P_n \setminus \{p_{n-2}\}}$.

The DFA A'_n accepts $L(A)^R$ and has $2^{n-2} + 1$ states. By the construction of A_n and since the equivalence classes of colorings partition the set $2^{P_n \setminus \{p_{n-2}\}}$, all states of A'_n are reachable.

In order to show that A'_n is minimal, first consider the states $\{p_{n-2}\}$ and $R_i \in S'_n$, for $0 \leq i \leq n - 3$. For $p_l \in R_i$ we have $\delta'_n(R_i, a^l) \in F'_n$ while $\delta'_n(\{p_{n-2}\}, a^l) \notin F'_n$.

Now let R_i and R_j be two different states from $S'_n \setminus \{p_{n-2}\}$ and let p_l be in their symmetric difference, say, $p_l \in R_i \setminus R_j$. Then $\delta'_n(R_i, a^l) \in F'_n$ while $\delta'_n(R_j, a^l) \notin F'_n$. Therefore, A'_n is minimal.   □

Next, we turn to the operation iteration. Here, we will obtain tight upper and lower bounds that lie strictly in between the bounds for unary regular and arbitrary regular languages. Roughly speaking, the bounds for unary regular languages are quadratic and for arbitrary regular languages exponential. The bound for strongly bounded regular languages turns out to be the sum of a quadratic and an exponential function, where the quadratic function depends on the number of states of the first part of the given DFA and the exponential function depends on the number of remaining states. The partitioning of the state set of a DFA accepting a strongly bounded language $L \subseteq a_1^* a_2^* \cdots a_k^*$ into two sets is done, roughly speaking, as follows: the first set is given by all states that are reachable with words from a_1^*. Since the DFA is deterministic, these states form a line or a line followed by a cycle in the state graph.

More precisely, let $A = \langle S, \Sigma, \delta, s_0, F \rangle$ be a minimal DFA accepting a strongly bounded language $L \subseteq a_1^* a_2^* \cdots a_k^*$ and let e denote the rejecting sink state of A if it exists. In the sequel, the set of states $q \in S$ with $q \neq e$ such that there exists a word v from $a_1^* a_2^* \cdots a_k^* \setminus a_1^*$ with $\delta(s_0, v) = q$ is denoted by S_2. The set $S \setminus (S_2 \cup \{e\})$ is denoted by S_1.

So, all states from S_1 are reachable only by words of the form a_1^*. For $k \geq 2$, we have $S = S_1 \cup S_2 \cup \{e\}$. The next theorem shows the upper bound for the iteration.

Theorem 8. *Let $n_1 \geq 2$ and $n_2 \geq 1$ be two integers and A be an $(n_1 + n_2 + 1)$-state DFA with state set S that accepts a strongly bounded language, so that $S = S_1 \cup S_2 \cup \{e\}$ with $|S_1| = n_1$ and $|S_2| = n_2$. Then $(n_1 - 1)^2 + 2^{n_2} + 2$ states are sufficient for a DFA to accept the language $L(A)^*$.*

In order to show a matching lower bound the coloring of cycles is exploited. Next, the construction of a $(2n + 1)$-state witness DFA is given. Let $n \geq 1$ be an integer. The DFA $B_n = \langle S, \Sigma, \delta_n, 0, \{n-1, 2n-1\} \rangle$ is constructed as follows (see Fig. 5): $S = \{0, 1, \ldots, 2n-1, e\}$ where e denotes the rejecting sink state, $\Sigma = \{a, b, a_1, a_2, \ldots, a_k\}$ with $k = \frac{1}{n} \left(\sum_{d|n} \varphi(d) \cdot 2^{\frac{n}{d}} \right) - n - 1$, and

$$\delta_n(i, a) = \begin{cases} (i+1) \bmod n & \text{for } 0 \leq i \leq n-1 \\ e & \text{otherwise} \end{cases},$$

$$\delta_n(i, b) = \begin{cases} i + n & \text{for } 0 \leq i \leq n-1 \\ (i+1) \bmod n & \text{for } n \leq i \leq 2n-1 \\ e & \text{otherwise} \end{cases}.$$

In order to complete the definition of the transition function we consider colorings of the cycle on input letter b, that is, of the states $\{n, n+1, \ldots, 2n-1\}$, whereby we disregard $gg \cdots g, wgg \cdots g, wwgg \cdots g, \cdots, ww \cdots wg$, and $ww \cdots w$. There remain $k = \frac{1}{n} \left(\sum_{d|n} \varphi(d) \cdot 2^{\frac{n}{d}} \right) - n - 1$ inequivalent colorings. From each equivalence class M_j one element m_j, $1 \leq j \leq k$, is chosen and identified by the states that are colored white. The states in m_j are denoted by $r_{0,j}, r_{1,j}, \ldots, r_{|m_j|-1,j}$. The definition of the transition function is completed by setting

$$\delta_n(i, a_j) = \begin{cases} r_{i,j} & \text{if } 0 \leq i \leq n-1 \text{ and } r_{i,j} \text{ is defined} \\ e & \text{otherwise} \end{cases}$$

for $1 \leq j \leq k$. The DFA B_n accepts the language

$$L(B_n) = (a^n)^* a^{n-1} \cup \bigcup_{j=0}^{n-1} (a^n)^* a^j b^{n-j} (b^n)^* \cup \bigcup_{j=1}^{k} \bigcup_{i=0}^{|m_j|-1} (a^n)^* a^i a_j b^{2n-1-r_{i,j}} (b^n)^*,$$

that is, $L(B_n) \subseteq a^* a_1^* a_2^* \cdots a_k^* b^*$.

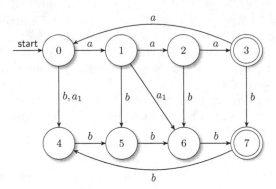

Fig. 5. The witness DFA B_4 for iteration. Edges to the rejecting sink state are omitted.

Example 9. From the six inequivalent possibilities to color the four-state cycle (see Example 6) of B_4 on input letter b only the sole equivalence class $M_1 = \{\{4,6\}, \{5,7\}\}$ remains. Choosing $m_1 = \{4,6\}$ yields $r_{0,1} = 4$ and $r_{1,1} = 6$. So, $\delta_n(0, a_1) = 4$ and $\delta(1, a_1) = 6$ are defined (see Fig. 5).   ∎

Theorem 10. *For any integers $n_1 = n_2 \geq 1$, there exists an $(n_1 + n_2 + 1)$-state DFA A that accepts a (strongly) bounded language such that any DFA accepting $L(A)^*$ needs at least $(n_1 - 1)^2 + 2^{n_2} + 2$ states.*

The final operation we consider is the concatenation. Again, the structure of (strongly) boundedness allows to reduce the descriptional complexity compared with the general case. As for iteration we obtain that the upper bound is described by the sum of a quadratic and an exponential function, where the number of states of the first DFA appear as quadratic resp. linear factor in both addends. As is done for iteration, the states of the second DFA are partitioned into two parts and the number of states of the first part appear as linear factor in the quadratic addend and as exponential factor in the other addend. The proof of the next theorem gives a detailed construction of the upper bound for the concatenation of two strongly bounded regular languages.

Theorem 11. *Let $m, n_1, n_2 \geq 1$ be integers, A be an m-state DFA, and A' be an $(n_1 + n_2 + 1)$-state DFA with state set S' that accept strongly bounded languages, so that $S' = S'_1 \cup S'_2 \cup \{e'\}$ with $|S'_1| = n_1$ and $|S'_2| = n_2$. Then in total $m^2 n_1 + (2m - 1)2^{n_2}$ states are sufficient for a DFA to accept the language $L(A)L(A')$.*

The currently best known lower bound for the concatenation of strongly bounded languages is derived from the concatenation of unary languages. In [21] it is shown that for any $m, n \geq 1$ with $\gcd(m, n) = 1$ there exist an m-state DFA A and an n-state DFA A' accepting unary (and thus strongly bounded) languages so that any DFA that accepts the concatenation $L(A)L(A')$ has at least mn states. The results on the deterministic state complexity obtained in this section are summarized in Table 2.

Table 2. Summary of the deterministic state complexity of the operations studied in this section. The upper and lower bounds for bounded regular languages are obtained in this section. The results for unary regular and arbitrary regular languages may be found, for example, in [6,21].

	Unary regular	Bounded regular	Regular
$L_1 \cup L_2$	$\leq mn$	$\leq mn$	mn
$L_1 \cap L_2$	$\geq mn$, if $\gcd(m,n) = 1$	$\geq mn$, if $\gcd(m,n) = 1$	
$\overline{L}$	m	m	m
$L_1 L_2$	$\leq mn$	$\leq m^2 n_1 + (2m - 1)2^{n_2}$	$(2m - 1)2^{n-1}$
	$\geq mn$, if $\gcd(m,n) = 1$	$\geq mn$, if $\gcd(m,n) = 1$	
L^*	$(m - 1)^2 + 1$	$(m_1 - 1)^2 + 2^{m_2} + 2$	$2^{m-1} + 2^{m-2}$
L^R	m	$\leq 2^{m-1}$	2^m
		$\geq 2^{m-2} + 1$	

5 Conclusions

In this paper, we have studied the descriptional complexity of (strongly) bounded regular languages. We have described a procedure for determinizing nondeterministic finite automata accepting strongly bounded regular languages. The obtained upper bound on the number of states is close to the known upper bound for the determinization of unary nondeterministic finite automata. Moreover, we have determined the deterministic state complexity of several operations on strongly bounded regular languages, in particular, of the operations reversal, iteration, and concatenation. The resulting upper and lower bounds are basically strictly in between the known bounds for unary and arbitrary regular languages. As interesting points for further research on the topic we would like to mention the improvement of the lower bound on concatenation, the study of additional operations, and the investigation of the nondeterministic state complexity of operations. Another interesting question is to look more closely at the size of the alphabets of the witness languages for the lower bounds. In the proofs given in this paper, the size is depending on the given number of states. It would clearly be of interest to study fixed alphabets or to consider the size of the alphabet as an additional parameter for upper and lower bounds.

References

1. Birget, J.C.: Intersection and union of regular languages and state complexity. Inform. Process. Lett. **43**, 185–190 (1992)
2. Bordihn, H., Holzer, M., Kutrib, M.: Determination of finite automata accepting subregular languages. Theor. Comput. Sci. **410**(35), 3209–3222 (2009)
3. Chrobak, M.: Finite automata and unary languages. Theoret. Comput. Sci. **47**(2), 149–158 (1986)

4. Chrobak, M.: Errata to "Finite automata and unary languages". Theoret. Comput. Sci. **302**, 497–498 (2003)
5. Erickson, M.J.: Introduction to Combinatorics. Wiley, New York (1996)
6. Gao, Y., Moreira, N., Reis, R., Yu, S.: A survey on operational state complexity. CoRR abs/1509.03254 (2015). http://arxiv.org/abs/1509.03254
7. Ginsburg, S., Spanier, E.H.: Bounded regular sets. Proc. Amer. Math. Soc. **17**(5), 1043–1049 (1966)
8. Ginsburg, S.: The Mathematical Theory of Context-Free Languages. McGraw Hill, New York (1966)
9. Glaister, I., Shallit, J.: A lower bound technique for the size of nondeterministic finite automata. Inform. Process. Lett. **59**, 75–77 (1996)
10. Hopcroft, J.E., Ullman, J.D.: Introduction to Automata Theory, Languages, and Computation. Addison-Wesley, Reading (1979)
11. Ibarra, O.H.: Reversal-bounded multicounter machines and their decision problems. J. ACM **25**(1), 116–133 (1978)
12. Ibarra, O.H., Ravikumar, B.: On bounded languages and reversal-bounded automata. Inf. Comput. **246**, 30–42 (2016)
13. Ibarra, O.H., Seki, S.: Characterizations of bounded semilinear languages by one-way and two-way deterministic machines. Int. J. Found. Comput. Sci. **23**(6), 1291–1306 (2012)
14. Leiss, E.L.: Succinct representation of regular languages by Boolean automata. Theoret. Comput. Sci. **13**, 323–330 (1981)
15. Malcher, A., Pighizzini, G.: Descriptional complexity of bounded context-free languages. Inf. Comput. **227**, 1–20 (2013)
16. Meyer, A.R., Fischer, M.J.: Economy of description by automata, grammars, and formal systems. In: SWAT 1971, pp. 188–191. IEEE (1971)
17. Pighizzini, G., Shallit, J.: Unary language operations, state complexity and Jacobsthal's function. Int. J. Found. Comput. Sci. **13**, 145–159 (2002)
18. Salomaa, K., Yu, S.: NFA to DFA transformation for finite languages over arbitrary alphabets. J. Autom. Lang. Comb. **2**, 177–186 (1997)
19. Valiant, L.G.: Regularity and related problems for deterministic pushdown automata. J. ACM **22**, 1–10 (1975)
20. Yu, S.: State complexity of regular languages. J. Autom. Lang. Comb. **6**, 221–234 (2001)
21. Yu, S., Zhuang, Q., Salomaa, K.: The state complexities of some basic operations on regular languages. Theoret. Comput. Sci. **125**(2), 315–328 (1994)

The Complexity of Languages Resulting from the Concatenation Operation

Galina Jirásková[1(✉)], Alexander Szabari[2], and Juraj Šebej[2]

[1] Mathematical Institute, Slovak Academy of Sciences, Grešákova 6,
040 01 Košice, Slovakia
jiraskov@saske.sk
[2] Faculty of Science, Institute of Computer Science, P.J. Šafárik University,
Jesenná 5, 040 01 Košice, Slovakia
alexander.szabari@gmail.com, juraj.sebej@gmail.com

Abstract. We prove that for all m, n, and α with $1 \leq \alpha \leq f(m,n)$, where $f(m,n)$ is the state complexity of the concatenation operation, there exist a minimal m-state DFA A and a minimal n-state DFA B, both defined over an alphabet Σ with $|\Sigma| \leq 2n+4$, such that the minimal DFA for the language $L(A)L(B)$ has exactly α states. This improves a similar result in the literature that uses an exponential alphabet.

1 Introduction

Iwama et al. [4] stated the question of whether there always exists a minimal nondeterministic finite automaton (NFA) of n states whose equivalent minimal deterministic finite automaton (DFA) has α states for all integers n and α satisfying $n \leqslant \alpha \leqslant 2^n$. The question was also considered by Iwama et al. [5], and answered positively in [9] for a ternary alphabet. However, in the unary case, the existence of holes, so called "magic numbers", was proved by Geffert [1]. The binary case is still open.

The same problem on sub-regular language families was studied by Holzer et al. [2]. It turned out that the existence of non-trivial magic numbers is rare, and that the ranges of possible complexities are usually contiguous. One interesting exception was obtained by Čevorová [18]. She studied the star operation on unary regular languages, and proved that there are two linear segments of magic numbers in the range from 1 to $(n-1)^2 + 1$, that is, of values that cannot be met by the state complexity of the star of a unary language accepted by a minimal n-state DFA. On the other hand, she proved that for the square operation in the unary case no magic numbers exist [19]. Another example of the existence of magic numbers for symmetric difference NFAs was presented by Zijl [17], but they could possibly be trivial.

A similar problem for the reversal, star, and concatenation operation was studied in [7,8], where it was shown that for all the three operations the whole

---

G. Jirásková — Research supported by grant VEGA 2/0084/15.
A. Szabari and J. Šebej— Research supported by grant VEGA 1/0142/15.

Published by Springer International Publishing Switzerland 2016. All Rights Reserved
C. Câmpeanu et al. (Eds.): DCFS 2016, LNCS 9777, pp. 153–167, 2016.
DOI: 10.1007/978-3-319-41114-9_12

range of possible complexities up to known upper bounds can be produced using an exponential alphabet.

The result for reversal and star was improved in [10,14] by showing that a linear alphabet is enough to produce the whole range of complexities.

In this paper we complement these results, and show that a linear alphabet can also be used for the concatenation operation. We prove that for all m, n, and α with $1 \leq \alpha \leq f(m,n)$, where $f(m,n)$ is the state complexity of the concatenation operation, there exist a minimal m-state DFA A and a minimal n-state DFA B, both defined over an alphabet Σ with $|\Sigma| \leq 2n + 4$, such that the minimal DFA for the language $L(A)L(B)$ has exactly α states.

To get this result, we describe three constructions, in which we are able to get m-state and $(n + 1)$-state DFAs A_i, B_i for $i = 1, 2, 3$ from m-state and n-state DFAs A and B, by adding a new state to B, and by adding the transitions on two new symbols. Moreover, if the state complexity of the concatenation of $L(A)$ and $L(B)$ is α, then the state complexity of the concatenation of $L(A_i)$ and $L(B_i)$, $i = 1, 2, 3$, is 2α, $2\alpha - 1$, and $\alpha + 1$, respectively. As a results, we get a contiguous range of complexities from $m + n + 1$ up to known upper bound for a linear alphabet. To get complexities from 1 to $m + n - 1$, we use a known result from [8]. We deal with the value $m + n$ separately, and use a binary alphabet here.

The paper is organized as follows. The next section contains some definitions and preliminary results. In Sect. 3, we recall known results concerning the state complexity of concatenation. In Sect. 4, we prove that the range of possible complexities for the languages resulting from the concatenation operation is contiguous from 1 up to known upper bound, and we show that a linear alphabet is enough for this. Section 5 contains some concluding remarks.

2 Preliminaries

In this section we give some basic definitions and preliminary results. For details, the reader may refer to [3,13,15].

Let Σ be a finite alphabet of symbols. Then Σ^* denotes the set of strings over Σ including the empty string ε. The length of a string w is denoted by $|w|$, and the number of occurrences of a symbol a in a string w is denoted by $\#_a(w)$. A language is any subset of Σ^*. The concatenation of languages K and L is the language $KL = \{uv \mid u \in K \text{ and } v \in L\}$. The cardinality of a finite set A is denoted by $|A|$, and its power-set by 2^A.

A *nondeterministic finite automaton* (NFA) is a quintuple $A = (Q, \Sigma, \cdot, I, F)$, where Q is a finite set of states, Σ is a finite alphabet, $\cdot : Q \times \Sigma \to 2^Q$ is the transition function which is extended to the domain $2^Q \times \Sigma^*$ in the natural way, $I \subseteq Q$ is the set of initial states, and $F \subseteq Q$ is the set of final states. The *language accepted by* A is the set $L(A) = \{w \in \Sigma^* \mid I \cdot w \cap F \neq \emptyset\}$. For a symbol a, we say that (p, a, q) is a transition in NFA A if $q \in p \cdot a$, and for a string w, we write $p \xrightarrow{w} q$ if $q \in p \cdot w$. We say that (p, a, q) is an in-transition going to state q.

An NFA A is *deterministic* (DFA) (and complete) if $|I| = 1$ and $|q \cdot a| = 1$ for each q in Q and each a in Σ. In such a case, we write $q \cdot a = q'$ instead of $q \cdot a = \{q'\}$.

The *state complexity* of a regular language L, $sc(L)$, is the smallest number of states in any DFA for L. The state complexity of a binary regular operation $\circ$ is defined as a function $f(m, n)$ given by

$$f(m, n) = \max\{sc(K \circ L) \mid K, L \subseteq \Sigma^*, sc(K) = m, sc(L) = n\}.$$

Every NFA $A = (Q, \Sigma, \cdot, I, F)$ can be converted to an equivalent DFA $A' = (2^Q, \Sigma, \cdot', I, F')$, where $R \cdot' a = R \cdot a$ and $F' = \{R \in 2^Q \mid R \cap F \neq \emptyset\}$ [12]. The DFA A' is called the *subset automaton* of the NFA A. The subset automaton may not be minimal since some of its states may be unreachable or equivalent to other states.

In the following proposition, we provide a sufficient condition for an NFA, which guarantees that the corresponding subset automaton does not have equivalent states.

Proposition 1. *Let $N = (Q, \Sigma, \cdot, I, F)$ be an NFA. Assume that for each state q in Q, there is a string w_q in Σ^* which is accepted by N only from the state q, that is, we have $q \cdot w_q \cap F \neq \emptyset$, and $p \cdot w_q \cap F = \emptyset$ if $p \neq q$. Then the subset automaton of N does not have equivalent states.*

Proof. Let S and T be two distinct subsets of the subset automaton. Then, without loss of generality, there is a state q with $q \in S \setminus T$. Then the string w_q is accepted by the subset automaton from the subset S, but it is rejected from T. $\qquad \square$

To describe string w_q accepted by an NFA only from state q, we usually use the next observation.

Proposition 2. *Let a string w_q be accepted by an NFA N only from state q. If (p, a, q) is the unique in-transition going to state q by symbol a, then the string aw_q is accepted by N only from state p.*

In what follows, we often need to show how the set of all the reachable subsets in a subset automaton looks like. To do this, the following observation is useful.

Proposition 3. *Let D be a subset automaton of an NFA $N = (Q, \Sigma, \cdot, I, F)$. Let $\mathcal{R}$ be a family of subsets of Q such that*

(1) each subset in $\mathcal{R}$ is reachable in D,
(2) $I \in \mathcal{R}$, and
(3) for each S in $\mathcal{R}$ and each symbol a in Σ, the set $S \cdot a$ is in $\mathcal{R}$.

Then $\mathcal{R}$ is the family of all reachable subsets of DFA D.

Proof. Each set in $\mathcal{R}$ is reachable in D by (1). Let S be a reachable subset of D. Then there is a string w in Σ^* such that $S = I \cdot w$. We prove the proposition by induction on $|w|$. If $|w| = 0$, then $w = \varepsilon$ and $S = I \cdot \varepsilon = I$, which is in $\mathcal{R}$ by (2). Now let $w = va$ for a string v and a symbol a. By the induction hypothesis, the set $S' = I \cdot v$ is in $\mathcal{R}$. Then $S = S' \cdot a$, so S is in $\mathcal{R}$ by (3). $\qquad \square$

3   State Complexity of Concatenation

Consider minimal DFAs A and B. Without loss of generality, we assume that the state set of A is $\{q_0, q_1, \ldots, q_{m-1}\}$ with the initial state q_0, and the state set of B is $\{0, 1, \ldots, n-1\}$ with the initial states 0. Moreover, in both A and B, let us denote the transition function by $\cdot$. This is not confusing since the state sets of A and B are disjoint. First, let us recall the construction of an NFA for the language $L(A)L(B)$.

Construction of NFA for concatenation:
(DFA A and DFA $B \rightarrow$ NFA N for $L(A)L(B)$)
Let $A = (\{q_0, q_1, \ldots, q_{m-1}\}, \Sigma, \cdot, q_0, F_A)$ and $B = (\{0, 1, \ldots, n-1\}, \Sigma, \cdot, 0, F_B)$ be DFAs. Construct NFA N from DFAs A and B as follows:

(a) for each symbol a and each state q_i with $q_i \cdot a \in F_A$, add transition $(q_i, a, 0)$;
(b) the set of initial states of N is $\{q_0\}$ if $q_0 \notin F_A$, and it is $\{q_0, 0\}$ otherwise;
(c) the set of final state of N is F_B.

In the subset automaton of NFA N constructed as above, each reachable subset is of the form $\{q_i\} \cup S$, where $S \subseteq \{0, 1, \ldots, n-1\}$ since A is deterministic and complete. Moreover, if q_i is a final state of A, then $0 \in S$ since N has the transition $(q, a, 0)$ whenever a state q of A goes to a final state q_i on a symbol a. It follows that the subset automaton of N has at most $(m-k)2^n + k2^{n-1}$ reachable states. Next we have $(m-k)2^n + k2^{n-1} = m2^n - k2^{n-1}$, which is maximal if $k = 1$ [11,16]. We write this upper bound as $(m-1)2^n + 2^{n-1}$. The bound is known to be tight if $m \geq 1$ and $n \geq 2$ [6,11,16]. If $m \geq 1$ and $n = 1$, then $L = \emptyset$ or $L = \Sigma^*$, so the tight upper bound in this case is m. Hence we get the following result.

Proposition 4 [6,11,16]. *Let $m, n \geq 1$ and $f(m, n)$ be the state complexity of the concatenation operation on languages over an alphabet of size at least two defined as $f(m, n) = \max\{\mathrm{sc}(KL) \mid K, L \subseteq \Sigma^*, |\Sigma| \geq 2, \mathrm{sc}(K) = m, \mathrm{sc}(L) = n\}$. Then we have*

$$f(m, n) = \begin{cases} m, & \text{if } n = 1; \\ (m-1)2^n + 2^{n-1}, & \text{if } n \geq 2. \end{cases}$$

4   The Range of Possible Complexities

The aim of this section is to show that the whole range of complexities from 1 to $f(m, n)$ for the concatenation operation can be produced using an alphabet that grows linearly with n.

To this aim consider minimal DFAs $A = (\{q_0, q_1, \ldots, q_{m-1}\}, \Sigma, \cdot, q_0, \{q_{m-1}\})$, and $B = (\{0, 1, \ldots, n-1\}, \Sigma, \cdot, 0, \{1\})$. Construct an NFA N for $L(A)L(B)$ as described in Sect. 3. Let D be the subset automaton of N, and $\mathcal{R}$ the family of all the reachable subsets in DFA D. We assume that A, B, N, D, and $\mathcal{R}$ satisfy the following conditions.

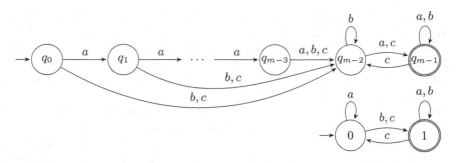

Fig. 1. Transitions on a, b, c in states in $\{q_0, q_1, \ldots, q_{m-1}\} \cup \{0, 1\}$.

(1) The transitions on symbols a, b, c in states in $\{q_0, q_1, \ldots, q_{m-1}\} \cup \{0, 1\}$ are defined as in Fig. 1.
(2) If (q_i, σ, q_0) is a transition in A for some σ in Σ, then $i = m - 1$.
(3) Each set in $\mathcal{R} \setminus \{\{q_0\}\}$ is reachable from $\{q_1\}$ in the subset automaton D.
(4) For each state q of NFA N, there exists a string w_q in Σ^* accepted by N only from state q. Moreover, we have

$$w_1 = \varepsilon,$$
$$w_0 = c,$$
$$w_{q_{m-1}} = bc,$$
$$w_{q_{m-2}} = cbc,$$
$$w_{q_{m-2-i}} = a^i cbc \text{ for } i = 1, 2, \ldots, m - 2, \text{ and}$$
$$w_j = a_j \text{ for } j = 2, 3, \ldots, n - 1.$$

Proposition 5. *Let A, B, N, D, and $\mathcal{R}$ satisfy conditions (1)–(4). Then*

(a) The sets $\{q_1\}$, $\{q_{m-1}, 0\}$, $\{q_{m-1}, 0, 1\}$, $\{q_{m-2}, 0, 1\}$ are in $\mathcal{R}$.
(b) The initial subset $\{q_0\}$ of the subset automaton D cannot be reached from any other reachable subset of D.
(c) The subset automaton D of NFA N does not have equivalent states, so $\mathrm{sc}(L(A)L(B)) = |\mathcal{R}|$.

Proof. (a) By (1), the transitions on a, b, c are as in Fig. 1. It follows that in the subset automaton D, we have

$$\{q_0\} \xrightarrow{a} \{q_1\} \xrightarrow{a^{m-2}} \{q_{m-1}, 0\} \xrightarrow{b} \{q_{m-1}, 0, 1\} \xrightarrow{c} \{q_{m-2}, 0, 1\}.$$

(b) Assume for a contradiction that there is a set S in $\mathcal{R}$ and a symbol σ such that $S \cdot \sigma = \{q_0\}$. Then we must have $q_{m-1} \in S$ by (2). It follows that the initial state 0 of B must be in S since q_{m-1} is final in A. However then $S \cdot \sigma \supseteq \{q_0, 0 \cdot \sigma\}$, a contradiction.

(c) By (4), the NFA N satisfies the condition in Proposition 1. Therefore the subset automaton D of N does not have equivalent states, and we have $\mathrm{sc}(L(A)L(B)) = |\mathcal{R}|$. $\qquad\square$

Now our goal is to construct a minimal m-state DFA A_i and a minimal $(n+1)$-state DFA B_i for $i = 1, 2, 3$ over the alphabet $\Sigma \cup \{a_n, b_n\}$ from automata A and B, such that $A, B, N, D, \mathcal{R}$ satisfy conditions (1)–(4), in such a way that A_i and B_i, the NFA N_i for $L(A_i)L(B_i)$, the subset automaton D_i of N_i and the family $\mathcal{R}_i$ of reachable states of D_i satisfy conditions (1)–(4). Moreover, if $\mathcal{R} = \alpha$, then $|\mathcal{R}_1| = 2\alpha$, $|\mathcal{R}_2| = 2\alpha - 1$, and $|\mathcal{R}_3| = \alpha + 1$.

We construct automata A_i and B_i from automata A and B by adding a new state n to DFA B, and by adding the transitions on two new symbols a_n and b_n. The transitions on a_n are the same in all the three constructions, and they guarantee that the string a_n is accepted by N_i only from state n. The transitions on b_n are used to reach the set $\{q_0, n\}$ in D_1, the set $\{q_1, n\}$ in D_2 and the set $\{q_{m-1}, 0, n\}$ in D_3. We have to be careful with condition (4), especially in the third construction.

Table 1. New transitions; $i \in \{0, 1, \ldots, m-1\}$, $j \in \{0, 1, \ldots, n-1\}$.

	C1	C2	C3
$\sigma \in \Sigma$	$n \to n$	$n \to n$	$n \xrightarrow{c} 0$
			$n \xrightarrow{\sigma} 0 \cdot \sigma$ if $\sigma \neq c$
a_n	$q_i \to q_{m-1}$	$q_i \to q_{m-1}$	$q_i \to q_{m-1}$
	$n \to 1$	$n \to 1$	$n \to 1$
	$j \to 0$	$j \to 0$	$j \to 0$
b_n	$q_{m-1} \to q_0$	$q_{m-1} \to q_1$	
	$q_i \to q_{m-1}$ if $i \neq m-1$	$q_i \to q_{m-1}$ if $i \neq m-1$	$q_i \to q_{m-1}$
	$n \to n$	$n \to n$	$n \to n$
	$j \to n$	$j \to n$	$j \to n$

Construction 1. $(\alpha \to 2\alpha)$

Construct DFAs A_1 and B_1 from DFAs A and B as follows:

(1) add a new state n to DFA B going to itself on each old symbol σ in Σ;

(2) add the transitions on two new symbols a_n and b_n as shown in Table 1 in column C1.

Construction 2. $(\alpha \to 2\alpha - 1)$

Construct DFAs A_2 and B_2 from DFAs A and B as follows:

(1) add a new state n to DFA B going to itself on each old symbol σ in Σ;

(2) add the transitions on two new symbols a_n and b_n as shown in Table 1 in column C2.

Construction 3. $(\alpha \to \alpha + 1)$

Construct DFAs A_3 and B_3 from DFAs A and B as follows:

(1) add a new state n to DFA B with $n \cdot c = 1$ and $n \cdot \sigma = 0 \cdot \sigma$ if $\sigma \in \Sigma \setminus \{c\}$;

(2) add the transitions on two new symbols a_n and b_n as shown in Table 1 in column C3.

Lemma 6. *Let $A, B, N, D, \mathcal{R}$ satisfy conditions (1)–(4). Let A_i, B_i for $i = 1, 2, 3$ be the DFAs resulting from Constructions 1, 2, 3, respectively. Let N_i be an NFA for $L(A_i)L(B_i)$ constructed as described in Sect. 3, D_i be the corresponding subset automaton, and $\mathcal{R}_i$ be the family of all the reachable subsets in DFA D_i. Then all these automata satisfy conditions (1)–(4). Moreover, if $|\mathcal{R}| = \alpha$, then $|\mathcal{R}_1| = 2\alpha$, $|\mathcal{R}_2| = 2\alpha - 1$, and $|\mathcal{R}_3| = \alpha + 1$.*

Proof. Since we do not change transitions on symbols in Σ on states of A and B, condition (1) is satisfied. Since the only new transition to q_0 is (q_{m-1}, b_n, q_0) in Construction 1, condition (2) is satisfied in each A_i.

In each N_i, the string a_n is accepted only from state n. Moreover, in B_1 and B_2, state n goes to itself on each symbol in Σ. It follows that condition (4) is satisfied for N_1 and N_2. In B_3, we have $n \cdot c = 0$ and $n \cdot b = 0 \cdot b = b$. It follows that $(0, c, 1)$ is the only transition on c going to state 1, and $(q_{m-1}, b, 0)$ is the only transition on b going to state 0. It follows that (4) is satisfied for N_3 as well.

Now consider the subset automata D_1, D_2, D_3. Since we did not change transitions on symbols in Σ on states in A and B, we have $\mathcal{R} \subseteq \mathcal{R}_i$ for $i = 1, 2, 3$. Let us show that

$\mathcal{R}_1 = \mathcal{R} \cup \{S \cup \{n\} \mid S \in \mathcal{R}\}$,
$\mathcal{R}_2 = \mathcal{R} \cup \{S \cup \{n\} \mid S \in \mathcal{R} \text{ and } S \neq \{q_0\}\}$,
$\mathcal{R}_3 = \mathcal{R} \cup \{\{q_{m-1}, 0, n\}\}$.

If S is in $\mathcal{R}$ then S is reachable in D, so S can be reached from the initial state $\{q_0\}$ by a string u_S over Σ. If moreover, $S \neq \{q_0\}$, then, by (3), S is reached from $\{q_1\}$ by a string v_S.

In D_1 we have $\{q_0\} \xrightarrow{a} \{q_1\} \xrightarrow{a^{m-2}} \{q_{m-1}, 0\} \xrightarrow{b_n} \{q_0, n\} \xrightarrow{u_S} S \cup \{n\}$. Thus $\mathcal{R} \cup \{S \cup \{n\} \mid S \in \mathcal{R}\} \subseteq \mathcal{R}_1$, and every new set $S \cup \{n\}$ can be reached from $\{q_1\}$. Let us show that no other set is reachable in D_1. For each set S in $\mathcal{R}$ and each σ in Σ, we have

$S \cdot \sigma \in \mathcal{R}$,
$S \cdot a_n \in \{\{q_{m-1}, 0\}, \{q_{m-1}, 0, 1\}\}$,
$S \cdot b_n \in \{\{q_0, n\}, \{q_{m-1}, 0\}, \{q_{m-1}, 0, n\}\}$,
$(S \cup \{n\}) \cdot \sigma = S \cdot \sigma \cup \{n\}$,
$(S \cup \{n\}) \cdot a_n = \{q_{m-1}, 0, 1\}$, and
$(S \cup \{n\}) \cdot b_n \in \{\{q_0, n\}, \{q_{m-1}, 0, n\}\}$.

Using Proposition 5(a), we get that all the resulting sets are in $\mathcal{R} \cup \{S \cup \{n\} \mid S \in \mathcal{R}\}$. By Proposition 3, we have $\mathcal{R}_1 = \mathcal{R} \cup \{S \cup \{n\} \mid S \in \mathcal{R}\}$, and, moreover, $\mathcal{R}_1$ satisfies condition (3).

Next, in D_2 we have $\{q_0\} \xrightarrow{a} \{q_1\} \xrightarrow{a^{m-2}} \{q_{m-1}, 0\} \xrightarrow{b_n} \{q_1, n\} \xrightarrow{v_S} S \cup \{n\}$ if $S \neq \{q_0\}$. So every new set $S \cup \{n\}$ is reached from $\{q_1\}$. The transitions on each σ in Σ and on a_n are the same as in Construction 1, and for each S in $\mathcal{R}$,

$S \cdot b_n \in \{\{q_1, n\}, \{q_{m-1}, 0\}, \{q_{m-1}, 0, n\}\}$, and
$(S \cup \{n\}) \cdot b_n \in \{\{q_1, n\}, \{q_{m-1}, 0, n\}\}$.

All the resulting sets are in $\mathcal{R} \cup \{S \cup \{n\} \mid S \in \mathcal{R} \text{ and } S \neq \{q_0\}\}$. Moreover, $\{q_0\}$ cannot be reached from any other subset in $\mathcal{R}$. By Proposition 3, we have $\mathcal{R}_2 = \mathcal{R} \cup \{S \cup \{n\} \mid S \in \mathcal{R} \text{ and } S \neq \{q_0\}\}$. Moreover, $\mathcal{R}_2$ satisfies condition (3).

Finally, in D_3 we have $\{q_0\} \xrightarrow{a} \{q_1\} \xrightarrow{a^{m-2}} \{q_{m-1}, 0\} \xrightarrow{b_n} \{q_{m-1}, 0, n\}$,
so the new set $\{q_{m-1}, 0, n\}$ is reached from $\{q_1\}$. The transitions on a_n are
the same as above, and for each S in $\mathcal{R}$ and each σ in Σ, we have $S \cdot b_n \in$
$\{\{q_{m-1}, 0\}, \{q_{m-1}, 0, n\}\}$. Next, for the new set $\{q_{m-1}, 0, n\}$, we have
$$\{q_{m-1}, 0, n\} \cdot c = \{q_{m-2}, 0, 1\},$$
$$\{q_{m-1}, 0, n\} \cdot \sigma = \{q_{m-1}, 0\} \cdot \sigma \text{ if } \sigma \in \Sigma \text{ and } \sigma \neq c;$$
$$\{q_{m-1}, 0, n\} \cdot b_n = \{q_{m-1}, 0, n\}.$$
All the resulting subsets are in $\mathcal{R} \cup \{\{q_{m-1}, 0, n\}\}$. By Proposition 3, we have
$\mathcal{R}_3 = \{\mathcal{R} \cup \{q_{m-1}, 0, n\}\}$, and again, $\mathcal{R}_3$ satisfies condition (3).   □

Recall that $f(m, n) = (m-1)2^n + 2^{n-1}$ is the state complexity of concatenation if $n \geq 2$. Our first aim is to show that each value in the range from $m+n+1$
to $f(m, n)$ may be attained by the state complexity of concatenation of m-state
and n-state DFA languages provided that $m \geq 3$. We show this by induction,
with the basis proved in the next lemma.

Lemma 7. *Let $m \geq 3$ and $n = 2$. For each α with $m + 3 \leq \alpha \leq f(m, 2) =$
$4m - 2$, there exist a minimal m-state DFA A and a minimal 2-state DFA B,
both defined over an alphabet Σ with $|\Sigma| \leq 7$, such that $\mathrm{sc}(L(A)L(B)) = \alpha$.
Moreover, the corresponding NFA N for $L(A)L(B)$, the subset automaton D of
N, and the set $\mathcal{R}$ of reachable states of D satisfy conditions (1)–(4) on page 5.*

Proof. We first consider the values $\alpha = i(m - 2) + 6$ for $i = 1, 2, 3, 4$. Then we
consider all the intermediate values of α. Finally we deal with the case $\alpha = m+3$.
    First let $i = 1$, so $\alpha = (m - 2) + 6 = m + 4$. Define a minimal m-state
DFA $A_{1,0} = (\{q_0, q_1, \ldots, q_{m-1}\}, \{a, b, c, d\}, \cdot, q_0, \{q_{m-1}\})$ where for each i in
$\{0, 1, \ldots, m - 1\}$,
    $q_i \cdot a = q_{i+1}$ if $i \neq m - 1$ and $q_{m-1} \cdot a = q_{m-1}$,
    $q_i \cdot b = q_{m-2}$ if $i \neq m - 1$ and $q_{m-1} \cdot b = q_{m-1}$,
    $q_i \cdot c = q_{m-2}$ if $i \neq m - 2$ and $q_{m-2} \cdot c = q_{m-1}$, and
    $q_i \cdot d = q_{m-2}$.
Define a minimal two-state DFA $B_{1,0} = (\{0, 1\}, \{a, b, c, d\}, \cdot, 0, \{1\})$ where
    $0 \cdot a = 0$, and $1 \cdot a = 1$,
    $0 \cdot b = 1$, and $1 \cdot b = 1$,
    $0 \cdot c = 1$, and $1 \cdot c = 0$,
    $0 \cdot d = 0$, and $1 \cdot d = 1$.
Construct NFA $N_{1,0}$ for $L(A_{1,0})L(B_{1,0})$, and let $D_{1,0}$ be the corresponding
subset automaton. Notice that (1), (2), and (4) are satisfied. Next, in $D_{1,0}$ we have
$$\{q_0\} \xrightarrow{a} \{q_1\} \xrightarrow{a^{i-1}} \{q_i\} \text{ for } i = 0, 1, \ldots, m - 2,$$
$$\{q_{m-2}\} \xrightarrow{a} \{q_{m-1}, 0\} \xrightarrow{b} \{q_{m-1}, 0, 1\} \xrightarrow{c} \{q_{m-2}, 0, 1\},$$
$$\{q_{m-1}, 0\} \xrightarrow{d} \{q_{m-2}, 0\} \xrightarrow{c} \{q_{m-2}, 1\}.$$
Thus the subset automaton has $m + 4$ reachable subsets. Next, notice that
each of these $m + 4$ subsets goes to some of them by each symbol in $\{a, b, c, d\}$.
By Proposition 3, no other set is reachable, so the complexity of $L(A_{1,0})L(B_{1,0})$
is $m + 4$. Notice that all the possible subsets containing states q_{m-1} and q_{m-2}
are reachable in $D_{1,0}$.

Now we construct appropriate DFAs from automata $A_{1,0}$ and $B_{1,0}$ by adding transitions on new symbols. Thus we do not change the transitions on symbols a, b, c, d, and therefore the conditions (1) and (4) are always satisfied. Moreover, for each new symbol, the new transition is defined in such a way that condition (2) is satisfied as well. Finally, notice that $\{q_{m-1}, 0\}$ is reachable from $\{q_1\}$ by a^{m-2} in the subset automaton $D_{1,0}$. In what follows, we always reach new subsets in the corresponding subset automata for concatenation from the subset $\{q_{m-1}, 0\}$. Hence condition (3) is always satisfied.

Next, let $\alpha = 2(m-2) + 6$. Construct DFAs $A_{2,0}, B_{2,0}$ from DFAs $A_{1,0}, B_{1,0}$ by adding the transitions on a new symbol e_0 as follows:

$q_{m-1} \cdot e_0 = q_0$ and $q_i \cdot e_0 = q_{m-1}$ for $i = 0, 1, \ldots, m-2$;
$0 \cdot e_0 = 0$ and $1 \cdot e_0 = 0$.

Construct the NFA $N_{2,0}$ for $L(A_{2,0})L(B_{2,0})$. In the subset automaton $D_{2,0}$, all the sets that were reachable in the subset automaton $D_{1,0}$ are reachable as well, since the transitions on the old symbols a, b, c, d are the same. For the same reason, the NFA $N_{2,0}$ satisfies (4), and therefore the subset automaton $D_{2,0}$ does not have equivalent states. Next, in $D_{2,0}$, we have

$$\{q_{m-1}, 0\} \xrightarrow{e_0} \{q_0, 0\} \xrightarrow{a^i} \{q_i, 0\} \text{ for } i = 1, 2, \ldots, m-3.$$

No other new set is reachable since each set $\{q_i, 0\}$ goes either to a set $\{q_j, 0\}$ or to a set containing q_{m-2} or q_{m-1} by each symbol in $\{a, b, c, d, e_0\}$, and moreover, by e_0, each set goes either to $\{q_0, 0\}$ or to a set containing q_{m-1}. Therefore the resulting complexity of the concatenation $L(A_{2,0})L(B_{2,0})$ is $2(m-2) + 6$.

In a similar way, we construct DFAs $A_{3,0}, B_{3,0}$ from $A_{2,0}, B_{2,0}$ by adding transitions on a new symbol e_{01} defined as follows:

$q_{m-1} \cdot e_{01} = q_0$ and $q_i \cdot e_{01} = q_{m-1}$ for $i = 0, 1, \ldots, m-2$;
$0 \cdot e_{01} = 0$ and $1 \cdot e_{01} = 1$.

This results in the reachability of $m-2$ new subsets $\{q_i, 0, 1\}$ in the subset automaton of $N_{3,0}$. Since no other new set is reachable, the complexity of $L(A_{3,0})L(B_{3,0})$ is $3(m-2) + 6$.

Finally, construct DFAs $A_{4,0}, B_{4,0}$ from $A_{3,0}, B_{3,0}$ by adding the transitions on a new symbol e_1 defined as

$q_{m-1} \cdot e_1 = q_0$ and $q_i \cdot e_1 = q_{m-1}$ for $i = 0, 1, \ldots, m-2$;
$0 \cdot e_1 = 1$ and $1 \cdot e_1 = 1$.

This results in the reachability of subsets $\{q_i, 1\}$ in the subset automaton of $N_{4,0}$, and the complexity of $L(A_{4,0})L(B_{4,0})$ is $4(m-2) + 6$.

Up to now we have defined appropriate automata $A_{i,0}$ and $B_{i,0}$ for the values $\alpha = i(m-2) + 6$ for $i = 1, 2, 3, 4$. Now let us consider an intermediate value $\alpha = i(m-2) + 6 + j$ where $1 \le i \le 3$ and $1 \le j \le m-3$. Construct DFAs $A_{i,j}$ and $B_{i,j}$ from automata $A_{i,0}$ and $B_{i,0}$ by adding the transitions on a new symbol f_1 as follows:

$q_{m-1} \cdot f_1 = q_{m-2-j}$ and $q_i \cdot f_1 = q_{m-1}$ for $i = 0, 1, \ldots, m-2$;
$0 \cdot f_1 = 1$ and $1 \cdot f_1 = 1$.

This results in the reachability of the following j new subsets in the subset automaton of $N_{i,j}$:

$$\{q_{m-1},0\} \xrightarrow{f_1} \{q_{m-2-j},1\} \xrightarrow{a} \{q_{m-2-j+1},1\} \xrightarrow{a} \cdots \xrightarrow{a} \{q_{m-3},1\}.$$

Recall that the subset automaton of $N_{i,0}$ has $i(m-2)+6$ reachable states, and since $i \le 3$, the subsets $\{q_i,1\}$ are unreachable in the subset automaton of $N_{i,0}$. Hence the resulting complexity of of $L(A_{i,j})L(B_{i,j})$ is $i(m-2)+6+j$ as desired. Moreover, all the automata satisfy conditions (1)–(4).

Finally notice that if A and B are DFAs over a,b,c shown in Fig. 1 then $\mathrm{sc}(L(A)L(B)) = m+3$. This concludes our proof.   □

Now we are ready to prove the main lemma. Recall that the state complexity of concatenation is $f(m,n) = (m-1)2^n + 2^{n-1}$ if $n \ge 2$. Moreover, notice that we have $f(m,n+1) = (m-1)2^{n+1} + 2^n = 2((m-1)2^n + 2^{n-1}) = 2f(m,n)$.

Lemma 8. *Let $m \ge 3$ and $n \ge 2$. For each α with $m+n+1 \le \alpha \le f(m,n)$, there exist a minimal m-state DFA A, and a minimal n-state DFA B, both defined over an alphabet Σ with $|\Sigma| \le 2n+4$, such that $\mathrm{sc}(L(A)L(B)) = \alpha$.*

Proof. We prove the claim by induction on n. Moreover, in the induction hypothesis, we assume that DFAs A and B, the corresponding NFA N for $L(A)L(B)$ constructed as in Sect. 3, the subset automaton D of N, and the set $\mathcal{R}$ of reachable states of D satisfy conditions (1)–(4) on page 5.

The basis, in which we have $m \ge 3$, $n = 2$, and $m+3 \le \alpha \le f(m,2) = 4m-2$, is proved in Lemma 7. Let $m \ge 3$, $n \ge 2$, and assume that for each β with $m+n+1 \le \beta \le f(m,n)$, there exist a minimal m-state DFA A and a minimal n-state DFA B, both defined over an alphabet Σ with $|\Sigma| \le 2n+4$, such that $\mathrm{sc}(L(A)L(B)) = \beta$. Moreover, assume that DFAs A and B, the NFA N for $L(A)L(B)$, the subset automaton D of N, and the set of reachable states $\mathcal{R}$ of D satisfy conditions (1)–(4) on page 5. Let us show that the claim holds for $n+1$. To this aim let α be an integer with $m+(n+1)+1 \le \alpha \le f(m,n+1)$.

First, let $2m+2n+2 \le \alpha \le f(m,n+1)$ and α be even. Let $\beta = \alpha/2$. Then $m+n+1 \le \beta \le f(m,n)$, and by the induction hypothesis, there exists a minimal m-state DFA A and a minimal n-state DFA B, both defined over an alphabet Σ with $|\Sigma| \le 2n+4$, such that $\mathrm{sc}(L(A)L(B)) = \beta$. Moreover, conditions (1)–(4) are satisfied for $A, B, N, D, \mathcal{R}$. We use Construction 1, in which we add a new state to DFA B and the transitions on two new symbols to get a minimal m-state A_1 and a minimal $(n+1)$-state DFA B_1. By Lemma 6, all conditions (1)–(4) are satisfied for A_1, B_1, N_1, D_1, and R_1. It follows that $\mathrm{sc}(L(A_1)L(B_1)) = 2\beta = \alpha$.

Now, let $2m+2n+1 \le \alpha \le f(m,n+1)-1$ and α be odd. Let $\beta = (\alpha+1)/2$. Then $m+n+1 \le \beta \le f(m,n)$, and we use the induction hypothesis and our Construction 2, to get automata A_2 and B_2 over $\Sigma \cup \{a_n, b_n\}$ satisfying (1)–(4) such that $\mathrm{sc}(L(A_2)L(B_2)) = 2\beta - 1 = \alpha$.

Finally, if $m+(n+1)+1 \le \alpha \le 2m+2n$, we set $\beta = \alpha - 1$. Then $m+n+1 \le \beta \le f(m,n)$ since we have $2m+2n-1 \le m2^n - 2^{n-1}$ if $m \ge 3$ and $n \ge 2$. We use the induction hypothesis and Construction 3, get appropriate automata A_3

and B_3, satisfying (1)–(4) such that $sc(L(A_3)L(B_3)) = \beta + 1 = \alpha$. Our proof is complete.   □

Now we consider the case of $m = 2$ and $n \geq 2$. In such a case, we only need to modify conditions (1)–(4). All the proofs are the same as above, except for the base case, which is a bit more complicated in this case.

Lemma 9. *Let $m = 2$, $n \geq 2$. For each α with $n+3 \leq \alpha \leq f(2,n) = 2^n + 2^{n-1}$, there exist a minimal 2-state DFA A, and a minimal n-state DFA B, both defined over an alphabet Σ with $|\Sigma| \leq 2n + 4$, and such that $sc(L(A)L(B)) = \alpha$.*

Proof. We modify conditions (1)–(4) as follows.

(1') The transitions on symbols a, b, c in states in $\{q_0, q_1\} \cup \{0, 1\}$ are defined as in Fig. 1 for $m = 2$. This means that the subsets $\{q_1, 0\}$, $\{q_1, 0, 1\}$, and $\{q_0, 0, 1\}$ are reachable in D, that is, they are in $\mathcal{R}$.

(2') If (q_i, a, q_0) is a transition in A, then $i = 1$

(3') Each set in $\mathcal{R} \setminus \{\{q_0\}\}$ is reachable from $\{q_1, 0\}$ in the subset automaton D.

(4') For each state q of NFA N, there exists a string w_q in Σ^* which is accepted by N only from state q. Moreover, we have

$$w_1 = \varepsilon,$$
$$w_0 = c,$$
$$w_{q_1} = bc,$$
$$w_{q_0} = cbc,$$
$$w_j = a_j \text{ for } j = 2, 3, \ldots, n - 1.$$

Now we continue with exactly the same constructions as in the case of $m \geq 3$, and, using induction on n, we get the lemma.   □

The case of $m = 1$ and $n \geq 3$ is slightly different, although, the main idea is the same.

Lemma 10. *Let $m = 1$ and $n \geq 3$. For each α with $n+1 \leq \alpha \leq f(1,n) = 2^{n-1}$, there exist a minimal 1-state DFA A, and a minimal n-state DFA B, both defined over an alphabet Σ with $|\Sigma| = n - 1$, and such that $sc(L(A)L(B)) = \alpha$.*

Proof (Proof Idea). Let A be a 1-state DFA accepting Σ^*. We prove the lemma again by induction on n, where we assume that the following conditions hold for DFA B, the NFA N for Σ^*B, constructed from B by adding a loop in the initial state 0 on each input symbol in Σ, for the subset automaton D of N, and the set $\mathcal{R}$ of reachable subsets in D:

(1") In DFA B, the transitions on a, b, c in states $0, 1, 2$ are as in Fig. 2.

(2") In DFA B, we have $0 \cdot \sigma \neq 0$ for each $\sigma \in \Sigma$.

(3") Each subset in $\mathcal{R}$, except for the initial subset $\{0\}$, can be reached from the subset $\{0, 1\}$.

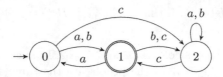

Fig. 2. Base case if $m = 1$.

(4") NFA N satisfies the condition in Proposition 1, that is, for each state j of N, there exists a string w_j in Σ^* which is accepted by N only from state j. Moreover, we have $w_0 = c$ and $w_1 = \varepsilon$.

The basis, in which we have $n = 3$ and $n + 1 = f(1,3) = 4$, holds true since the 3-state DFA B shown in Fig. 2 satisfies (1")–(4").

For the induction step, we again describe three constructions: We construct $(n + 1)$-state DFAs B_1, B_2, B_3 from DFA B by adding a new state n, and by adding transitions on new symbol a_n, b_n, as shown in Table 2 in columns C1, C2, and C3, respectively.

We can show that all the resulting automata satisfy conditions (1")–(4"), and, moreover, if $|\mathcal{R}| = \beta$, then $|\mathcal{R}_1| = 2\beta$, $|\mathcal{R}_2| = 2\beta - 1$, $|\mathcal{R}_3| = \beta + 1$. Since N and N_i satisfy (4"), we have $\mathrm{sc}(L(A)L(B)) = |\mathcal{R}|$ and $\mathrm{sc}(L(A_i)L(B_i)) = |\mathcal{R}_i|$. This proves the lemma by induction. □

Table 2. The three constructions in the case of $m = 1$.

	C1	C2	C3
$\sigma \in \Sigma$	$n \to n$	$n \to n$	$n \to 0$
a_n	$n \to 1$	$n \to 2$	$n \to 1$
	$j \to n$	$0 \to 1$	$j \to 2$
		$1 \to n$	
		$j \to 0$ if $j \geq 2$	
b_n	—	—	$n \to n$
			$j \to n$
w_n	a_n	$a_n c$	a_n

Up to now we have considered the complexities in the range from $m + n + 1$ to $f(m,n)$. The complexities from 1 to $m + n - 1$ are covered by the following result from [8]. Notice that this lemma also covers the case of $m = 1$ and $n = 2$, since then $f(1,2) = 2^1 = 2 = m + n - 1$.

Lemma 11 ([8, **Lemma 5**]). *Let $m, n \geq 1$. For each α with $1 \leq \alpha \leq m+n-1$, there exist a minimal m-state DFA A and a minimal n-state DFA B, both defined over an alphabet of at most two symbols, such that $\mathrm{sc}(L(A)L(B)) = \alpha$.*

The next lemma shows that the complexity $m + n$ can be produced. Then we consider the case of $n = 1$.

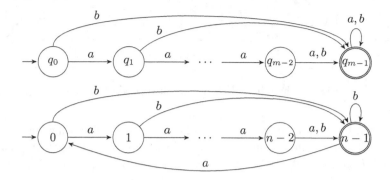

Fig. 3. The minimal DFAs A and B with $\mathrm{sc}(L(A)L(B)) = m + n$.

Lemma 12. *Let* $m \geq 2, n \geq 2$. *There exist binary regular languages* K *and* L *with* $\mathrm{sc}(K) = m$ *and* $\mathrm{sc}(L) = n$ *such that* $\mathrm{sc}(KL) = m + n$.

Proof. Let K and L be the binary languages accepted by minimal DFAs A and B shown in Fig. 3, where for each i in $\{0, 1, \ldots, m-1\}$ and j in $\{0, 1, \ldots, n-1\}$, we have

$q_i \cdot a = q_{i+1}$ if $i \neq m-1$, $q_{m-1} \cdot a = q_{m-1}$ and $q_i \cdot b = q_{m-1}$;
$j \cdot a = j+1$ if $j \neq n-1$, $(n-1) \cdot a = 0$, and $j \cdot b = n-1$.

Construct an NFA N from DFAs A and B by adding transitions $(q_{m-2}, a, 0)$, $(q_{m-1}, a, 0)$, and $(q_i, b, 0)$ for each i; the initial state of N is q_0, and the set of final states is $\{n-1\}$. In the corresponding subset automaton, the initial subset is $\{q_0\}$, and we have

$$\{q_0\} \xrightarrow{a^i} \{q_i\} \text{ for } i = 1, 2, \ldots, m-2,$$

$$\{q_{m-2}\} \xrightarrow{a} \{q_{m-1}, 0\} \xrightarrow{a^j} \{q_{m-1}, 0, 1, \ldots, j\} \text{ for } j = 1, 2, \ldots, n-1, \text{ and}$$

$$\{q_{m-1}, 0\} \xrightarrow{b} \{q_{m-1}, 0, n-1\}.$$

It follows that the subset automaton has $m + n$ reachable subsets. Notice that each of these $m + n$ subsets goes to some of them by a, and each of them goes to $\{q_{m-1}, 0\}$ or to $\{q_{m-1}, 0, n-1\}$ by b. By Proposition 3, no other set is reachable.

To prove distinguishability, let $\{q_i\} \cup S$ and $\{q_j\} \cup T$ be two distinct reachable subsets. Since NFA N accepts the string a^{n-1-t} only from state t ($0 \leq t \leq n-1$), the two subsets are distinguishable if $S \neq T$. Next, if $i < j$, then we have

$$\{q_j\} \xrightarrow{a^{m-1-j}} \{q_{m-1}, 0\} \text{ and } \{q_i\} \xrightarrow{a^{m-1-j}} \{q_{m-1-(j-i)}\},$$

where the resulting subsets are distinguishable since they differ in a state of B. This proves distinguishability and concludes the proof. □

Lemma 13. *Let* $m \geq 1$ *and* $n = 1$. *For each integer* α *with* $1 \leq \alpha \leq f(m, 1) = m$, *there exist a minimal* m-state *DFA* A *and a minimal* 1-state *DFA* B, *both defined over a unary alphabet, such that* $\mathrm{sc}(L(A)L(B)) = \alpha$.

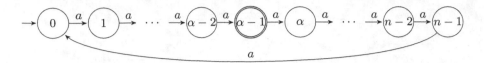

Fig. 4. The minimal DFA A with $\mathrm{sc}(L(A)\, \Sigma^*) = \alpha$.

Proof. Let A be a minimal m-state DFA shown in Fig. 4 accepting the language $a^{\alpha-1}(a^m)^*$. Let B be the minimal 1-state DFA accepting the unary language a^*. Then $L(A)L(B) = a^{\alpha-1}(a^m)^*a^* = \{a^k \mid k \le \alpha - 1\}$, so $\mathrm{sc}(L(A)L(B)) = \alpha$. □

The next theorem summarizes our results, and shows that the whole range of complexities for the concatenation operation can be produced using an alphabet which grows linearly with n. Recall that $f(m,n)$ is the state complexity of the concatenation operation on languages over an alphabet of size at least two and we have $f(m,1) = m$ and $f(m,n) = (m-1)2^n + 2^{n-1}$ if $n \ge 2$.

Theorem 14. *Let $m, n \ge 1$. For each α with $1 \le \alpha \le f(m,n)$, there exist regular languages K and L defined over an alphabet Σ with $|\Sigma| \le 2n + 4$ such that $\mathrm{sc}(K) = m$, $\mathrm{sc}(L) = n$, and $\mathrm{sc}(KL) = \alpha$.*

Proof. In each of the following six cases, we refer to the corresponding lemma dealing with this case:

(1) If $n = 1$, then $f(m,n) = m$, and the theorem follows by Lemma 13.
(2) If $n \ge 2$ and $1 \le \alpha \le m + n - 1$, then the theorem follows by Lemma 11.
(3) If $m = 1$ and $n = 2$, then $f(1,2) = 2 = m + n - 1$, so this case is covered by Lemma 11 as well.
(4) If $m = 1$, $n \ge 3$, and $m + n = n + 1 \le \alpha \le f(1,n) = 2^{n-1}$, then the theorem follows by Lemma 10.
(5) The case of $m \ge 2, n \ge 2$, and $\alpha = m + n$ follows by Lemma 12.
(6) Finally, if $m \ge 2, n \ge 2$, and $m + n + 1 \le \alpha \le f(m,n)$, then the theorem follows by Lemma 9 if $m = 2$, and by Lemma 8 if $m \ge 3$.

This covers all the possible cases, and proves the theorem. □

5   Conclusions

We investigated the state complexity of languages resulting from the concatenation operation. We proved that for all m, n, α with $m, n \ge 1$ and $1 \le \alpha \le f(m,n)$, where $f(m,n)$ is the state complexity of the concatenation operation, there exist regular languages K and L defined over an alphabet of size at most $2n + 4$ such that $\mathrm{sc}(K) = m$, $\mathrm{sc}(L) = n$, and $\mathrm{sc}(KL) = \alpha$. This improves the result from [8], where an alphabet of size growing exponentially with n is used to produce the whole range of complexities for the concatenation operation. Our result complements similar results from [10,14], where a linear alphabet is used to get the whole range of complexities for the reversal and Kleene closure operations.

A similar problem for the square operation, defined as $L^2 = LL$, remains open even for an exponential alphabet.

References

1. Geffert, V.: Magic numbers in the state hierarchy of finite automata. Inform. Comput. **205**, 1652–1670 (2007)
2. Holzer, M., Jakobi, S., Kutrib, M.: The magic number problem for subregular language families. Internat. J. Found. Comput. Sci. **23**, 115–131 (2012)
3. Hopcroft, J.E., Ullman, J.D.: Introduction to Automata Theory and Computation. Addison-Wesley Publishing Company, Reading (1979)
4. Iwama, K., Kambayashi, Y., Takaki, K.: Tight bounds on the number of states of DFAs that are equivalent to n-state NFAs. Theoret. Comput. Sci. **237**, 485–494 (2000)
5. Iwama, K., Matsuura, A., Paterson, M.: A family of NFAs which need $2^n - \alpha$ deterministic states. Theoret. Comput. Sci. **301**, 451–462 (2003)
6. Jirásek, J., Jirásková, G., Szabari, A.: State complexity of concatenation and complementation. Internat. J. Found. Comput. Sci. **16**, 511–529 (2005)
7. Jirásková, G.: On the state complexity of complements, stars, and reversals of regular languages. In: Ito, M., Toyama, M. (eds.) DLT 2008. LNCS, vol. 5257, pp. 431–442. Springer, Heidelberg (2008)
8. Jirásková, G.: Concatenation of regular languages and descriptional complexity. Theory Comput. Syst. **49**, 306–318 (2011)
9. Jirásková, G.: Magic numbers and ternary alphabet. Internat. J. Found. Comput. Sci. **22**, 331–344 (2011)
10. Jirásková, G., Palmovský, M., Šebej, J.: Kleene closure on regular and prefix-free languages. In: Holzer, M., Kutrib, M. (eds.) CIAA 2014. LNCS, vol. 8587, pp. 226–237. Springer, Heidelberg (2014)
11. Maslov, A.N.: Estimates of the number of states of finite automata. Soviet Math. Doklady **11**, 1373–1375 (1970)
12. Rabin, M., Scott, D.: Finite automata and their decision problems. IBM Res. Develop. **3**, 114–129 (1959)
13. Sipser, M.: Introduction to the Theory of Computation. PWS Publishing Company, Boston (1997)
14. Šebej, J.: Reversal on regular languages and descriptional complexity. In: Jurgensen, H., Reis, R. (eds.) DCFS 2013. LNCS, vol. 8031, pp. 265–276. Springer, Heidelberg (2013)
15. Yu, S.: Regular languages, Chap. 2. In: Rozenberg, G., Salomaa, A. (eds.) Handbook of Formal Languages, vol. I, pp. 41–110. Springer, Heidelberg (1997)
16. Yu, S., Zhuang, Q., Salomaa, K.: The state complexity of some basic operations on regular languages. Theoret. Comput. Sci. **125**, 315–328 (1994)
17. Zijl, L.: Magic numbers for symmetric difference NFAs. Internat. J. Found. Comput. Sci. **16**, 1027 (2005)
18. Čevorová, K.: Kleene star on unary regular languages. In: Jurgensen, H., Reis, R. (eds.) DCFS 2013. LNCS, vol. 8031, pp. 277–288. Springer, Heidelberg (2013)
19. Čevorová, K., Jirásková, G., Krajňáková, I.: On the square of regular languages. In: Holzer, M., Kutrib, M. (eds.) CIAA 2014. LNCS, vol. 8587, pp. 136–147. Springer, Heidelberg (2014)

Minimal and Reduced Reversible Automata

Giovanna J. Lavado, Giovanni Pighizzini$^{(\boxtimes)}$, and Luca Prigioniero

Dipartimento di Informatica, Università degli Studi di Milano, Milan, Italy
{lavado,pighizzini}@di.unimi.it, luca.prigioniero@studenti.unimi.it

Abstract. A condition characterizing the class of regular languages which have several nonisomorphic minimal reversible automata is presented. The condition concerns the structure of the minimum automaton accepting the language under consideration. It is also observed that there exist reduced reversible automata which are not minimal, in the sense that all the automata obtained by merging some of their equivalent states are irreversible. Furthermore, it is proved that if the minimum deterministic automaton accepting a reversible language contains a loop in the "irreversible part" then it is always possible to construct infinitely many reduced reversible automata accepting such a language.

1 Introduction

A device is said to be *reversible* when each configuration has exactly one predecessor, thus implying that there is no loss of information during the computation. On the other hand, as observed by Landauer, logical irreversibility is associated with physical irreversibility and implies a certain amount of heat generation [8]. In order to avoid such a power dissipation and, hence, to reduce the overall power consumption of computational devices, the possibility of realizing reversible machines looks appealing.

A lot of work has been done to study reversibility in different computational devices. Just to give a few examples in the case of general devices as Turing machines, Bennet proved that each machine can be simulated by a reversible one [2], while Lange, McKenzie, and Tapp proved that each deterministic machine can be simulated by a reversible machine which uses the same amount of space [9]. As a corollary, in the case of a constant amount of space, this implies that each regular language is accepted by a *reversible two-way deterministic finite automaton*. Actually, this result was already proved by Kondacs and Watrous [5].

However, in the case of *one-way* automata, the situation is different. In fact, as shown by Pin, the regular language a^*b^* cannot be accepted by any reversible automaton [11].[1] So the class of languages accepted by *reversible automata* is a proper subclass of the class of regular languages. Actually, there are some different notions of reversible automata in literature. In 1982, Angluin introduced

---

[1] From now on, we will consider only *one-way automata*. Hence we will omit to specify "one-way" all the times.

© IFIP International Federation for Information Processing 2016
Published by Springer International Publishing Switzerland 2016. All Rights Reserved
C. Câmpeanu et al. (Eds.): DCFS 2016, LNCS 9777, pp. 168–179, 2016.
DOI: 10.1007/978-3-319-41114-9_13

reversible automata in algorithmic learning theory, considering devices having only one initial and only one final state [1]. On the other hand, the devices considered in [11], besides a set of final states, can have multiple initial states, hence they can take a nondeterministic decision at the beginning of the computation. An extension which allows to consider nondeterministic transitions, without changing the class of accepted languages, has been considered by Lombardy [10], introducing and investigating *quasi reversible automata*. Classical automata, namely automata with a single initial state and a set of final states, have been considered in the works by Holzer, Jakobi, and Kutrib [3,6,7]. In particular, in [3] the authors gave a characterization of regular languages which are accepted by reversible automata. This characterization is given in terms of the structure of the minimum deterministic automaton. Furthermore, they provide an algorithm that, in the case the language is acceptable by a reversible automaton, allows to transform the minimum automaton into an equivalent reversible automaton, which in the worst case is exponentially larger than the given minimum automaton. In spite of that, the resulting automaton is minimal, namely there are no reversible automata accepting the same language with a smaller number of states. However, it is not necessarily unique, in fact there could exist different reversible automata with the same number of states accepting the same language.

In this paper we continue the investigation of minimality in reversible automata. Our first result is a condition that characterizes languages having several different minimal reversible automata. Even this condition is on the structure of the transition graph of the minimum automaton accepting the language under consideration. As a special case, we show that each time the "irreversible part" of the minimum automaton contains a loop, it is possible to construct at least two different minimal reversible automata.

We also observe that there exist reversible automata which are not minimal but they are reduced, in the sense that when we try to merge some of their equivalent states we always obtain an irreversible automaton. Investigating this phenomenon more into details, we were able to find a language for which there exist arbitrarily large, and hence infinitely many, reduced reversible automata. In the paper, we present a general construction that allows to obtain arbitrarily large reversible automata for each language accepted by a minimum deterministic automaton satisfying the structural condition given in [3] and such that the "irreversible part" contains a loop. We know that this is also possible in other situations, namely that our condition is not necessary. We leave as an open problem, to find a characterization of the class of the languages having infinitely many reduced reversible automata.

2 Preliminaries

In this section we recall some basic definitions and results useful in the paper. We assume the reader is familiar with standard notions from automata and formal language theory (see, e.g., [4]). Given a set S, let us denote by $\#S$ its cardinality and by 2^S the family of all its subsets. Given an alphabet Σ, $|w|$ denotes the length of a string $w \in \Sigma^*$ and ε the empty string.

A *deterministic finite automaton* (DFA for short) is a tuple $A = (Q, \Sigma, \delta, q_I, F)$, where Q is the finite set of *states*, Σ is the *input alphabet*, $q_I \in Q$ is the *initial state*, $F \subseteq Q$ is the set of *accepting states*, and $\delta : Q \times \Sigma \to Q$ is the partial *transition function*. The *language accepted* by A is $L(A) = \{w \in \Sigma^* \mid \delta(q_I, w) \in F\}$. The *reverse* transition function of A is a function $\delta^R : Q \times \Sigma \to 2^Q$, with $\delta^R(p, a) = \{q \in Q \mid \delta(q, a) = p\}$. A state $p \in Q$ is *useful* if p is *reachable*, i.e., there is $w \in \Sigma^*$ such that $\delta(q_I, w) = p$, and *productive*, i.e., if there is $w \in \Sigma^*$ such that $\delta(p, w) \in F$. In this paper we only consider automata with all useful states.

We say that two states $p, q \in Q$ are *equivalent* if and only if for all $w \in \Sigma^*$, $\delta(p, w) \in F$ exactly when $\delta(q, w) \in F$. When $p \neq q$ are equivalent states, we can reduce the size of the automaton by "merging" p and q. This would imply to merge all the states reachable from p and q by reading a same string, namely the states $\delta(p, w)$ and $\delta(q, w)$, for $w \in \Sigma^*$.

Let $A' = (Q', \Sigma, \delta', q_I', F')$ be another DFA. A *morphism* φ from A to A', in symbols $\varphi : A \to A'$, is a function $\varphi : Q \to Q'$ such that $\varphi(q_I) = q_I'$, for each $q \in Q$, $a \in \Sigma$, $\varphi(\delta(q, a)) = \delta'(\varphi(q), a)$, and $q \in F$ if and only if $\varphi(q) \in F'$. Notice that if there exists a morphism $\varphi : A \to A'$ then it is unique and, for $x, y \in \Sigma^*$, $\delta(q_I, x) = \delta(q_I, y)$ implies $\delta'(q_I', x) = \delta'(q_I', y)$. We can observe that since in all automata we are considering all the states are useful, there exists the morphism $\varphi : A \to A'$ if and only if the automaton A' can be obtained from A after merging all pairs of states p, q of A, with $\varphi(p) = \varphi(q)$ (and possibly renaming the states). Hence, the number of states of A' cannot exceed that of A. Hence $\varphi^{-1}(s)$ denotes the set of states of A which are merged in the state s of A'. Two automata A and A' are said to be *equivalent* if they accept the same language, i.e., $L(A) = L(A')$.

Let $\mathcal{C}$ be a family of DFAs and $A \in \mathcal{C}$. We consider the following notions:

- The automaton A is *reduced* in $\mathcal{C}$ if for each morphism $\varphi : A \to A'$, the automaton A' does not belong to $\mathcal{C}$, i.e., every automaton obtained from A by merging some equivalent states does not belong to $\mathcal{C}$.
- The automaton A is *minimal* in $\mathcal{C}$ if and only if each automaton in $\mathcal{C}$ has at least as many states as A.
- The automaton A is the *minimum* in $\mathcal{C}$ if and only if it is the unique (up to an isomorphism, i.e., a renaming of the states) minimal automaton in $\mathcal{C}$.

Notice that each minimal automaton in a family $\mathcal{C}$ is reduced. Furthermore, if $\mathcal{C}$ contains a minimum automaton M, then M is also the only minimal and the only reduced automaton in $\mathcal{C}$. This happens, for instance, when $\mathcal{C}$ is the family of all DFAs accepting a given regular language L. However, a family $\mathcal{C}$ which does not have a minimum automaton, could contain reduced automata which are not minimal, as in the cases that will be presented in the paper.

A *strongly connected component* (SCC) C of a DFA $A = (Q, \Sigma, \delta, q_I, F)$ is a maximal subset of Q such that in the transition graph of A there exists a path between every pair of states in C. A SCC consisting of a single state q, *without* a looping transition, is said to be *trivial*. Otherwise C is *nontrivial* and, for each state in $q \in C$, there is a string $w \in \Sigma^* \setminus \{\varepsilon\}$ such that $\delta(q, w) = q$.

We introduce a partial order $\preceq$ on the set of SCCs of M, such that, for two such components C_1 and C_2, $C_1 \preceq C_2$ when no state in C_1 can be reached from a state in C_2, but a state in C_2 is reachable from a state in C_1. We write $C_1 \not\preceq C_2$ when $C_1 \preceq C_2$ is false, namely, $C_1 \neq C_2$ and either $C_2 \preceq C_1$ or C_1 and C_2 are incomparable.

Given a DFA $A = (Q, \Sigma, \delta, q_I, F)$, a state $r \in Q$ is said to be *irreversible* when $\#\delta^R(r, a) \geq 2$ for some $a \in \Sigma$, i.e., there are at least two transitions on the same letter entering r, otherwise r is said to be *reversible*. The DFA A is said to be *irreversible* if it contains at least one irreversible state, otherwise A is *reversible* (REV-DFA for short). As pointed out in [7], the notion of reversibility for a language is related to the computational model under consideration. In this paper we only consider DFAs. Hence, by saying that a language L is *reversible*, we refer to this model, namely we mean that there exists a REV-DFA accepting L.

The following result presents a characterization of reversible languages:

Theorem 1. [3] *Let L be a regular language and $M = (Q, \Sigma, \delta, q_I, F)$ be the minimum DFA accepting a language L. L is accepted by a REV-DFA if and only if there do not exist useful states $p, q \in Q$, a letter $a \in \Sigma$, and a string $w \in \Sigma^*$ such that $p \neq q$, $\delta(p, a) = \delta(q, a)$, and $\delta(q, aw) = q$.*

According to Theorem 1, a language L is reversible exactly when the minimum DFA accepting it does not contain the "forbidden pattern" consisting of two transitions on a same letter a entering in a same state r, with one of these transitions arriving from a state in the same SCC as r. Notice that, since transitions entering the initial state q_I can only arrive from states in the same SCC of q_I, if the language L is reversible, then the initial state q_I of M should be reversible.

An algorithm to convert a minimum DFA M into an equivalent REV-DFA, if any, was obtained in [3]. Furthermore, the resulting REV-DFA is minimal. We present an outline of it. The algorithm builds a REV-DFA A in the following way. At the beginning A is a copy of M. Then, the algorithm considers a minimal (with respect to $\preceq$) SCC C that contains an irreversible state and replace it with a number of copies which is equal to the maximum number of transitions on a same letter incoming in a state of C. This process is iterated until all the states in A are reversible.

3   Minimal Reversible Automata

In [3] it has been observed that there are reversible languages having several nonisomorphic minimal REV-DFAs. In this section we deepen that investigation by presenting a characterization of the languages having a unique minimal REV-DFA. (Notice that it could be different from the minimum DFA accepting the language.) To prove it we make use of a series of preliminary results. Hence, from now on, let us fix a reversible language L and the minimum DFA $M = (Q, \Sigma, \delta, q_I, F)$ accepting it.

Lemma 2. *Let $A' = (Q', \Sigma, \delta', q'_I, F')$ be a* REV-DFA *and $A'' = (Q'', \Sigma, \delta'', q''_I, F'')$ be a minimal* REV-DFA *both accepting L. Given the morphisms $\varphi' : A' \to M$ and $\varphi'' : A'' \to M$, it holds that $\#\varphi'^{-1}(s) \geq \#\varphi''^{-1}(s)$, for each $s \in Q$.*

Proof. By contradiction, suppose $\#\varphi'^{-1}(q) < \#\varphi''^{-1}(q)$ for some state q.

Let us partition Q in the set $Q_L = \{p \mid \exists w \in \Sigma^* \; \delta(p, w) = q\}$ of the states from which q is reachable and the set Q_R of remaining states. The sets Q' and Q'' are partitioned in a similar way, by defining $Q'_L = \varphi'^{-1}(Q_L)$, $Q'_R = \varphi'^{-1}(Q_R)$, $Q''_L = \varphi''^{-1}(Q_L)$, $Q''_R = \varphi''^{-1}(Q_R)$.

First, let us suppose $\#\varphi'^{-1}(p) \leq \#\varphi''^{-1}(p)$ for each $p \in Q_L$. We build another automaton $A''' = (Q''', \Sigma, \delta''', q'''_I, F''')$, which starts the computation by simulating A' using the states in Q'_L and, at some point, continues by simulating A'' using the states in Q''_R. In particular:

- $Q''' = Q'_L \cup Q''_R$
- The transitions are defined as follows:
  - For $s \in Q''_R$, $a \in \Sigma$: $\delta'''(s, a) = \delta''(s, a)$;
  - For $s \in Q'_L$, $a \in \Sigma$, such that $\delta'(s, a) \in Q'_L$: $\delta'''(s, a) = \delta'(s, a)$;
  - The remaining transitions, i.e., $\delta'''(s, a)$, in the case $s \in Q'_L$, $a \in \Sigma$, and $\delta'(s, a) \in Q'_R$, are obtained in the following way:
    Let us consider set of states $\{s_1, s_2, \ldots, s_k\}$ which are equivalent to s in A', i.e., $\varphi'(s_i) = \varphi'(s)$ for $i = 1, \ldots, k$ (notice that $s = s_h$ for some $h \in \{1, \ldots, k\}$), and the set of states $\{r_1, r_2, \ldots, r_j\}$ which are equivalent to s in A'', i.e., $\varphi''(r_i) = \varphi'(s)$ for $i = 1, \ldots, j$. Since $j \geq k$ we can safely define $\delta'''(s_i, a) = \delta''(r_i, a)$, for $i = 1, \ldots, k$.

The resulting automaton A''' still recognizes the language L, it is reversible and it has $\#Q'_L + \#Q''_R$ states. From $\#\varphi'^{-1}(p) \leq \#\varphi''^{-1}(p)$, for each $p \in Q_L$, and $\#\varphi'^{-1}(q) < \#\varphi''^{-1}(q)$, it follows that $\#Q'_L < \#Q''_L$, thus implying that the number of states of A''' is smaller than the one of A'', which is a contradiction.

In case $\#\varphi'^{-1}(p) > \#\varphi''^{-1}(p)$ for some $p \in Q_L$, we can apply the same construction, after switching the role of A' and A'', so producing an equivalent REV-DFA $\hat{A}'$ which is smaller than A' and still verifies $\#\hat{\varphi}'^{-1}(q) < \#\varphi''^{-1}(q)$, for the morphism $\hat{\varphi}' : \hat{A}' \to M$. Then, we iterate the proof on the two REV-DFAs $\hat{A}'$ and A''.

Hence, we can conclude that $\#\varphi'^{-1}(s) \geq \#\varphi''^{-1}(s)$, for each $s \in Q$. □

Lemma 2 allows to associate with each reversible language L and the minimum DFA $M = (Q, \Sigma, \delta, q_I, F)$ accepting it, the function $c : Q \to \mathbf{N}$ such that, for $q \in Q$, $c(q)$ is the number of states equivalent to q in any minimal REV-DFA A equivalent to M, i.e., $c(q) = \#\varphi^{-1}(q)$ for the morphism $\varphi : A \to M$. Notice that $c(q_I) = 1$. Furthermore, each REV-DFA accepting L should contain at least $c(q)$ states equivalent to q. These facts are summarized in the following result, where we also show that $c(q)$ has the same value for all states belonging to the same SCC of M.

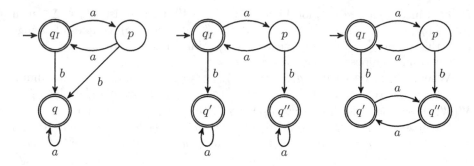

Fig. 1. A minimum DFA accepting the language $L = (aa)^* + a^*ba^*$, with two minimal nonisomorphic REV-DFAs

Lemma 3. *Let A be a* REV-DFA *accepting L, with the morphism $\varphi : A \to M$. If two states p, q of M belong to the same* SCC *of M then $\#\varphi^{-1}(p) = \#\varphi^{-1}(q) \geq c(p)$. Furthermore, if A is minimal then $c(p) = c(q) = \#\varphi^{-1}(p)$.*

Proof. Observe that since p, q belong to the same SCC there exists $x \in \Sigma^*$ such that $\delta(q, x) = p$. Let $\{q_1, q_2, \ldots, q_k\} = \varphi^{-1}(q)$ and $\{p_1, p_2, \ldots, p_j\} = \varphi^{-1}(p)$ be the sets of states in A which are equivalent to q and p, respectively. We are going to prove that $k = j$.

For each q_i, there exists p_{h_i} such that $\delta(q_i, x) = p_{h_i}$. Suppose $j < k$. In this case there are two indices i', i'' such that $p_{h_{i'}} = p_{h_{i''}}$ and then $\delta(q_{i'}, x) = \delta(q_{i''}, x) = p_{h_{i'}}$, implying that the state $p_{h_{i'}}$ is irreversible, which is a contradiction. This means that $j \geq k$. In the same way, by interchanging the roles of p and q, we can prove that $k \geq j$, which leads to the conclusion $j = k$.

The facts that $\#\varphi^{-1}(p) \geq c(p)$ and, for A minimal, $\#\varphi^{-1}(p) = c(p)$, follow from Lemma 2. □

In the following, for each SCC C of the transition graph of M, we use $c(C)$ to denote the value $c(q)$, for $q \in C$. Considering the algorithm outlined at the end of Sect. 2, we can observe that if C' is another SCC, then $C \preceq C'$ implies $c(C) \leq c(C')$.

As a consequence of Lemma 3, all the minimal REV-DFAs accepting L have the same "state structure", in the sense that they should contain exactly $c(q)$ states equivalent to the state q of M. However, they could differ in the transitions (see Fig. 1 for an example).

Lemma 4. *Let $A' = (Q', \Sigma, \delta', q'_I, F')$ and $A'' = (Q'', \Sigma, \delta'', q''_I, F'')$ be two* REV-DFAs *accepting L. If there are no morphisms $\varphi : A' \to A''$ then there exists a state $p \in Q$ with $\#\varphi''^{-1}(p) \geq 2$ such that either $p = q_I$, or*

$$\delta^R(p, a) \neq \emptyset \text{ and } \delta^R(p, b) \neq \emptyset$$

for two symbols $a, b \in \Sigma$, with $a \neq b$, and the morphism $\varphi'' : A'' \to M$.

Proof. Since there are no morphisms $\varphi : A' \to A''$, there exist $x, y \in \Sigma^*$ such that $\delta'(q_I', x) = \delta'(q_I', y)$ and $\delta''(q_I'', x) \neq \delta''(q_I'', y)$. Among all couples of strings with this property we choose one with $|xy|$ minimal. Furthermore, we observe that it cannot be possible that $x = y = \varepsilon$.

When $x = \varepsilon$, we have $\delta'(q_I', \varepsilon) = \delta'(q_I', y) = q_I'$ and, since M is minimum, $\delta(q_I, y) = q_I$. Hence, $\varphi''(\delta''(q_I'', y)) = \varphi''(q_I'') = q_I$. From $\delta''(q_I'', y) \neq q_I'' = \delta''(q_I'', \varepsilon)$, we conclude that $\#\varphi''^{-1}(q_I) \geq 2$. The case $y = \varepsilon$ is similar.

We now consider $x \neq \varepsilon$ and $y \neq \varepsilon$, i.e., $x = ua$, $y = vb$ for some $u, v \in \Sigma^*$ and $a, b \in \Sigma$. Let $\delta'(q_I', u) = q'$, $\delta'(q_I', v) = r'$, $\delta'(q', a) = \delta'(r', b) = \bar{p}$, $\delta''(q_I'', u) = q''$, $\delta''(q_I'', v) = r''$, $\delta''(q'', a) = s$, and $\delta''(r'', b) = t$, for states $q', r', \bar{p} \in Q'$, $q'', r'', s, t \in Q''$, with $s \neq t$.

Suppose $a = b$. Since A' is reversible from $\delta'(q', a) = \delta'(r', a) = \bar{p}$ we get $q' = r'$. Furthermore $q'' \neq r''$, otherwise A'' would be nondeterministic. Hence, on the strings u, v the automaton A' reaches the same state, while A'' reaches two different states, against the minimality of $|xy|$. Thus $a \neq b$.

Given the morphism $\varphi' : A' \to M$, let $p = \varphi'(\bar{p})$. Since M is minimum, it turns out that $\varphi''(s) = \varphi''(t) = \varphi'(\bar{p}) = p$. From $s \neq t$, we conclude that $\#\varphi''^{-1}(p) \geq 2$. Furthermore, from the previous discussion, the reader can observe that there are transitions on symbols a and b entering in p.   □

We are now able to prove the following:

Theorem 5. *Let* $M = (Q, \Sigma, \delta, q_I, F)$ *be the minimum* DFA *accepting a reversible language* L. *The following statements are equivalent:*

1. *There exists a state* $p \in Q$ *such that* $c(p) \geq 2$, $\delta^R(p, a) \neq \emptyset$, $\delta^R(p, b) \neq \emptyset$, *for two symbols* $a, b \in \Sigma$, *with* $a \neq b$.
2. *There exist at least two minimal nonisomorphic* REV-DFAs *accepting* L.

Proof. (2) implies (1): By Lemma 4, given two minimal nonisomorphic REV-DFAs A' and A'' accepting L, there is a state p such that $c(p) = \#\varphi''^{-1}(p) \geq 2$. Furthermore, since $c(q_I) = 1$, $p \neq q_I$. Hence, $\delta^R(p, a) \neq \emptyset$, $\delta^R(p, b) \neq \emptyset$, for two symbols $a, b \in \Sigma$, with $a \neq b$.

(1) implies (2): Let $w \in \Sigma^*$ be a string of minimal length such that $\delta(q_I, w) = p$, $a \in \Sigma$ be its last symbol, i.e., $w = xa$, with $x \in \Sigma^*$. Let $b \in \Sigma$ be a symbol with $b \neq a$ and $\delta^R(p, b) \neq \emptyset$. Given a minimal REV-DFA $A' = (Q', \Sigma, \delta', q_I', F')$ accepting L and the morphism $\varphi : A' \to M$, we consider the state $\hat{p} = \delta'(q_I', w)$. Then $\varphi'(\hat{p}) = p$.

We show how to build a minimal REV-DFA A'' nonisomorphic to A'. The idea is to use the set of states Q' as in A' and to modify only the transitions which simulates the transitions that in M enter the state p with the letter b. There are different cases.

When $\delta'^R(\hat{p}, b) = \emptyset$, it should exist $\tilde{p} \in \varphi'^{-1}(p)$ such that $\tilde{p} \neq \hat{p}$ and $\delta'(\tilde{q}, b) = \tilde{p}$, for some $\tilde{q} \in Q'$. The automaton A'' is defined as A', with the only difference that the transition $\delta'(\tilde{q}, b) = \tilde{p}$ is replaced by $\delta''(\tilde{q}, b) = \hat{p}$. To prove that it is nonisomorphic to A', we consider a string $y \in \Sigma^*$ of minimal length such that $\delta'(q_I', y) = \tilde{q}$. Then $\delta'(q_I', yb) = \tilde{p} \neq \delta'(q_I', w) = \hat{p}$, while $\delta''(q_I', yb) = \hat{p} = \delta''(q_I', w)$.

When $\delta'^R(\hat{p}, b) \neq \emptyset$ we can use one of the following possibilities:

- If there exists $\tilde{p} \neq \hat{p}$ such that $\delta'^R(\tilde{p}, b) \neq \emptyset$, then it should also exist $\tilde{q}, \hat{q} \in Q'$ with $\tilde{q} \neq \hat{q}$ such that $\delta'(\tilde{q}, b) = \tilde{p}$ and $\delta'(\hat{q}, b) = \hat{p}$. The automaton A'' is defined by switching the destinations of these two transitions, namely by replacing them by $\delta''(\tilde{q}, b) = \hat{p}$ and $\delta''(\hat{q}, b) = \tilde{p}$. The proof that A' and A'' are non isomorphic is exactly the same as in the previous case.
- If there exists $\tilde{p} \neq \hat{p}$ such that $\delta'^R(\tilde{p}, b) = \emptyset$, then we can consider $\hat{q}$ such that $\delta'(\hat{q}, b) = \hat{p}$, and define A'' by replacing this transition by $\delta''(\hat{q}, b) = \tilde{p}$. Let $y \in \Sigma^*$ be a string of minimal length such that $\delta'(q_I', y) = \hat{q}$. Then $\delta'(q_I', yb) = \hat{p} = \delta'(q_I', w)$. On the other hand $\delta''(q_I', yb) = \tilde{p} \neq \hat{p} = \delta''(q_I', w)$. Hence, A' and A'' are nonisomorphic.

Finally, we observe that in all cases, the automaton A'' has the same number of states as A'. Furthermore, the construction preserves reversibility.   □

As a consequence of Theorem 5 we obtain the following characterization of reversible languages having a unique minimal (hence a minimum) REV-DFA:

Corollary 6. *Let L be a reversible language and $M = (Q, \Sigma, \delta, q_I, F)$ be the minimum DFA accepting it. There exists a unique (up to isomorphism) minimal REV-DFA accepting L if and only if for each state $p \in Q$ with $c(p) \geq 2$, all the transitions entering in p are on the same symbol.*

When the minimum DFA accepting a reversible language contains a loop in the irreversible part, i.e., in the part "after" an irreversible state, the condition in Corollary 6 is always false, hence there exist at least two minimal nonisomorphic REV-DFAs. This is proved in the following result:

Theorem 7. *Let $M = (Q, \Sigma, \delta, q_I, F)$ be the minimum DFA accepting a reversible language L. If there exists an irreversible state $q \in Q$ such that the language accepted by computations starting in q is infinite, then there exists a state $p \in Q$ such that $c(p) \geq 2$, $\delta^R(p, a) \neq \emptyset$ and $\delta^R(p, b) \neq \emptyset$, for two symbols $a, b \in \Sigma$, with $a \neq b$.*

Proof. Let $p \in Q$ be a state reachable from q which belongs to a nontrivial SCC C. Hence $c(p) \geq 2$. Among all possibilities, we choose p in such a way that all the other states on a fixed path from q to p does not belong to C. Since C is nontrivial, it should exist a transition from a state of C, which enters in p. Let $a \in \Sigma$ be the symbol of such transition. Furthermore, it should exist another transition which enters in p from a state which does not belong to C. (If $p \neq q$ then we can take the last transition on the fixed path. Otherwise, since the initial state is always reversible, we have $q \neq q_I$, and so we can take the last transition entering in q on a path from q_I.) Let b the symbol of such transition. If $a = b$ the automaton M would contain the forbidden pattern (cfr. Theorem 1), thus implying that L is not reversible. Hence, we conclude $a \neq b$.   □

As a consequence of Theorem 7, considering Corollary 6 we can observe that when a reversible language has a unique minimal REV-DFA, all the loops in the minimum DFA accepting it should be in the reversible part. However, the converse does not hold, namely there are languages whose minimum DFA does not contain any loop in the irreversible part, which does not have a unique minimal REV-DFA. Indeed, in [3] an example with a finite language is presented.

4   Reduced Reversible Automata

In the section we show that there exist REV-DFAs which are reduced but not minimal, namely they have more states than equivalent minimal REV-DFAs, but merging some of their equivalent states would produce an irreversible automaton. Furthermore, we will prove that there exist reversible languages having arbitrarily large reduced REV-DFAs and, hence, infinitely many reduced REV-DFAs.

In Fig. 2 a reduced REV-DFA equivalent to the DFAs in Fig. 1 is depicted. If we try to merge two states in the loop, then the loop collapses to unique state, so producing the minimum DFA, which is irreversible. Actually, this example can be modified by using a loop of N states: if (and only if) N is prime, we get a reduced automaton. This is a special case of the construction which we are now going to present:

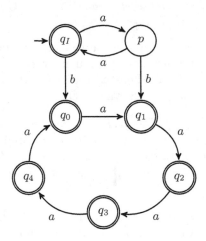

Fig. 2. A reduced REV-DFA

Theorem 8. *Let* $M = (Q, \Sigma, \delta, q_I, F)$ *be the minimum DFA accepting a reversible language* L. *If* M *contains a state* q *such that* $c(q) \geq 2$ *and the language accepted by computations starting in* q *is infinite, then there exist infinitely many nonisomorphic reduced REV-DFAs accepting* L.

Proof. Without loss of generality, we assume that the SCC C_q containing q is nontrivial. In fact, if this is not the case, we can find a state $\bar{q}$ which is reachable from q, and so $c(\bar{q}) \geq c(q) \geq 2$, and which belongs to a nontrivial SCC. Then, we can give the proof replacing q by $\bar{q}$.

Let A be a minimal REV-DFA A accepting L, obtained applying the algorithm outlined in Sect. 2, and $N \geq c(q)$ an integer. The idea is to modify A by replacing the part corresponding to the SCC C_q, with N copies of each state in C_q and arranging the transitions in such a way that all the states in these N copies form one SCC, without changing the accepted language. Furthermore, all the SCC that follow C_q will be replicated a certain number of times. More precisely, we build a DFA $A_N = (Q_N, \Sigma, \delta_N, q_{IN}, F_N)$ using the following steps:

(i) We put in A_N all the states of A which correspond to SCCs C of M with $C_q \not\preceq C$ and all the transitions between these states.

(ii) We add N copies of the states in C_q to the set of states of A_N. Given a state $r \in C_q$, let us denote its copies as $r_0, r_1, \ldots, r_{N-1}$.

(iii) We fix a transition $\delta(q, a) = q'$ of M, with $q, q' \in C_q$. For $i = 0, \ldots, N-1$, we define $\delta_N(q_i, a) = q'_{(i+1) \bmod N}$, and for the remaining transitions, namely $\delta(r, b) = r'$ with $(r, b) \neq (q, a)$, we define $\delta_N(r_i, b) = r'_i$. In this way in A_N we have N copies of the SCC C_q, modified in such a way that the transition from s_i on a in copy i leads to the state $q'_{(i+1) \bmod N}$ in copy $(i+1) \bmod N$.

(iv) We add to A_N each transition that in A leads from a state added in (i) to one state in the first $c(q)$ copies of C_q added in (iii). (We remind the reader that A should contain $c(q)$ copies of the SCC C_q. Hence, in A_N we keep exactly the same connections as in A from the states at point (i) to the states in these copies.)

(v) We complete the construction of A_N by adding a suitable number of copies of the remaining SCCs of M and suitable transitions, in order to derive a REV-DFA. This can be done just following the steps of the algorithm described in Sect. 2.

By construction, the automaton A_N so obtained is reversible and it accepts L. We are going to show that when N is a prime number then A_N is reduced. To this aim we shall prove that if we try to merge two equivalent states p', p'' of A_N then we obtain an irreversible automaton. The proof is divided in three cases:

– p', p'' are equivalent to a state p of M with $C_q \not\preceq C_p$, where C_p denotes the SCC containing p.

  These states have been added at step (i), copying them from the minimal REV-DFA A. By Lemma 3, A contains exactly $c(p)$ states equivalent to p. Hence, merging p' and p'', the resulting automaton would contain less than $c(p)$ states equivalent to p and, hence, it cannot be reversible.

– p', p'' are equivalent to a state p of M belonging to C_q.

  First, suppose $p' = q_0$ and $p'' = q_j$, $0 < j < n$. Considering step (iii), we observe that there is a string z such that $\delta(q', z) = q$, then $\delta(q, w) = q$ and $\delta_N(q_i, w) = q_{(i+1) \bmod N}$, where $w = az$. Thus, for each $k \geq 0$,

$\delta(q_0, w^{k(N-j)}) = q_{k(N-j) \bmod N}$ and $\delta(q_j, w^{k(N-j)}) = q_{j+k(N-j) \bmod N} = q_{(k-1)(N-j) \bmod N}$. Hence, merging q_0 and q_j would imply merging all the states whose indices are in the set $\{k(N-j) \bmod N \mid k \geq 0\}$, which, being N prime, coincides with $\{0, \ldots, N-1\}$. As a consequence, all the states q_i, should collapse in a unique state. However, since $c(q) \geq 2$, by Lemma 3 this implies that the resulting automaton is not reversible.

If $p' \neq q_0$, then we can always find a string y such that $\delta_N(p', y) = q_0$. Using the transitions introduced at step (iii), we get that $\delta_N(p'', y) = q_j$, for some $0 < j < N$. Hence, merging p' and p'' would imply merging q_0 and q_j, so reducing to the previous case.

- p', p'' are equivalent to a state p of M, such that $C_p \neq C_q$ and $C_q \preceq C_p$.

  Let $w \in \Sigma^*$ be such that $\delta(q, w) = p$ and $p' = \delta_N(q', w)$, $p'' = \delta_N(q'', w)$. From $p' \neq p''$, using the fact that A_N is reversible, we obtain $q' \neq q''$. So, to keep reversibility, merging p' and p'' would imply merging q' and q'', which are equivalent to q, so reducing to the previous case.

In summary, for each prime number $N \geq c(q)$ we obtained a reduced REV-DFA A_N with more than N states accepting the language L. Hence, we can conclude that there exist infinitely many nonisomorphic reduced REV-DFAs accepting L.   □

In Theorem 8 we gave a sufficient condition for the existence of infinitely many reduced REV-DFAs accepting a given language. This condition is not necessary. In fact, even if the minimum DFA does not contain any loop in the irreversible part, it could be possible to construct infinitely many reduced REV-DFAs. For instance, by modifying the construction given to prove Theorem 8, we can show that if the minimum DFA for a language L has a state p in the irreversible part, which is entered by transitions on at least two different letters (cfr. Theorem 7) and those transitions are used to recognize infinitely many strings, then there are infinitely many reduced REV-DFAs accepting L.

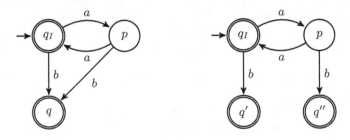

Fig. 3. The minimum DFA and the minimum REV-DFA accepting the language $L = (aa)^* + a^*b$

5   Conclusion

In this paper we studied the existence of minimal and reduced REV-DFAs. In some cases the minimum DFA accepting a language is already reversible, so assuring that the language is reversible. However, in general a minimum DFA does not need to be reversible, although the accepted language could be reversible. Using Theorem 1 and the construction from [3] outlined in Sect. 2, in the case the language is reversible, from a given minimum DFA we can obtain a minimal REV-DFA. Minimal REV-DFAs are not necessarily unique (see Fig. 1 for an example, while Fig. 3 shows a case with a unique minimal, and hence minimum, REV-DFA). In Sect. 3 we gave a characterization of the languages having a unique minimum REV-DFA, in terms of the structure of the minimum DFA.

Here we wanted to go beyond the investigation of minimal REV-DFAs studying reduced REV-DFAs. We observed the existence of reduced REV-DFAs which are not minimal and we gave a sufficient condition for the existence of infinitely many reduced REV-DFAs accepting a same reversible language.

References

1. Angluin, D.: Inference of reversible languages. J. ACM **29**(3), 741–765 (1982)
2. Bennett, C.: Logical reversibility of computation. IBM J. Res. Dev. **17**(6), 525–532 (1973)
3. Holzer, M., Jakobi, S., Kutrib, M.: Minimal reversible deterministic finite automata. In: Potapov, I. (ed.) DLT 2015. LNCS, vol. 9168, pp. 276–287. Springer, Heidelberg (2015)
4. Hopcroft, J.E., Ullman, J.D.: Introduction to Automata Theory, Languages and Computation. Addison-Wesley, Reading (1979)
5. Kondacs, A., Watrous, J.: On the power of quantum finite state automata. In: FOCS, pp. 66–75. IEEE Computer Society (1997)
6. Kutrib, M.: Aspects of reversibility for classical automata. In: Calude, C.S., Freivalds, R., Kazuo, I. (eds.) Computing with New Resources. LNCS, vol. 8808, pp. 83–98. Springer, Heidelberg (2014)
7. Kutrib, M.: Reversible and irreversible computations of deterministic finite-state devices. In: Italiano, G.F., Pighizzini, G., Sannella, D.T. (eds.) MFCS 2015, Part I. LNCS, vol. 9234, pp. 38–52. Springer, Heidelberg (2015)
8. Landauer, R.: Irreversibility and heat generation in the computing process. IBM J. Res. Dev. **5**(3), 183–191 (1961)
9. Lange, K., McKenzie, P., Tapp, A.: Reversible space equals deterministic space. J. Comput. Syst. Sci. **60**(2), 354–367 (2000)
10. Lombardy, S.: On the construction of reversible automata for reversible languages. In: Widmayer, P., Triguero, F., Morales, R., Hennessy, M., Eidenbenz, S., Conejo, R. (eds.) ICALP 2002. LNCS, vol. 2380, pp. 170–182. Springer, Heidelberg (2002)
11. Pin, J.-E.: On reversible automata. In: Simon, I. (ed.) LATIN 1992. LNCS, vol. 583, pp. 401–416. Springer, Heidelberg (1992)

Unary Self-verifying Symmetric Difference Automata

Laurette Marais[1,2] and Lynette van Zijl[1(✉)]

[1] Department of Computer Science, Stellenbosch University,
Stellenbosch, South Africa
lvzijl@sun.ac.za
[2] Meraka Institute, CSIR, Pretoria, South Africa
laurette.p@gmail.com

Abstract. We investigate self-verifying nondeterministic finite automata, in the case of unary symmetric difference nondeterministic finite automata (SV-XNFA). We show that there is a family of languages $\mathcal{L}_{n\geq 2}$ which can always be represented non-trivially by unary SV-XNFA. We also consider the descriptional complexity of unary SV-XNFA, giving an upper and lower bound for state complexity.

1 Introduction

Any nondeterministic finite automaton (NFA) has an equivalent deterministic finite automaton (DFA) which can by found by applying the subset construction [1]. This subset construction uses the union set operation. Symmetric difference NFA (XNFA), on the other hand, employ the symmetric difference set operation [2] during the determinisation process with the subset construction. XNFA may also be considered as a special case of weighted automata over GF(2) [3]. XNFA, even in the unary case, are interesting because of the different descriptional complexity when compared to traditional NFA. For example, an n-state unary XNFA may have an equivalent minimal DFA with $2^n - 1$ states, whereas the bound is $e^{\Theta\sqrt{n \ \ln n}}$ in the case of NFA [2]. In this work, we consider self-verification for XNFA.

Self-verifying NFA (SV-NFA) [4–6] are automata with two kinds of final states, namely, accept states and reject states. Each path in the automaton may reach either an "Accept", "Reject" or "I do not know" state. Once a path has been found that either accepts or rejects, it is guaranteed that no other path with the same label will reach the opposite answer. Furthermore, every word is guaranteed one path that reaches either an accept or a reject state. Consequently, unlike with NFA, rejection is the result of reaching a reject state, and not the result of a failure to reach an accept state.

Assent and Seibert [4] showed that any n-state SV-NFA has an equivalent DFA with $O(2^n/\sqrt{n})$ states. Jirásková and Pighizzini [6] improved their result, and showed a tight upper bound $h(n)$, where $h(n)$ grows like $3^{\frac{n}{3}}$, for an SV-NFA

© IFIP International Federation for Information Processing 2016
Published by Springer International Publishing Switzerland 2016. All Rights Reserved
C. Câmpeanu et al. (Eds.): DCFS 2016, LNCS 9777, pp. 180–191, 2016.
DOI: 10.1007/978-3-319-41114-9_14

with a binary alphabet. In the unary case, it was shown that the upper bound of $e^{\Theta\sqrt{n \ \ln n}}$ is not tight for unary SV-NFA.

In this article, we define self-verifying XNFA (SV-XNFA), and consider the case of unary SV-XNFA. We show the existence of a family of languages accepted by unary SV-XNFA, and point out some conditions for the existence of n-state unary SV-XNFA. We also give an upper bound and lower bound for the state complexity.

2   Preliminaries

An NFA N is a five-tuple $N = (Q, \Sigma, \delta, Q_0, F)$, where Q is a finite set of states, Σ is a finite alphabet, $\delta : Q \times \Sigma \to 2^Q$ is a transition function (here, 2^Q indicates the power set of Q), $Q_0 \subseteq Q$ is a set of initial states, and $F \subseteq Q$ is the set of final (acceptance) states. The transition function δ can be extended to strings in the Kleene closure Σ^* of the alphabet:

$$\delta'(q, w_0 w_1 \ldots w_k) = \delta(\delta(\ldots \delta(q, w_0), w_1), \ldots, w_k).$$

For convenience, we write $\delta(q, w)$ to mean $\delta'(q, w)$.

An NFA N is said to accept a string $w \in \Sigma^*$ if $q_0 \in Q_0$ and $\delta(q_0, w) \in F$, and the set of all strings (also called words) accepted by N is the language $\mathcal{L}(N)$ accepted by N. Any NFA has an equivalent DFA which accepts the same language. The DFA equivalent to a given NFA can be found by the subset construction [1]. In essence, the subset construction keeps track of all the states that the NFA may be in at the same time, and forms the states of the equivalent DFA by grouping of the states of the DFA. In short,

$$\delta(A, \sigma) = \bigcup_{q \in A} \delta(q, \sigma)$$

for any $A \subseteq Q$ and $\sigma \in \Sigma$.

An XNFA $M = (Q, \Sigma, \delta, Q_0, F)$ is defined similarly to an NFA, with the difference that the XNFA accepts a string $w \in \Sigma^*$ if $q_0 \in Q_0$, and $|\delta(q_0, w) \cap F|$ is odd. This acceptance condition reflects the parity nature of the XNFA, so that a string is accepted when there is an odd number of paths which lead to final states [7]. This definition of acceptance ensures that an XNFA can be seen as a special case of a weighted automaton [3]. When the subset construction is applied to find the DFA equivalent to the XNFA, the symmetric difference (in the set theoretic sense) is used to reflect the parity of the paths. That is,

$$\delta(A, \sigma) = \bigoplus_{q \in A} \delta(q, \sigma)$$

for any $A \subseteq Q$ and $\sigma \in \Sigma$.

For clarity, the DFA equivalent to an XNFA N is termed an XDFA and denoted with N_D (with corresponding Q_D, δ_D etc.).

It was shown (amongst others) in [2,7] that XNFA can be investigated by considering them as linear machines over the Galois field GF(2). We also use that approach in this work. Consider the transition table of a unary XNFA $N = (Q, \Sigma, \delta, Q_0, F)$, where each row represents a mapping from a state $q \in Q$ to a set of states $P \in 2^Q$. Then P can be written as a vector with a one in position i if $q_i \in P$, and a zero in position i if $q_i \notin P$. Hence, the transition table can be represented as a matrix of zeroes and ones (see Example 1). This is known as the characteristic or transition matrix of the XNFA.

Initial and final states can be represented by vectors, and appropriate vector and matrix multiplications over GF(2) represent the behaviour of the XNFA[1]. For more detail, see for example [3]. For the purposes of this work, we consider only unary XNFA with one alphabet symbol. In general, for larger alphabets, there is a matrix associated with each alphabet symbol.

Let M be the characteristic matrix of N. The characteristic polynomial $c(X)$ of M is given by $\det(M - IX)$, and $c(X)$ is said to be the characteristic polynomial of N.

Note that the characteristic matrix of an XNFA does not contain information about the choice of initial and final states, so in fact any such matrix represents a set of XNFA sharing a transition graph but differing in choice of initial and final states. A characteristic polynomial is associated with the matrix, but many matrices may share the same polynomial, so a polynomial over GF(2) represents a set of characteristic matrices. A useful result from linear field theory [8] states that any monic polynomial $c(X) = X^n + c^{n-1}X^{n-1} + \ldots c_2X^2 + c_1X + c_0$ over GF(2) has a so-called companion matrix (also called a normal form matrix) M of the form

$$M = \begin{bmatrix} 0 & 0 & \ldots & 0 & c_0 \\ 1 & 0 & \ldots & 0 & c_1 \\ 0 & 1 & \ldots & 0 & c_2 \\ \vdots & \vdots & & \vdots & \vdots \\ 0 & 0 & \ldots & 1 & c_{n-1} \end{bmatrix}.$$

Thus, given a polynomial over GF(2), it is possible to construct its companion matrix directly, and then construct an XNFA from the companion matrix. Such an XNFA will have the transition function $\delta(q_i, a) = q_{i+1}$ for $0 \leq i < n - 1$, and $q_j \in \delta(q_{n-1}, a)$ for all j such that $c_j \neq 0$.

Finally, each $c(X)$ over GF(2) is associated with a certain cycle structure. Specifically, given a unary XNFA N, the properties of its characteristic polynomial $c(X)$ allow conclusions about the possible length of the cycle of states of the equivalent XDFA N_D (see for example [2,8,9]).

Theorem 1. [8] *Let $c(X)$ be a polynomial of degree n over GF(2) that does not have X as a factor.*

– *If $c(X)$ is a primitive irreducible polynomial over GF(2), then $c(X)$ has a single cycle of length $2^n - 1$.*

---

[1] In GF(2), $1 + 1 = 0$.

– If $c(X)$ is an irreducible but not primitive polynomial over GF(2), then $c(X)$ has $(2^n - 1)/b$ cycles of length b, where b is a factor of $2^n - 1$.
– If $c(X)$ is a reducible polynomial over GF(2), consider its factors. For each cycle of length k_i induced by factor $\phi_i(X)$ and for each cycle of length k_j induced by factor $\phi_j(X)$, $c(X)$ has $\gcd(k_i, k_j)$ cycles of length $\mathrm{lcm}(k_i, k_j)$.

The choice of initial states for N determines which cycle in the cycle structure of $c(X)$ represents the equivalent XDFA N_D. We give an example of an XNFA to illustrate the discussion above.

Example 1. Let N be an XNFA where $Q = \{q_0, q_1, q_2, q_3\}$, $\Sigma = \{a\}$, $Q_0 = \{q_0\}$, $F = \{q_1, q_3\}$ and δ is defined in Table 1 (start states are indicated by $\rightarrow$, and final states by $\leftarrow$). This corresponds to the matrix M below and characteristic polynomial $c(X) = X^4 + X^3 + X + 1$.

$$M = \begin{bmatrix} 0 & 0 & 0 & 1 \\ 1 & 0 & 0 & 1 \\ 0 & 1 & 0 & 0 \\ 0 & 0 & 1 & 1 \end{bmatrix}$$

Table 1. Transition function of N

δ	a
$\rightarrow$ q_0	q_1
$\leftarrow$ q_1	q_2
q_2	q_3
$\leftarrow$ q_3	q_0, q_1, q_3

Table 2. Transition function of N_D

δ_D	a
$\rightarrow$ $[q_0]$	$[q_1]$
$\leftarrow$ $[q_1]$	$[q_2]$
$[q_2]$	$[q_3]$
$\leftarrow$ $[q_3]$	$[q_0, q_1, q_3]$
$[q_0, q_1, q_3]$	$[q_0, q_2, q_3]$
$\leftarrow$ $[q_0, q_2, q_3]$	$[q_0]$

The transition function δ_D of the equivalent XDFA N_D is shown in Table 2 and N_D is shown in Fig. 1. Note that $[q_0, q_1, q_3] \notin F_D$, since it contains an even number of states from F.

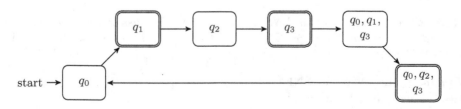

Fig. 1. Example 1: N_D

$\square$

We now recap the definition of SV-NFA:

Definition 1. [4,6] *A self-verifying nondeterministic finite automaton (SV-NFA) is a 6-tuple $N = (Q, \Sigma, \delta, Q_0, F^a, F^r)$, where Q, Σ, δ and Q_0 are defined as for standard NFA. Here, $F^a \subseteq Q$ and $F^r \subseteq Q$ are the sets of accept and reject states, respectively. The remaining states, that is, the states belonging to $Q \setminus (F^a \cup F^r)$, are called neutral states. For each input string w in Σ^*, it is required that there exists at least one path ending in either an accept or a reject state; that is, $\delta(q_0, w) \cap (F^a \cup F^r) \neq \emptyset$ for any $q_0 \in Q_0$, and there are no strings w such that both $\delta(q_0, w) \cap F^a$ and $\delta(q_1, w) \cap F^r$ for any $q_0, q_1 \in Q_0$ are nonempty.*

Unlike an NFA, an SV-NFA leads to an explicit answer state for any string $w \in \Sigma^*$. Hence, its equivalent DFA must do so too. The path for each w in a DFA is unique, so each state in the DFA is an accept or reject state. Hence, for any DFA state d, there is some SV-NFA state $q_r \in d$ such that $q_r \in F^a$ so that $d \in F_D^a$ or $q_r \in F^r$ so that $d \in F_D^r$. Since each state in the DFA is a subset of states of the SV-NFA, accept and reject states cannot occur together in a DFA state. That is, if d is a DFA state, then for any $p, q \in d$, if $p \in F^a$ then $q \notin F^r$ and vice versa.

Combining the notions of SV-NFA and XNFA, we now define SV-XNFA.

Definition 2. *A self-verifying symmetric difference finite automaton (SV-XNFA) is a 6-tuple $N = (Q, \Sigma, \delta, Q_0, F^a, F^r)$, where Q, Σ, δ and Q_0 are defined as for XNFA, and F^a and F^r are defined as for SV-XNFA. That is, each state in the SV-XDFA equivalent to N must contain an odd number of states from either F^a or F^r, but not both.*

Note that the acceptance condition for SV-XNFA (or the SV condition) implies that if a state in the SV-XDFA of an SV-XNFA N contains an odd number of states from F^a, it may also contain an even number of states from F^r, and so belongs to F_D^a, and vice versa. Parity is not applied to neutral states, so that any state in the XDFA may contain any number of neutral states from N.

The choice of F^a and F^r for a given SV-XNFA N is called an *SV-assignment* of N. An SV-assignment where either F^a or F^r is empty, is called a *trivial SV-assignment*. Otherwise, if both F^a and F^r are nonempty, the SV-assignment is non-trivial.

Definition 3. *Let N be an XNFA. A non-trivial SV-assignment for N such that $\mathcal{L}(N) \neq \emptyset$ and $\mathcal{L}(N) \neq \Sigma^*$, is called an interesting SV-assignment. An SV-XNFA with an interesting SV-assignment is called an interesting SV-XNFA.*

Example 2. Let N be an XNFA where $Q = \{q_0, q_1, q_2, q_3, q_4\}$, $\Sigma = \{a\}$, $Q_0 = \{q_0, q_1\}$ and δ is defined in Table 3.

The transition function δ_D of the equivalent XDFA is shown in Table 4. Then $F^a = \{q_2, q_4\}$ and $F^r = \{q_0\}$ is an interesting SV-assignment. The resulting SV-XDFA N_D is shown in Fig. 2. We see that $F_D^a = \{[q_1, q_2], [q_2, q_3], [q_3, q_4], [q_1, q_4]\}$, since these states each contain one state from F^a. Similarly, it holds that

Table 3. Transition function of N

δ		a
$r \rightleftarrows$	q_0	q_1
$\rightarrow$	q_1	q_2
$a \leftarrow$	q_2	q_3
	q_3	q_4
$a \leftarrow$	q_4	q_0, q_1, q_2

Table 4. Transition function of N_D

δ_D		a
$r \rightleftarrows$	$[q_0, q_1]$	$[q_1, q_2]$
$a \leftarrow$	$[q_1, q_2]$	$[q_2, q_3]$
$a \leftarrow$	$[q_2, q_3]$	$[q_3, q_4]$
$a \leftarrow$	$[q_3, q_4]$	$[q_0, q_1, q_2, q_4]$
$r \leftarrow$	$[q_0, q_1, q_2, q_4]$	$[q_0, q_3]$
$r \leftarrow$	$[q_0, q_3]$	$[q_1, q_4]$
$a \leftarrow$	$[q_1, q_4]$	$[q_0, q_1]$

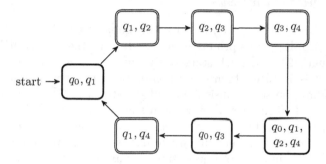

Fig. 2. Example 2: N_D

$F_D^r = \{[q_0, q_1], [q_0, q_1, q_2, q_4], [q_0, q_3]\}$, since each state contains q_0. Note that $[q_0, q_1, q_2, q_4]$ contains an even number of states from F^a.   □

We now investigate when interesting SV-assignments are possible for unary XNFA.

3   Unary SV-XNFA

Consider any unary XNFA N and its corresponding transition matrix M over GF(2). Then M can be either singular, or non-singular. If M is singular, it is known [8] that the XDFA N_D equivalent to N forms a state graph with a transient head followed by a cycle. If M is non-singular, then N_D forms a cycle. In the rest of this article, we only consider unary XNFA whose transition matrices are non-singular. By Lemma 1 below, this means we only consider polynomials over GF(2) that do not have X as a factor.

Noting the correspondence between a given XNFA, its matrix representation over GF(2) and the corresponding characteristic polynomial $c(X)$, we investigate whether there are properties of polynomials that guarantee the existence or non-existence of SV-assignments for XNFA with certain characteristic polynomials. We are specifically interested in finding properties that will guarantee the existence of SV-XNFA with n states that accept languages that require $n_D > n$ states in the equivalent SV-XDFA. This implies that we focus on interesting

SV-assignments when determining the existence of SV-XNFA for certain polynomials.

Lemma 1. *The companion matrix of a polynomial over GF(2) is singular if
and only if X is a factor of the polynomial.*

Proof. Let $c(X)$ be some polynomial over GF(2) of degree n. If and only if X
is a factor of $c(X)$, then the coefficient of X^0 is zero, and so in the companion
matrix, $M_{0,n-1} = 0$. Then $det(M) = 0$, which implies that M is singular [9]. $\square$

Theorem 2. *There is no n-state SV-XNFA such that its characteristic polynomial is primitive and irreducible.*

Proof. Let $N = (Q, \Sigma, \delta, Q_0, F)$ be an n-state unary XNFA with characteristic
polynomial $c(X)$. If $c(X)$ is primitive and irreducible, then the XDFA N_D forms
a cycle of length $2^n - 1$. Each state in the cycle is a non-empty subset of Q.
If there are $2^n - 1$ states, then every non-empty subset of Q is a state in the
XDFA, including the state consisting of all the states in Q.

Since each $q \in Q$ appears as a state in the cycle, each q must either be an
accept or reject state. There are two cases to consider. If n is even, then the
state consisting of all the states in Q contains either an even number of accept
states and an even number of reject states, or an odd number of accept states
and an odd number of reject states. In either case the SV condition is violated,
since each SV-XDFA state must contain an odd number of either F^a or F^r, but
not both.

On the other hand, if n is odd, then – in order for the state consisting of all
the states in Q to be either accepting or rejecting – some $A \subset Q$ where $|A|$ is odd
must contain, say, the accepting states, while $Q \setminus A$ contains the rejecting states
and $|Q \setminus A|$ is even. But the XDFA also contains a state consisting of $Q \setminus A$,
that is a state consisting entirely of an even number of reject states. Hence if
n is odd, this necessarily results in a neutral state in the XDFA, which again
violates the SV condition. Therefore, no SV-XNFA is possible. $\square$

Note that Theorem 2 excludes all SV-assignments for primitive polynomials,
including trivial or uninteresting SV-assignments. On the other hand, we now
prove that for a certain family of polynomials, interesting SV-assignments are
always possible.

Theorem 3. *Let $c(X) = X^n + X^{n-1} + X + 1$, with companion matrix M, and
let N be an XNFA with transition matrix M and $Q_0 = \{q_0\}$. Then N has an
interesting SV-assignment, and the equivalent XDFA N_D forms a cycle of length
$2n - 2$.*

Proof.   The transition function of N is given in Table 5. Since $\delta(q_i, a) =$
q_{i+1} for all $i < n - 1$, the XDFA N_D contains the states $[q_0], [q_1], ..., [q_{n-1}]$
in its cycle. Also, $\delta(q_{n-1}, a) = \{q_0, q_1, q_{n-1}\}$. Now, $\delta(\{q_0, q_i, q_{n-1}\}, a) =$
$\{q_0, q_{i+1}, q_{n-1}\}$ for $1 \leq i \leq n - 3$, since $\delta(\{q_0, q_i, q_{n-1}\}, a) = \{q_1\} \oplus \{q_{i+1}\} \oplus$

Table 5. Transition function of N with $c(X) = X^n + X^{n-1} + X + 1$

δ	a
q_0	q_1
q_1	q_2
$\vdots$	$\vdots$
q_{n-2}	q_{n-1}
q_{n-1}	q_0, q_1, q_{n-1}

$\{q_0, q_1, q_{n-1}\} = \{q_0, q_{i+1}, q_{n-1}\}$, as $q_{i+1} \neq q_{n-1}$ for $1 \leq i \leq n-3$. However, $\delta(\{q_0, q_{n-2}, q_{n-1}\}, a) = \{q_1\} \oplus \{q_{n-2+1}\} \oplus \{q_0, q_1, q_{n-1}\} = \{q_0\}$. Therefore, N_D contains $[q_0], [q_1], ..., [q_{n-1}]$ and $[q_0, q_i, q_{n-1}]$ for $1 \leq i \leq n-2$, and hence has $n + n - 2 = 2n - 2$ states.

Now, since every state in N_D has odd size, any choice of F^a and F^r so that $F^a \cup F^r = Q$ and $F^a \cap F^r = \emptyset$ with F^a and F^r non-empty will guarantee that each state in the XDFA contains an odd number of states from either F^a or F^r and zero or an even number of states from the other, and hence will be a non-trivial SV-assignment. Since $[q_0], [q_1], ..., [q_{n-1}] \in Q_D$, it will also necessarily be an interesting SV-assignment. $\qquad \square$

3.1   Languages for Unary SV-XNFA

Given the existence of SV-XNFA for certain $c(X)$ as shown above, we may now consider whether there is a family of languages $\mathcal{L}_{n \geq 2}$ such that each $\mathcal{L}_i$ may be represented by an SV-XNFA in a non-trivial way. That is, we consider whether there are languages that may be represented by SV-XNFA with n states that require $n_D > n$ states in their equivalent SV-XDFA. The next theorem presents such a language family.

Theorem 4. *For any integer $n \geq 2$, let $\mathcal{L}_n = a^{(2n-2)i+j}$, for $i \geq 0$ and $0 \leq j < n - 1$, and $\mathcal{L}_n^c = a^{(2n-2)i+j}$, for $i > 0$ and $n - 1 \leq j < 2n - 2$. Then there exists a pair of SV-XNFA with n states and the same transition graph that accept $\mathcal{L}_n$ and $\mathcal{L}_n^c$ respectively. Moreover, these languages each require an SV-XDFA with $2n - 2$ states.*

Proof. Using the construction given in the proof of Theorem 3, we construct N with n states so that N_D has $2n-2$ states. The states in the SV-XDFA are given in Fig. 3.

Then $F^a = \{q_i | 0 \leq i \leq n - 2\}$ and $F^r = \{q_{n-1}\}$ is an interesting SV-assignment. Consequently, $d_0, d_1, ..., d_{n-2} \in F_D^a$ and $d_{n-1}, d_n, ..., d_{2n-3} \in F_D^r$, so the language accepted by N is $\mathcal{L}_n = a^{(2n-2)i+j}$, for $i \geq 0$ and $0 \leq j < n - 1$. Since this pattern of $n - 1$ accept states followed by $n - 1$ reject states requires $2n - 2$ states, N_D is the minimal SV-XDFA that accepts $\mathcal{L}_n$.

$$d_0 = [q_0]$$
$$d_1 = [q_1]$$
$$\vdots$$
$$d_{n-2} = [q_{n-2}]$$
$$d_{n-1} = [q_{n-1}]$$
$$d_n = [q_0, q_1, q_{n-1}]$$
$$d_{n+1} = [q_0, q_2, q_{n-1}]$$
$$\vdots$$
$$d_{2n-3} = [q_0, q_{n-2}, q_{n-1}]$$

Fig. 3. Theorem 4: states in the SV-XDFA

Also, $F^r = \{q_i | 0 \leq i \leq n-2\}$ and $F^a = \{q_{n-1}\}$ is an interesting SV-assignment that would cause $d_0, d_1, ..., d_{n-2} \in F_D^r$ and $d_{n-1}, d_n, ..., d_{2n-3} \in F_D^a$ so that N accepts $\mathcal{L}_n^c = a^{(2n-2)i+j}$, for $i > 0$ and $n-1 \leq j < 2n-2$. Similarly as for $\mathcal{L}_n$, $2n-2$ states are required to accept $\mathcal{L}_n^c$.

This leads to a pair of SV-XNFA with n states and the same transition graphs that accept $\mathcal{L}_n$ and $\mathcal{L}_n^c$ respectively, while in the deterministic case, an SV-XDFA with at least $2n-2$ states is required.　　□

The ability of self-verifying automata to represent complementary pairs of languages using the same transition graph is discussed in [10].

3.2　Descriptional Complexity of Unary SV-XNFA

We now turn to the question of state complexity for SV-XNFA. By Theorem 1, the maximum cycle length for any $c(X)$ of degree n is $2^n - 1$, and therefore this is an upper bound for the number of states in the equivalent XDFA of any XNFA with n states. However, it is not a tight upper bound for SV-XNFA, because this cycle length is only achieved if $c(X)$ is primitive and from Theorem 2 it is clear such XDFA cannot have SV-assignments. Instead, we show in this section that for certain $c(X)$ of degree n, there exist SV-XNFA with characteristic polynomial $c(X)$ for which the equivalent SV-XDFA have at least $2^{n-1} - 1$ states, and that for any $n \geq 2$, there is a language $\mathcal{L}_n$ that can be represented by an n-state SV-XNFA while requiring an $(2^{n-1} - 1)$-state SV-XDFA.

Lemma 2. *Let* $c(X) = (X+1)\phi(X)$ *be a polynomial of degree n with non-singular companion matrix M, and let N be an XNFA with transition matrix M and $Q_0 = \{q_0\}$. Then the equivalent XDFA N_D has the following properties:*

1. $|Q_D| > n$
2. $|d|$ is odd for $d \in Q_D$

3. $[q_0], [q_1], ..., [q_{n-1}] \in Q_D$

Proof. Since $X + 1$ is a factor, 1 is a root of the polynomial, and so $c(X)$ must have an even number of terms, including X^0. The companion matrix M of $c(X)$ is

$$M = \begin{bmatrix} 0 & 0 & 0 & \cdots & 0 & 0 & c_0 \\ 1 & 0 & 0 & \cdots & 0 & 0 & c_1 \\ 0 & 1 & 0 & \cdots & 0 & 0 & c_2 \\ \vdots & & \ddots & & & & \vdots \\ \vdots & & & \ddots & & & \vdots \\ 0 & 0 & 0 & \cdots & 1 & 0 & c_{n-2} \\ 0 & 0 & 0 & \cdots & 0 & 1 & c_{n-1} \end{bmatrix}.$$

The last column contains an odd number of 1's, representing an odd number of transitions from state q_{n-1}. That is, $\delta(q_{n-1}, a) = \{q_c | c \in C\} = Q_c$ where $|C|$ is odd.

Since $\delta(q_i, a) = q_{i+1}$ for all $i < n - 1$, N_D contains the states $[q_0], [q_1], ..., [q_{n-1}]$, as well as $[Q_c]$, and therefore forms a cycle with at least $n + 1$ states. These states all have odd size, so it only remains to show that all other states in the cycle must have odd size as well.

Let $P = \{q_{i_0}, q_{i_1}, ..., q_{i_k}\} \subseteq Q$ where k is even and so $|P|$ is odd. Then if $i_j < n - 1$ for all $0 \leq j \leq k$, then $\delta(P, a) = \{q_{i_0+1}, q_{i_1+1}, ..., q_{i_k+1}\}$, and so $|\delta(P, a)|$ must be odd as well. However, suppose $q_{n-1} \in P$. We may assume that $q_{n-1} = q_{i_k}$. Let $P' = \{q_{i_0+1}, q_{i_1+1}, ..., q_{i_k}\}$, so $|P'| = |P| - 1$ and therefore even. Then $\delta(P, a) = P' \oplus Q_c$. Let $m = |P' \cap Q_c|$. Then $|\delta(P, a)| = |P'| + |Q_c| - 2m$. Since $|P'|$ is even, $|Q_c|$ is odd and $2m$ is even, it follows that $|\delta(P, a)|$ is odd.

Therefore, any state with odd size in N_D transitions to a state with odd size, and so all the states in the XDFA cycle have odd size. □

Theorem 5. *Let* $c(X) = (X + 1)\phi(X)$ *be a polynomial of degree n with non-singular companion matrix M. Then there is an XNFA N with transition matrix M and $Q = \{q_0\}$ for which there is an interesting SV-assignment.*

Proof. From Lemma 2 it follows that the XNFA N whose transition matrix is the companion matrix of $c(X)$ has a cycle with length greater than n in which each state has odd size. Furthermore, $[q_0], [q_1], ..., [q_{n-1}]$ are all states in Q_D, so $q_0, q_1, ..., q_{n-1}$ must all be in either F^a or F^r.

Therefore, any choice of F^a and F^r so that $F^a \cup F^r = Q$ and $F^a \cap F^r = \emptyset$ with F^a and F^r non-empty will guarantee that each state in the XDFA contains an odd number of states from either F^a or F^r and zero or an even number of states from the other, and hence will be an interesting SV-assignment. □

Lemma 3. *Let* $c(X) = (X + 1)\phi(X)$ *be a polynomial of degree n with non-singular companion matrix M and where $\phi(X)$ is a primitive polynomial. Let N be an XNFA with transition matrix M and $Q_0 = \{q_0\}$, then N_D forms a cycle of length $2^{n-1} - 1$.*

Proof. We calculate the number and lengths of all cycles for $c(X)$. By Theorem 1, factors $X + 1$ and $\phi(X)$ each induce a single cycle of length $2^m - 1$ with $m = 1$ and $m = n - 1$ respectively, as well as a single cycle each of length 1, which is the so-called empty cycle ε. Therefore $c(X)$ has the following cycles:

- ε_{X+1} and $X + 1$: $gcd(1, 1)$ cycle(s) of length $lcm(1, 1)$

- $\varepsilon_{\phi(X)}$ and $X + 1$: $gcd(1, 1)$ cycle(s) of length $lcm(1, 1)$

- ε_{X+1} and $\phi(X)$: $gcd(1, 2^{n-1} - 1)$ cycle(s) of length $lcm(1, 2^{n-1} - 1)$

- $\varepsilon_{\phi(X)}$ and $\phi(X)$: $gcd(1, 2^{n-1} - 1)$ cycle(s) of length $lcm(1, 2^{n-1} - 1)$

Therefore, $c(X)$ has two cycles of length 1, one of which is $\varepsilon_{c(X)}$, and two cycles of length $2^{n-1} - 1$. By Lemma 2, N_D must be a cycle with length greater than n, so it must have length $2^{n-1} - 1$. $\qquad \square$

Theorem 6. *For any $n \geq 2$, there is an interesting SV-XNFA N whose equivalent N_D has $2^{n-1} - 1$ states.*

Proof. Let $c(X) = (X + 1)\phi(X)$ be a polynomial of degree n, where $\phi(X)$ is a primitive polynomial, and let M be its non-singular companion matrix. Let N be an XNFA with transition matrix M and let $Q_0 = \{q_0\}$. By Theorem 5, N has an interesting SV-assignment, and by Lemma 3, the equivalent N_D has $2^{n-1} - 1$ states. $\qquad \square$

The following theorem shows that, for any $n \geq 2$, there exists an n-state SV-XNFA that accepts a language requiring at least $2^{n-1} - 1$ states in an equivalent SV-XDFA.

Theorem 7. *For any $n \geq 2$, there is a language $\mathcal{L}_n$ so that some n-state SV-XNFA accepts $\mathcal{L}_n$ and the minimal SV-XDFA that accepts $\mathcal{L}_n$ has $2^{n-1} - 1$ states.*

Proof. Let $c(X) = (X + 1)\phi(X)$ where $\phi(X)$ is a primitive polynomial and let $c(X)$ have degree n. We construct an SV-XNFA N with n states whose equivalent N_D has $2^{n-1} - 1$ states as in Theorem 6, and let $F^a = \{q_0\}$ and $F^r = Q \setminus F^a$. Then $\mathcal{L} = a^{(2^{n-1}-1)i+j}$ for $i \geq 0$ and $j \in J$, where J is some set of integers. Now, from the transition matrix of N it follows that $0, n \in J$, while $1, 2, ..., n - 1 \notin J$, since $q_0 \in \delta(q_0, a^n)$ and $q_0 \notin \delta(q_0, a^m)$ for $m < n$.

If there is an N_D' with fewer than $2^{n-1} - 1$ states that accepts $\mathcal{L}$, then there must be some $d_j \neq \{q_0\} \in Q_D$ such that $q_0 \in d_j$, $q_0 \in \delta(d_j, a^n)$ and there is no $m < n$ so that $q_0 \in \delta(d_j, a^m)$.

Let d_k be any state in N_D such that $d_k \neq \{q_0\}$. Let $max(d_k)$ be the largest subscript of any SV-XNFA state in d_k. Then $max(d_k) > 0$. Let $m = n - max(d_k)$, so $m < n$, then from the transition matrix of N it follows that $q_0 \in \delta(d_k, a^m)$. That is, for any d_k there is an $m < n$ so that $q_0 \in \delta(d_k, a^m)$.

Therefore, there is no N_D' with fewer than $2^{n-1} - 1$ states that accepts $\mathcal{L}$. $\square$

This gives a lower bound of $2^{n-1} - 1$ for the state complexity of unary SV-XNFA.

4   Conclusion

We introduced the notion of unary self-verifying symmetric difference automata, and showed that for certain polynomials, interesting SV-XNFA exist. We also showed that for primitive polynomials, no SV-assignments for unary XNFA are possible. This provides an upper bound of $2^n - 1$ on the state complexity of unary SV-XNFA that is known not to be tight. Furthermore, we demonstrated that $2^{n-1} - 1$ is a lower bound for unary SV-XNFA.

Directions for future work include determining a tight bound, as well as providing a more detailed exposition of the properties of polynomials over GF(2) that lead to SV-assignments, and especially interesting SV-assignments. Also, further consideration may be given to the question of which languages can be represented succinctly by SV-XNFA.

References

1. Hopcroft, J.E., Ullman, J.D.: Introduction to Automata Theory, Languages, and Computation, 1st edn. Addison-Wesley Longman Publishing Co., Inc., Boston (1990)
2. Van Zijl, L.: Generalized nondeterminism and the succinct representation of regular languages. Ph.D. thesis, University of Stellenbosch (1997). http://www.cs.sun.ac.za/~lvzijl/publications/boek.ps.gz
3. Van der Merwe, B., Tamm, H., Van Zijl, L.: Minimal DFA for symmetric difference NFA. In: Kutrib, M., Moreira, N., Reis, R. (eds.) DCFS 2012. LNCS, vol. 7386, pp. 307–318. Springer, Heidelberg (2012)
4. Assent, I., Seibert, S.: An upper bound for transforming self-verifying automata into deterministic ones. RAIRO-Theoretical Informatics and Applications-Informatique Théorique et Applications 41(3), 261–265 (2007)
5. Hromkovič, J., Schnitger, G.: On the power of Las Vegas II. Two-way finite automata. In: Wiedermann, J., Van Emde Boas, P., Nielsen, M. (eds.) ICALP 1999. LNCS, vol. 1644, pp. 433–442. Springer, Heidelberg (1999)
6. Jiráskova, G., Pighizzini, G.: Optimal simulation of self-verifying automata by deterministic automata. Inf. Comput. 209(3), 528–535 (2011). Special Issue: 3rd International Conference on Language and Automata Theory and Applications (LATA 2009)
7. Vuillemin, J., Gama, N.: Compact normal form for regular languages as Xor automata. In: Maneth, S. (ed.) CIAA 2009. LNCS, vol. 5642, pp. 24–33. Springer, Heidelberg (2009)
8. Stone, H.S.: Discrete Mathematical Structures and their Applications. Science Research Associates, Chicago (1973)
9. Dornhoff, L.L., Hohn, F.E.: Applied Modern Algebra. Macmillan Publishing Co., Inc., Collier Macmillan Publishers, New York, London (1978)
10. Geffert, V., Pighizzini, G.: Pairs of complementary unary languages with "balanced" nondeterministic automata. Algorithmica 63(3), 571–587 (2010)

State Complexity of Prefix Distance
of Subregular Languages

Timothy Ng, David Rappaport, and Kai Salomaa[⊠]

School of Computing, Queen's University, Kingston, ON K7L 3N6, Canada
{ng,daver,ksalomaa}@cs.queensu.ca

Abstract. The neighbourhood of a regular language of constant radius with respect to the prefix distance is always regular. We give upper bounds and matching lower bounds for the size of the minimal deterministic finite automaton (DFA) needed for the radius k prefix distance neighbourhood of an n state DFA that recognizes, respectively, a finite, a prefix-closed and a prefix-free language. For prefix-closed languages the lower bound automata are defined over a binary alphabet. For finite and prefix-free regular languages the lower bound constructions use an alphabet that depends on the size of the DFA and it is shown that the size of the alphabet is optimal.

1 Introduction

The neighbourhood of radius r of a language L consists of all strings that are within distance at most r from some string of L. A distance measure d is said to be regularity preserving if the neighbourhood of any regular language with respect to d is regular. Calude et al. [2] have shown that *additive distances* are regularity preserving. Additivity requires, roughly speaking, that the distance is compatible with concatenation of words in a certain sense and best known examples of additive distances include the Levenshtein distance and the Hamming distance [2,5].

The prefix distance of two words u and v is the sum of the lengths of the suffixes of u and v that begin after the longest common prefix of u and v. The suffix distance and the factor distance are defined analogously in terms of the longest common suffix (respectively, factor) of two words. It is known that the prefix, suffix and factor distance preserve regularity [4].

By the state complexity of a regularity preserving distance we mean the worst-case size of the minimal deterministic finite automaton (DFA) needed to recognize radius r neighbourhood of an n state DFA language (as a function of n and r). Tight bounds for the state complexity of prefix distance were recently obtained by the authors [14].

Worst-case state complexity bounds for general regular languages typically cannot be matched by finite languages, as first observed by Câmpeanu et al. [3], and the same holds for other proper sub-families of the regular languages.

© IFIP International Federation for Information Processing 2016
Published by Springer International Publishing Switzerland 2016. All Rights Reserved
C. Câmpeanu et al. (Eds.): DCFS 2016, LNCS 9777, pp. 192–204, 2016.
DOI: 10.1007/978-3-319-41114-9_15

Relations between different sub-regular language families have been investigated recently by Holzer and Truthe [11]. Bordihn et al. [1] have studied the state complexity of determinization of automata for the different sub-regular language families and further recent work on the state complexity of sub-regular language families has been done by Holzer et al. [8, 10].

Here we study the state complexity of prefix distance for finite languages. Additionally, we concentrate on the classes of prefix-closed and prefix-free regular languages because their corresponding restricting properties can be viewed to be related to the definition of the prefix distance measure. We give tight state complexity bounds for the prefix distance of finite, prefix-closed and prefix-free regular languages. In the case of finite languages and prefix-free languages the lower bound construction uses an alphabet that depends linearly on the size of the DFA. We establish that the general upper bound cannot be matched by languages defined over an alphabet of smaller size.

2   Preliminaries

We briefly recall some definitions and notation used in the paper. For all unexplained notions on finite automata and regular languages the reader may consult the textbook by Shallit [15] or the survey by Yu [16]. A survey of distances is given by Deza and Deza [5]. Recent surveys on descriptional complexity of regular languages include [6, 9, 13].

In the following Σ is always a finite alphabet, the set of strings over Σ is Σ^* and ε is the empty string. The reversal of a string $x \in \Sigma^*$ is x^R. The set of nonnegative integers is $\mathbb{N}_0$. The cardinality of a finite set S is denoted $|S|$ and the powerset of S is 2^S. A string $w \in \Sigma^*$ is a *substring* or *factor* of x if there exist strings $u, v \in \Sigma^*$ such that $x = uwv$. If $u = \varepsilon$, then w is a *prefix* of x. If $v = \varepsilon$, then w is a *suffix* of x.

A *nondeterministic finite automaton* (NFA) is a 5-tuple $A = (Q, \Sigma, \delta, Q_0, F)$ where Q is a finite set of states, Σ is an alphabet, δ is a multi-valued transition function $\delta : Q \times \Sigma \to 2^Q$, $Q_0 \subseteq Q$ is a set of initial states, and $F \subseteq Q$ is a set of final states. We extend the transition function δ to a function $Q \times \Sigma^* \to 2^Q$ in the usual way. A string $w \in \Sigma^*$ is *accepted* by A if, for some $q_0 \in Q_0$, $\delta(q_0, w) \cap F \neq \emptyset$ and the language recognized by A consists of all strings accepted by A. An ε-NFA is an extension of an NFA where transitions can be labeled by the empty string ε [15, 16], i.e., δ is a function $Q \times (\Sigma \cup \{\varepsilon\}) \to 2^Q$. It is known that every ε-NFA A has an equivalent NFA without ε-transitions and with the same number of states as A. An NFA $A = (Q, \Sigma, \delta, Q_0, F)$ is a *deterministic finite automaton* (DFA) if $|Q_0| = 1$ and, for all $q \in Q$ and $a \in \Sigma$, $\delta(q, a)$ either consists of one state or is undefined. Two states p and q of a DFA A are equivalent if $\delta(p, w) \in F$ if and only if $\delta(q, w) \in F$ for every string $w \in \Sigma^*$. A DFA A is *minimal* if each state $q \in Q$ is reachable from the initial state, a final state is reachable from each state q, and no two states are equivalent.

Note that our definition of a DFA allows some transitions to be undefined, that is, by a DFA we mean an incomplete DFA. It is well known that, for a regular language L, the sizes of the minimal incomplete and complete DFAs differ by at most one. The constructions used in this paper are more convenient to formulate using incomplete DFAs but our results would not change in any significant way if we were to require that all DFAs are complete. The (incomplete deterministic) *state complexity* of a regular language L, sc(L), is the size of the minimal DFA recognizing L.

We define pref(L) to be the language of all prefixes of words belonging to L,

$$\text{pref}(L) = \{u \in \Sigma^* \mid (\exists v \in \Sigma^*)\, uv \in L\}.$$

A language L is *prefix-closed* if $L = \text{pref}(L)$. A language L is *prefix-free* if no word $u \in L$ is a proper prefix of any other word in L. A DFA A is *non-exiting* if a final state of A has no outgoing transitions. The minimal DFAs recognizing a prefix-free language have always the following property.

Lemma 1 ([7]). *If A is minimal and $L(A)$ is prefix-free, then A is non-exiting.*

To conclude this section, we recall definitions of the distance measures used in the following. Generally, a function $d : \Sigma^* \times \Sigma^* \to [0, \infty)$ is a *distance* if it satisfies for all $x, y, z \in \Sigma^*$, the conditions $d(x,y) = 0$ if and only if $x = y$, $d(x,y) = d(y,x)$, and $d(x,z) \leq d(x,y) + d(y,z)$. The *neighbourhood* of a language L of radius k with respect to a distance d is the set

$$E(L, d, k) = \{w \in \Sigma^* \mid (\exists x \in L)\, d(w,x) \leq k\}.$$

Let $x, y \in \Sigma^*$. The *prefix distance* of x and y counts the number of symbols which do not belong to the longest common prefix of x and y [4]. Formally, it is defined by

$$d_p(x,y) = |x| + |y| - 2 \cdot \max_{z \in \Sigma^*}\{|z| \mid x, y \in z\Sigma^*\}.$$

The state complexity of prefix distance was established in [14].

Theorem 1 ([14]). *For $n > k \geq 0$, if sc$(L) = n$ then*

$$\text{sc}(E(L, d_p, k)) \leq n \cdot (k+1) - \frac{k(k+1)}{2}$$

and this bound can be reached in the worst case.

To conclude this section we recall from [14] the construction of a DFA that recognizes the prefix-distance neighbourhood of a regular language.

Let $A = (Q, \Sigma, \delta, q_0, F)$ be a DFA and $\varphi_A : Q \to \mathbb{N}_0$ be a function defined by

$$\varphi_A(q) = \min_{w \in \Sigma^*}\{|w| \mid \delta(q, w) \in F\}$$

The function $\varphi_A(q)$ gives the length of the shortest path from a state q to the closest reachable final state. Note that if $q \in F$, then $\varphi_A(q) = 0$.

We construct a DFA $A' = (Q', \Sigma, \delta', q_0', F')$ for the neighbourhood $E(L(A), d_p, k)$, $k \in \mathbb{N}$, as follows. We define the state set

$$Q' = ((Q - F) \times \{1, \ldots, k+1\}) \cup F \cup \{p_1, \ldots, p_k\}. \tag{1}$$

The initial state q_0' is defined by

$$q_0' = \begin{cases} q_0, & \text{if } q_0 \in F; \\ (q_0, \varphi_A(q_0)) & \text{if } q_0 \notin F \text{ and } \varphi_A(q_0) \le k; \\ (q_0, k+1) & \text{if } q_0 \notin F \text{ and } \varphi_A(q_0) > k. \end{cases}$$

The set of final states is given by

$$F' = ((Q - F) \times \{1, \ldots, k\}) \cup F \cup \{p_1, \ldots, p_k\}.$$

Let $q_{i,a} = \delta(i, a)$ for $i \in Q$ and $a \in \Sigma$, if $\delta(i, a)$ is defined. Then for all $a \in \Sigma$, the transition function δ' is defined for states $i \in F$ by

$$\delta'(i, a) = \begin{cases} (q_{i,a}, 1), & \text{if } q_{i,a} \in Q - F; \\ q_{i,a}, & \text{if } q_{i,a} \in F; \\ p_1, & \text{if } \delta(i, a) \text{ is undefined.} \end{cases}$$

For states $(i, j) \in Q - F \times \{1, \ldots, k+1\}$, δ' is defined

$$\delta'((i, j), a) = \begin{cases} q_{i,a}, & \text{if } q_{i,a} \in F; \\ (q_{i,a}, \min\{j+1, \varphi_A(q_{i,a})\}), & \text{if } \varphi_A(q_{i,a}) \text{ or } j+1 \le k; \\ (q_{i,a}, k+1), & \text{if } \varphi_A(q_{i,a}) \text{ and } j+1 > k; \\ p_{j+1}, & \text{if } \delta(i, a) \text{ is undefined.} \end{cases}$$

Finally, we define δ' for states p_ℓ for $\ell = 1, \ldots, k-1$ by $\delta'(p_\ell, a) = p_{\ell+1}$.

The following Proposition 1 follows from the proof of Proposition 2 of [14]. Note that Proposition 2 of [14] establishes a stronger claim and the statement of the below proposition includes only the parts that we need in the later sections.

Proposition 1 ([14]).   (a) The DFA A' recognizes the neighbourhood $E(L(A), d_p, k)$.

(b) The elements of the set $S_{ur} = \{(q, j) \mid q \in Q - F, 1 \le j \le k+1, j > \varphi_A(q)\}$ are all unreachable as states of the DFA A'.

3   Neighbourhoods of Finite Languages

We first consider the state complexity of neighbourhoods of finite languages with respect to the prefix distance.

Proposition 2. Let L be a finite language recognized by a minimal DFA $A = (Q, \Sigma, \delta, q_0, F)$ with n states. Then

$$\mathrm{sc}(E(L, d_p, k)) \le (n-2) \cdot (k+1) - k^2 + 2.$$

Proof. We know that the neighbourhood of L of radius k with respect to the prefix distance is recognized by a DFA $A' = (Q', \Sigma, \delta', q_0'.F')$ obtained from A as in Proposition 1 where, furthermore, all elements of the set S_{ur} are unreachable. We show that there are more unreachable states in the case of finite languages.

Since A is acyclic, the number and length of words that reach each state $q \in Q$ is bounded. For $q \in Q$, let w_q denote the longest word that reaches q from the initial state q_0 without passing through a final state. Then for all states q with $|w_q| \leq k$, the states $(q, j) \in Q'$ with $j > |w_q|$ are unreachable as states of A' (where the set of states of A' is as in (1). That is, all states in the set

$$R_{ur} = \{(q, j) \mid q \in Q - F, 1 \leq j \leq k + 1, j > |w_q|\}$$

are unreachable in A'. By Proposition 1 (b) all elements of the set $S_{ur} = \{(q, j) \mid q \in Q - F, 1 \leq j \leq k + 1, j > \varphi_A(q)\}$ are also unreachable in A'. We note that increasing the number of final states of A by one decreases the cardinality of Q' by k and decreases the cardinality of S_{ur} and R_{ur} by at most k. However, we observe that A must have at least two final states to reach the bound. The last state of A, with no outgoing transitions, must be a final state since, otherwise, there are useless states. But this cannot be the only final state, since otherwise, for every state $q \in Q$ with $\varphi_A(q) > k$, only $(q, k + 1)$ is reachable. Thus, the initial state q_0 must also be a final state.

As in [14], we note that the cardinality of S_{ur} is minimized when exactly one non-final state has a shortest path of length i that reaches q_f. From the above it then follows that reaching the upper bound requires exactly two final states, one of which must be the initial state and the other which must have no outgoing transitions. Since A is acyclic, the initial state cannot have any incoming transitions, so the states in S_{ur} consist of those that can reach the non-initial final state, giving $\frac{k(k+1)}{2}$ unreachable states. Similarly, the cardinality of R_{ur} is minimized when exactly one non-final state has a longest word of length i which reaches it from q_0, giving $\frac{k(k+1)}{2}$ unreachable states.

Thus, the number of states of the minimal DFA for $E(L, d_p, k)$ is upper bounded by

$$(n - 2)(k + 1) + 2 + k - 2 \cdot \frac{k(k + 1)}{2} = (n - 2)(k + 1) - k^2 + 2.$$

$\square$

Next we give a lower bound construction that matches the upper bound of Proposition 2.

Lemma 2. *There exists a finite language recognized by a DFA with n states such that $E(L(A), d_p, k)$ requires at least $(n - 2)(k + 1) - k^2 + 2$ states.*

Proof. Let $A_n = (Q_n, \Sigma_n, \delta_n, q_0, F_n)$ where $Q_n = \{0, \ldots, n - 1\}$, $\Sigma_n = \{a_1, \ldots, a_{n-3}\}$, $q_0 = 0$, $F_n = \{0, n-1\}$, and the transition function is defined by

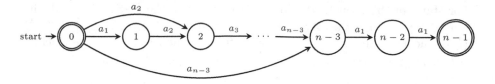

Fig. 1. The DFA A_n.

- $\delta_n(0, a_i) = i$ for $1 < j \leq n - 3$,
- $\delta_n(i, a_{i+1}) = i + 1$ for $0 \leq i < n - 3$,
- $\delta_n(i, a_1) = i + 1$ for $i = n - 3, n - 2$.

The DFA A_n is depicted in Fig. 1.

Let $A'_n = (Q'_n, \Sigma_n, \delta'_n, q'_0, F'_n)$ be the DFA constructed from A_n as in Proposition 1. First, we show that $(n - 2)(k + 1) - k^2 + 2$ states are reachable. States of the form p_i with $1 \leq i \leq k$ are reachable from states $0 \leq i \leq k$ on symbols a_j with $j \neq i + 1$. For states of the form $(i, j) \in (Q_n - F_n) \times \{1, \ldots, k + 1\}$, with $\varphi_{A_n}(i) > k$ and $j \leq i$, each (i, j) is reachable on the word $a_{i-j} a_{i-j+1} \cdots a_i$. However, states (i, j) with $j > \varphi_{A_n}(i)$ are unreachable by definition of A'_n and states (i, j) with $i < j \leq k$ are unreachable. Thus the number of unreachable states in $(Q_n - F_n) \times \{1, \ldots, k + 1\}$ is

$$\sum_{i=n-k}^{n-1} |\{i\} \times \{\varphi_{A_n}(i) + 1, \ldots, k + 1\}| + \sum_{i=1}^{k} |\{i + 1, \ldots, k + 1\}|$$

$$= 2 \cdot \sum_{i=1}^{k} |\{i = 1, \ldots, k + 1\}| = 2 \cdot \sum_{i=1}^{k} i = 2 \cdot \frac{k(k+1)}{2}.$$

Thus the number of reachable states is

$$(n - 2)(k + 1) - 2 + k - 2 \cdot \frac{k(k + 1)}{2} = (n - 2)(k + 1) - k^2 + 2.$$

Now, we show that all reachable states are pairwise inequivalent.

- For states of the form p_i and p_j, $i < j$, the word a_1^{k-i} takes the machine from state p_i to p_k and is accepted. However, from state p_j, the word a_1^{k-i} reaches state p_k on the prefix a_1^{k-j} with no further transitions to read a_1^{j-i} and thus, the word is not accepted.
- For states of the form (i, j) and p_ℓ with $\ell < k$, we consider the word $z = w_i a_2^k$ with

$$w_i = a_{n-i+1} a_{n-i+2} \cdots a_{n-3} a_1 a_1.$$

The prefix w_i takes the machine from state (i, j) to state $n - 1$ and on the rest of the word a_2^k, the machine moves from $n - 1$ to p_k and is accepted. However, from state p_ℓ, the computation on z reaches p_k before all of z is read, since $|z| = n - i + k > k - \ell$ and it is rejected.

- For states of the form (i, j) and (i', j') with $i < i'$ the states can be distinguished by $z = w_i a_2^k$ as above. For $i = i'$ and $j < j'$, let $z = a_i a_1^{k-j}$. From (i, j), the machine reads a_i and is taken to p_j, while from (i, j'), the machine is taken to $p_{j'}$. From above, p_j and $p_{j'}$ are distinguishable by a_1^{k-j}.

Thus, we have shown that there are $(n - 2)(k + 1) - k^2 + 2$ reachable states and that all reachable states are pairwise inequivalent.

$\square$

Proposition 2 and Lemma 2 now yield a tight state complexity bound for the prefix distance neighbourhoods of regular languages.

Theorem 2. *Let L be a finite language. For $n > 2k \geq 0$, if $\mathrm{sc}(L) = n$, then*

$$\mathrm{sc}(E(L, d_p, k)) \leq (n - 2) \cdot (k + 1) - k^2 + 2,$$

and this bound can be reached in the worst case.

The lower bound construction of Lemma 2 uses, for a DFA with n states, an alphabet of cardinality $n - 3$. To conclude this section we show that the construction is optimal in the sense that the upper bound of Theorem 2 cannot be reached with an alphabet of cardinality less than $n - 3$.

Proposition 3. *Let A be a DFA recognizing a finite language with n states. If the state complexity of $E(L(A), d_p, k)$ equals $(n - 2)(k + 1) - k^2 + 2$, then the alphabet of A needs at least $n - 3$ letters.*

Proof. Let $A = (Q, \Sigma, \delta, q_0, F)$ with $|Q| = n$. Let $A' = (Q', \Sigma, \delta', q_0' F')$ be the DFA recognizing $E(L(A), d_p, k)$ constructed in Proposition 1. Recall from the proof of Proposition 2 that in order for A' to have the maximal number of states $(n-2)(k+1) - k^2 + 2$, a necessary condition is that $F = \{q_0, q_f\}$ and that there can be only one state q_1 with $\varphi_A(q_1) = 1$.

Now for all $q \in Q - \{q_0, q_f, q_1\}$, $\varphi_A(q) \geq 2$. By definition of the transition function δ', if $\varphi_A(q) \geq 2$, the state $(q, 1)$ can only be reached by a direct transition from a final state. Since q_f does not have any outgoing transitions, q_0 must have $n - 3$ outgoing transitions—one for each state q.

Furthermore, since A contains a final state q_f with no outgoing transitions, no additional symbols are required to reach p_1, as it can be reached from q_f via a direct transition on any symbol.

Since A is a DFA and q_0 has at least $n-3$ outgoing transitions, the cardinality of the alphabet must be at least $n - 3$. $\square$

4 Neighbourhoods of Prefix-Closed and Prefix-Free Languages

Next, we consider the state complexity of neighbourhoods of prefix-closed and prefix-free regular languages with respect to the prefix distance.

Theorem 3. *Let L be a prefix-closed regular language recognized by an n-state DFA A. Then there is a DFA A' that recognizes the neighbourhood $E(L, d_p, k)$ with at most $n + k$ states and this bound is reachable.*

Proof. Since L is prefix-closed, every state of A must be an accepting state [12]. If A has n states, this means that the DFA A' constructed in Proposition 1 for the radius k neighbourhood has $n + k$ states.

We now define a prefix-closed regular language L_n such that a DFA recognizing $E(L_n, d_p, k)$ requires at least $n + k$ states. Let $L_n = \{a^i \mid 0 \le i \le n\}$. Then we define $A_n = (Q_n, \{a, b\}, \delta_n, q_0, F_n)$ where $Q_n = F_n = \{0, \dots, n - 1\}$, $q_0 = 0$, and the transition function δ_n is defined by $\delta_n(i, a) = i + 1$ for $0 \le i \le n - 1$.

Then we define the DFA recognizing $E(L_n, d_p, k)$ by $A' = (Q'_n, \{a, b\}, \delta'_n, q_0, F'_n)$ where $Q'_n = F'_n = Q_n \cup \{p_1, \dots, p_k\}$ and the transition function defined by

- $\delta'_n(i, a) = i + 1$ for $0 \le i < n - 1$,
- $\delta'_n(n - 1, a) = p_1$,
- $\delta'_n(i, b) = p_1$ for $0 \le i < n - 1$,
- $\delta'_n(p_i, a) = \delta'_n(p_i, b) = p_{i+1}$ for $1 \le i < k$.

Every state i, $0 \le i \le n - 1$, is reachable on the word a^i and every state p_i, $1 \le i \le k$ is reachable on the word b^i. The states $0 \le i, i' \le n-1$ are distinguished by the word b^{k-i} and the states p_i, p'_i, $1 \le i, i' \le k$ are also distinguished by the word b^{k-i}. The states i, $0 \le i \le n-1$ and p_j, $1 \le j \le k$ are distinguished by the word $a^{n-j}b^k$. Thus, there are $n + k$ reachable states and they are all pairwise distinguishable. $\square$

Proposition 4. *Let L be a prefix-free regular language recognized by a minimal n-state DFA $A = (Q, \Sigma, \delta, q_0, F)$. Then there is a DFA B with at most $(n - 1)k + 2 - \frac{k(k-1)}{2}$ states that recognizes the neighbourhood $E(L, d_p, k)$.*

Proof. Let $A' = (Q', \Sigma, \delta, q'_0, F')$ be the DFA constructed for the neighbourhood $E(L, d_p, k)$ as in Proposition 1. Since L is prefix-free, A must be non-exiting. That is, A has a single final state with no outgoing transitions. This property creates additional unreachable states in the DFA A' for $E(L, d_p, k)$.

For all non-final states $q \in Q - F$, the state $(q, 1)$ is reachable only if either $\varphi_A(q) = 1$ or there is a transition from a final state to q. However, since A is non-exiting, no final states may have any outgoing transitions, so the only states q where $(q, 1)$ is reachable are those with $\varphi_A(q) = 1$. However, for all such states q, the states (q, i) with $2 \le i \le k + 1$ are unreachable. Thus, to reach the upper bound on the number of states, the number of states q with $\varphi_A(q) = 1$ must be minimized if $k \ge 2$. If $k = 1$, then for each state $q \in Q - F$, either $(q, 1)$ is reachable or $(q, k + 1)$ is reachable, so the number of states with $\varphi_A(q) = 1$ need not be minimized.

By Proposition 1 (b) elements of the set $S_{ur} = \{(q, j) \mid q \in Q - F, 2 \le j \le k + 1, j > \varphi_A(q)\}$ are unreachable as states of A' (even without assuming that $L(A)$ is prefix-free. Let q_f be the sole final state of A. The set S_{ur} is minimized when exactly one non-final state q_i in the DFA A for each $1 \le i \le k$ has a shortest path of length i that reaches q_f. In this case, we have $|S_{ur}| = \frac{k(k-1)}{2}$.

Thus, in order to maximize the number of reachable states of A', the DFA A has a single final state and a single state q_1 with $\varphi_A(q_1) = 1$ if $k \geq 2$, giving us at most $(n-2)k + k + 2 - \frac{k(k-1)}{2} = (n-1)k + 2 - \frac{k(k-1)}{2}$ states of A' which are reachable.   □

Next we present a lower bound construction that matches the bound of Proposition 4.

Lemma 3. *There exists a DFA A with n states recognizing a prefix-free regular language such that a DFA recognizing the neighbourhood $E(L(A), d_p, k)$ requires at least $(n-1)k + 2 - \frac{k(k-1)}{2}$ states.*

Proof. We define a DFA $A_n = (Q_n, \Sigma_n, \delta_n, q_0, F)$, shown in Fig. 2, by choosing

$$Q_n = \{0, \dots, n-1\}, \Sigma_n = \{a_1, \dots, a_{n-3}, b\},$$

$q_0 = 0$, $F = \{n-1\}$, and the transition function δ_n is given by

– $\delta_n(0, a_i) = i$ for $i = 1, \dots, n-3$,
– $\delta_n(i, a_i) = i$ for $i = 1, \dots, n-3$,
– $\delta_n(i, a_{i+1}) = i+1$ for $i = 1, \dots, n-4$,
– $\delta_n(n-3, b) = n-2$, $\delta_n(n-2, b) = 0$, $\delta_n(0, b) = n-1$.

We transform A_n into the DFA $A'_n = (Q'_n, \Sigma_n, \delta'_n, q'_0, F')$ via the construction from Proposition 1. To determine the reachable states of Q'_n, we first note that the state $(0, 1)$ is reachable as it is the initial state. Note that the initial state

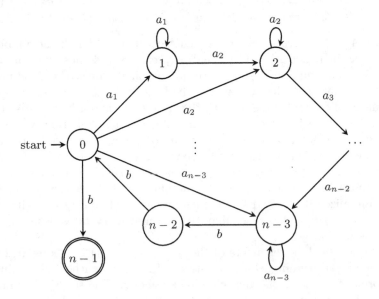

Fig. 2. The DFA A_n.

is $(0, 1)$ since $\varphi_{A_n}(0) = 1$. The final state $n - 1$ is reachable on the word b. Now consider states $p_1, \ldots, p_k$. The state p_ℓ is reachable on the word $b^{\ell+1}$ by first reading b to reach the final state and b^ℓ to reach the state p_ℓ.

Now consider states of the form $(i, j) \in (Q_n - \{0, n-1\}) \times \{2, \ldots, k+1\}$. Recall that states $(i, 1)$ are unreachable for any state $i \in Q_n$ with $\varphi_{A_n} > 1$. Then for states $i \in Q_n$ with $\varphi_{A_n} > k$ and each $2 \le j \le k+1$, we can reach state (i, j) from $(0, 1)$ via the word a_i^{j-1}. For states $i \in Q_n$ with $\varphi_{A_n} \le k$, we can reach state (i, j) via the word a_i^{j-1} for $j = 2, \ldots, \varphi_{A_n}(i)$ and states (i, j) with $j > \varphi_{A_n}(i)$ are unreachable by definition of A'_n.

Finally, we can reach state $(n-2, 2)$ via the word $a_{n-3}b$ and states $(n-2, j)$ are unreachable for $j > 2$ since $\varphi_{A_n}(n-2) = 2$. Thus the number of unreachable states in $(Q_n - \{0, n-1\}) \times \{2, \ldots, k+1\}$ is

$$\sum_{i=n-k}^{n-2} |\{i\} \times \{\varphi_{A_n}(i)+1, \ldots, k+1\}| = \sum_{i=1}^{k} |\{i+1, \ldots, k+1\}| = \sum_{i=1}^{k} i = \frac{k(k-1)}{2}.$$

Thus, the number of reachable states is

$$(n-2) \cdot k + 2 - \frac{k(k-1)}{2} + k = (n-1) \cdot k + 2 - \frac{k(k-1)}{2}.$$

Now, we show that all reachable states are pairwise inequivalent. First, note that as a final state of A, $n-1$ is not equivalent to a state of the form (i, j) in A'. Next, we distinguish states of the form (i, j) from states of the form p_ℓ. For each $1 \le i \le n-3$, reading the word a_i^k from state (i, j) takes the machine to state $(i, \min\{\varphi_A(i), k+1\})$. Then subsequently reading $a_{i+1}a_{i+2}\cdots a_{n-3}bbb$ takes the machine to the final state $n-1$. However, for every state p_ℓ, reading a_i^k forces the machine beyond state p_k, after which there are no transitions defined. The state $(n-2, 2)$ is distinguished from all p_ℓ by the word b^{2+k}, $(0, 1)$ by b^{1+k}, and $n-1$ by b^k.

Next, without loss of generality, let $\ell < \ell'$ and consider states p_ℓ and $p_{\ell'}$. Choose $z = b^{k-\ell}$. The string z takes state p_ℓ to the state p_k, where it is accepted. However, the computation on string z from state $p_{\ell'}$ is undefined since $\ell' + k - \ell > k$.

Finally, we consider states of the form (i, j). Let $i < i'$ and consider states (i, j) and (i', j'). Let $z = a_{i+1}a_{i+2}\cdots a_{n-3}bbbb^k$. From state (i, j), the word z goes to state $n-1$ on $a_{i+1}\cdots a_{n-3}bbb$. Then by reading b^k from state $n-1$, we reach state p_k, an accepting state. However, when reading z from state (i', j'), we immediately reach state $p_{j'+1}$ on a_{i+1}, since the transition on a_{i+1} is defined only for states $(0, 1)$ and (i, j). Since the rest of the word z is of length greater than k, reading it takes us to state p_k with no further defined transitions for the rest of the word.

Next, consider the state (i, j) and (i, j'), where $j < j'$. First, consider the case when $\varphi_{A_n}(i) > k$. Then let $z = a_i^{k-j}$. Reading z from (i, j) takes us to state (i, k), which is a final state. However, from (i, j'), reading z brings us to state $(i, k+1)$ and so the computation is rejected.

Now, consider the case when $\varphi_{A_n}(i) \leq k$. Let $z = bb^{k-j-1}$. From state (i,j), reading b takes the machine to state p_{j+1} and reading b^{k-j-1} puts the machine in the accepting state p_k. However, reading z from (i, j') takes us to state p_k with $b^{j'-j}$ still unread since $j' + k - j - 1 > k$ and thus, with no further transitions available, the computation is rejected.

Thus, we have shown that there are $(n-1) \cdot k + 2 - \frac{k(k-1)}{2}$ reachable states and that all reachable states are pairwise inequivalent. □

Combining Proposition 4 and Lemma 3 we have:

Theorem 4. *Let L be a prefix-free regular language. For $n > k \geq 0$, if $sc(L) = n$, then*

$$sc(E(L, d_p, k)) \leq (n-1) \cdot k + 2 - \frac{k(k-1)}{2},$$

and this bound can be reached in the worst case.

The construction of Lemma 3 that establishes the lower bound for Theorem 4 uses an alphabet of size $n - 2$, where n is the number of states of the DFA. The below result establishes that the size of the alphabet cannot be reduced.

Proposition 5. *Let A be a DFA recognizing a prefix-free regular language with n states. If the state complexity of $E(L(A), d_p, k)$ equals $(n-1)k + 2 - \frac{k(k-1)}{2}$, then the alphabet of A needs at least $n - 2$ letters.*

Proof. Let $A = (Q, \Sigma, \delta, q_0, F)$ with $|Q| = n$. Let $A' = (Q', \Sigma, \delta', q_0' F')$ be the DFA recognizing $E(L(A), d_p, k)$ constructed in Proposition 1. Recall that as an automaton recognizing a prefix-free regular language A must be non-exiting. That is, A has a single final state q_f and it cannot have any outgoing transitions. Recall also from the proof of Proposition 4 that in order for A' to have the maximal number of states $(n-1)k + 2 - \frac{k(k-1)}{2}$, a necessary condition is that there can be only one state q_1 with $\varphi_A(q_1) = 1$ and one state q_2 with $\varphi_A(q_2) = 2$.

Now for all $q \in Q - \{q_f, q_1, q_2\}$, $\varphi_A(q) \geq 3$. Recall that since the sole final state q_f has no outgoing transitions, states $(q, 1)$ are reachable only if $\varphi_A(q) = 1$. Then by definition of the transition function δ', if $\varphi_A(q) \geq 3$, the state $(q, 2)$ can only be reached by a direct transition from a state q with $\varphi_A(q) = 1$. Thus, q_1 must have $n - 2$ outgoing transitions—one for each state q with $\varphi_A(q) \geq 3$ and one additional transition to the final state q_f. Note that q_2 requires no direct transition from q_1 since $\varphi_A(q_2) = 2$ and thus $(q_2, 2)$ is the only reachable state of the form (q_2, j).

Furthermore, since A contains a final state q_f with no outgoing transitions, no additional symbols are required to reach p_1, as it can be reached from q_f via a direct transition on any symbol.

Since A is a DFA and q_1 has at least $n-2$ outgoing transitions, the cardinality of the alphabet must be at least $n - 2$. □

5   Conclusion

We have given tight state complexity bounds for the prefix-distance neighbour-hood of, respectively, finite, prefix-closed, and prefix-free languages. As can, per-haps, be expected the bound for prefix-closed languages is relatively easier to obtain and the matching lower bound construction uses a binary alphabet. The upper bound constructions for the finite and the prefix-free languages are more involved and the lower bound constructions use a variable size alphabet. Fur-thermore, we have shown that, in both cases, the alphabet size is optimal.

Since the reversal of a DFA is not, in general, deterministic, the state com-plexity bounds for suffix-distance (or factor-distance) neighbourhoods differ sig-nificantly from the corresponding bounds for prefix-distance neighbourhoods. Tight lower bounds are not known for suffix-distance neighbourhoods of gen-eral regular languages [14] or for various sub-regular language families. Such questions can be a topic for further research.

References

1. Bordihn, H., Holzer, M., Kutrib, M.: Determination of finite automata accepting subregular languages. Theor. Comput. Sci. **410**(35), 3209–3222 (2009)
2. Calude, C.S., Salomaa, K., Yu, S.: Additive distances and quasi-distances between words. J. Univ. Comput. Sci. **8**(2), 141–152 (2002)
3. Câmpeanu, C., Culik II, K., Salomaa, K., Yu, S.: State complexity of basic oper-ations on finite languages. In: Boldt, O., Jürgensen, H. (eds.) WIA 1999. LNCS, vol. 2214, p. 60. Springer, Heidelberg (2001)
4. Choffrut, C., Pighizzini, G.: Distances between languages and reflexivity of rela-tions. Theor. Comput. Sci. **286**(1), 117–138 (2002)
5. Deza, M.M., Deza, E.: Encyclopedia of Distances. Springer, Heidelberg (2009)
6. Gao, Y., Moreira, N., Reis, R., Yu, S.: A review on state complexity of individual operations. Faculdade de Ciencias, Universidade do Porto, Technical report DCC-2011-8. To appear in Computer Science Review. www.dcc.fc.up.pt/dcc/Pubs/TReports/TR11/dcc-2011-08.pdf
7. Han, Y.S., Salomaa, K., Wood, D.: State complexity of prefix-free regular lan-guages. In: Proceedings of the 8th International Workshop on Descriptive Com-plexity of Formal Systems. pp. 165–176 (2006)
8. Holzer, M., Jakobi, S., Kutrib, M.: The magic number problem for subregular language families. Int. J. Found. Comput. Sci. **23**(1), 115–131 (2012)
9. Holzer, M., Kutrib, M.: Descriptional and computational complexity of finite automata – a survey. Inform. Comput. **209**, 456–470 (2011)
10. Holzer, M., Kutrib, M., Meckel, K.: Nondeterministic state complexity of star-free languages. In: Bouchou-Markhoff, B., Caron, P., Champarnaud, J.-M., Maurel, D. (eds.) CIAA 2011. LNCS, vol. 6807, pp. 178–189. Springer, Heidelberg (2011)
11. Holzer, M., Truthe, B.: On relations between some subregular language families. Proc. NCMA **2015**, 109–124 (2015)
12. Kao, J.Y., Rampersad, N., Shallit, J.: On NFAs where all states are final, initial, or both. Theor. Comput. Sci. **410**(47–49), 5010–5021 (2009)
13. Kutrib, M., Pighizzini, G.: Recent trends in descriptional complexity of formal languages. Bull. EATCS **111**, 70–86 (2013)

14. Ng, T., Rappaport, D., Salomaa, K.: State complexity of prefix distance. In: Drewes, F. (ed.) CIAA 2015. LNCS, vol. 9223, pp. 238–249. Springer, Heidelberg (2015)
15. Shallit, J.: A Second Course in Formal Languages and Automata Theory. Cambridge University Press, Cambridge (2009)
16. Yu, S.: Regular languages. In: Rozenberg, G., Salomaa, A. (eds.) Handbook of Formal Languages, pp. 41–110. Springer, Heidelberg (1997)

Two Results on Discontinuous Input Processing

Vojtěch Vorel[✉]

Faculty of Mathematics and Physics, Charles University,
Malostranské nám. 25, Prague, Czech Republic
vorel@ktiml.mff.cuni.cz

Abstract. First, we show that universality and other properties of general jumping finite automata are undecidable, which answers questions asked by Meduna and Zemek in 2012 [12]. Second, we close a study started by Černo and Mráz in 2010 [3] by proving that a clearing restarting automaton using contexts of length two can accept a binary non-context-free language.

1 Introduction

In 2012, Meduna and Zemek [12,13] introduced *general jumping finite automata* as a model of discontinuous information processing in modern software. A general jumping finite automaton (GJFA) is described by a finite set Q of states, a finite alphabet Σ, a finite set R of *rules* from $Q \times \Sigma^* \times Q$, an initial state $q_0 \in Q$, and a set $F \subseteq Q$ of final states. In a step of computation, the automaton switches from a state r to a state s using a rule $(r, v, s) \in R$ and deletes a factor equal to v from any part of the input word. A rule (r, v, s) and an occurrence of the factor v are chosen nondeterministically (in other words, the read head can *jump* to any position). A word $w \in \Sigma^*$ is accepted if the GJFA can reduce w to the empty word while passing from the initial state to an accepting state. The boldface term **GJFA** refers to the class of languages accepted by GJFA. The initial work [12,13] deals mainly with closure properties of **GJFA** and its relations to classical language classes (the publications [12,13] contain flaws, see [17]). It turns out that the class **GJFA** is not closed under operations related to continuous processing (concatenation, Kleene star, homomorphism, inverse homomorphism, shuffle) nor some Boolean closure operations (complementation, intersection). The class is incomparable with both regular and context-free languages. It is a proper subclass of both context-sensitive languages and of the class NP, while there exist NP-complete **GJFA**languages (see [5], which is an extended version of [6]).

On the other hand, the concept of *restarting automata* [10,14] is motivated by reduction analysis and grammar checking of natural language sentences. In 2010, Černo and Mráz [3] introduced a subclass named *clearing restarting automata*

Research supported by the Czech Science Foundation grant GA14-10799S and the GAUK grant No. 52215.

(cl-RA) in order to describe systems that use only very basic types of reduction rules (see also [2]). Clearing restarting automata may delete factors according to contexts and endmarks, but, unlike GJFA and classical restarting automata, they are not controlled by states and rules. A key property of a cl-RA is the maximum length k of context used. For $k \geq 0$, a k-*clearing restarting automaton* (k-cl-RA) is described by a finite alphabet Σ and a finite set I of instructions of the form (u_L, v, u_R), where $v \in \Sigma^*$, $u_L \in \Sigma^k \cup \mathbb{c}\Sigma^{k-1}$, and $u_R \in \Sigma^k \cup \Sigma^{k-1}\$$. The words u_L, u_R specify the left and right context for consuming a factor v, while $\mathbb{c}$ and $\$$ stand for the left and right end of input, respectively. A word is accepted by a cl-RA if it may be completely consumed using a series of instructions. The class of languages accepted by cl-RA is not closed under complementation, intersection, or union [3]. It forms a superset of regular languages, a subset of context-sensitive languages, and is incomparable with context-free languages [3].

Tough both the formalisms are defined as acceptors, they may be equivalently treated as generative systems. Moreover, they share important properties with *insertion systems* [16] (possibly *graph-controlled* [1]) and semi-contextual grammars [15] (possibly using *regular control without appearance checking* [11]), as we briefly discuss in the conclusion. The present paper consists of two main parts:

In Sect. 3 we show that, given a GJFA M with an alphabet Σ, it is undecidable whether M accepts the universal language Σ^*. In other words, *universality* of GJFA is undecidable. As a direct consequence, the more general problems of *equivalence* and *inclusion* are undecidable for GJFA as well. Decidability of these tasks was listed as an open problem in [12,13].

In Sect. 4 we deal with expressive power of cl-RA with short contexts and small alphabets, as it was addressed in [3]. The authors showed that a language accepted by a 2-cl-RA may not be context-free, but the example automata required at least six-letter alphabets, so they asked what is the least sufficient alphabet size. We provide a binary example, which forms a tight bound.

2   Preliminaries

We use the notion of *insertion* as it was defined, e.g., in [4,7,9]:

Definition 1. *Let* $K, L \subseteq \Sigma^*$ *be languages. The* insertion *of K to L is*

$$L \leftarrow K = \{u_1 v u_2 \mid u_1 u_2 \in L, v \in K\}.$$

More generally, for each $k \geq 1$ we denote

$$L \leftarrow^k K = \left(L \leftarrow^{k-1} K\right) \leftarrow K,$$
$$L \leftarrow^* K = \bigcup_{i \geq 0} L \leftarrow^i K,$$

where $L \leftarrow^0 K$ stands for L. In expressions with $\leftarrow$ and $\leftarrow^*$, a singleton set $\{w\}$ may be replaced by w.

A chain $L_1 \leftarrow L_2 \leftarrow \cdots \leftarrow L_d$ of insertions is evaluated from the left, e.g., $L_1 \leftarrow L_2 \leftarrow L_3$ means $(L_1 \leftarrow L_2) \leftarrow L_3$. The empty word is denoted by ϵ.

As described above, a GJFA is a quintuple $M = (Q, \Sigma, R, q_0, F)$. For a rule $(r, v, s) \in R$ with $r, s \in Q$, the word $v \in \Sigma^*$ is called the *label* of the rule. A sequence

$$(r_1, v_1, s_1), (r_2, v_2, s_2), \ldots, (r_k, v_k, s_k)$$

of rules from R is a *path* if $k \geq 1$ and $s_i = r_{i+1}$ for $1 \leq i \leq k - 1$. The sequence $v_1, v_2, \ldots, v_k$ is the *labeling* of the path. The path is *accepting* if $r_1 = q_0$ and $s_k \in F$. The original definition [12,13] of the language $L(M)$ accepted by M is based on *configurations* that specify positions of the read head (i.e., starting positions of the factor to be erased in the next step). For our proofs, this type of configurations is useless, whence we directly use the following generative characterization [17, Corollary 1] of $L(M)$ as a definition:

Definition 2. *Let* $M = (Q, \Sigma, R, s, F)$ *be a GJFA and* $w \in \Sigma^*$. *Then* $w \in L(M)$ *if and only if* $w = \epsilon$ *and* $s \in F$, *or*

$$w \in \epsilon \leftarrow v_d \leftarrow v_{d-1} \leftarrow \cdots \leftarrow v_2 \leftarrow v_1, \tag{1}$$

where $d \geq 1$ *and* $v_1, v_2, \ldots, v_d$ *is a labeling of an accepting path in* M.

If a GJFA $M = (Q, \Sigma, R, s, F)$ is clear, we write $(r, w) \curvearrowright (s, u)$ for $r, s \in Q$ and $u, v \in \Sigma^*$ if $w \in u \leftarrow v$ for some $(r, v, s) \in R$.

In the case of clearing restarting automata we include the original definition, which builds on *context rewriting systems* [3]:

Definition 3. *For* $k \geq 0$, *a* k-*context rewriting system is a tuple* $M = (\Sigma, \Gamma, I)$, *where* Σ *is an input alphabet,* $\Gamma \supseteq \Sigma$ *is a working alphabet not containing the special symbols* $\mathcal{c}$ *and* $\$$, *called* sentinels, *and* I *is a finite set of instructions of the form*

$$(u_L, v \rightarrow t, u_R),$$

where u_L *is a left context,* $u_L \in \Gamma^k \cup \mathcal{c}\Gamma^{k-1}$, u_R *is a right context,* $u_R \in \Gamma^k \cup \Gamma^{k-1}\$$, *and* $v \rightarrow t$ *is a rule,* $v, t \in \Gamma^*$. *A word* $w = u_1 v u_2$ *can be rewritten into* $u_1 t u_2$ *(denoted by* $u_1 v u_2 \rightarrow_M u_1 t u_2$*) if and only if there exists an instruction* $(u_L, v \rightarrow t, u_R) \in I$ *such that* u_L *is a suffix of* $\mathcal{c}u_1$ *and* u_R *is a prefix of* $u_2\$$.

We use the star in $\curvearrowright^*, \rightarrow^*, \dashv^*$ and other symbols to denote reflexive-transitive closures of binary relations.

Definition 4. *For* $k \geq 0$, *a* k-*clearing restarting automaton (*k-cl-RA*) is a system* $M = (\Sigma, I)$, *where* $M' = (\Sigma, \Sigma, I)$ *is a* k-*context rewriting system such that for each* $\mathbf{i} = (u_L, v \rightarrow t, u_R) \in I$ *it holds that* $v \in \Sigma^+$ *and* $t = \epsilon$. *Since* t *is always the empty word, the notation* $\mathbf{i} = (u_L, v, u_R)$ *is used. A* k-cl-RA M *accepts the language*

$$L(M) = \{w \in \Sigma^* \mid w \vdash_M^* \epsilon\},$$

where $\vdash_M$ *denotes the rewriting relation* $\rightarrow_{M'}$ *of* M'. *The term* $\mathcal{L}(k$-cl-RA$)$ *denotes the class of languages accepted by* k-cl-RA.

The generative approach is formalized by writing $w_2 \dashv w_1$ instead of $w_1 \vdash w_2$.

3   Undecidability in General Jumping Finite Automata

Theorem 5. *Given a GJFA $M = (Q, \Sigma, R, s, F)$, it is undecidable whether $L(M) = \Sigma^*$.*

Let us prove the theorem. Given a context-free grammar G with terminal alphabet Σ_T, it is undecidable whether $L(G) = \Sigma_T^*$ [8]. We present a reduction from this problem to the universality of GJFA. Assume that the given grammar G

- has non-terminal alphabet Σ_N and a start symbol $A_S \in \Sigma_N$,
- accepts the empty word ϵ, and
- is given in Greibach normal form [8], i.e., the rules are $A_S \to \epsilon$ and $A_i \to u_i$, where $A_i \in \Sigma_N$ and $u_i \in \Sigma_T \Sigma_N^*$ for $i \in \{1, \ldots, m\}$, $m \geq 0$.

Note that any context-free grammar that accepts ϵ can be algorithmically converted to the form above. Next, we construct a GJFA $M_G = (Q, \Gamma, R, s, F)$ as follows, denoting $\Sigma_B = \{b_1, \ldots, b_m\}$:

$$Q = \{q_0, q_1, q_2, q_3, q_4\},$$
$$\Gamma = \Sigma_T \cup \Sigma_N \cup \Sigma_B,$$

$s = q_0$, $F = \{q_2, q_4\}$. The set R of rules is defined in Fig. 1. In this figure, each arrow labeled with a finite set $S \subseteq \Gamma^*$ stands for $|S|$ rules, each labeled with a word $v \in S$. The following finite sets are used:

$$P_{BU} = \{b_i u_i \mid i = 1, \ldots, m\}, \qquad P_C = \{xA_1 \mid x \in \Sigma_T\}$$
$$P_{NB} = \{A_i b_i \mid i = 1, \ldots, m\}, \qquad \cup \{A_i b_i \mid i = 1, \ldots, m\}$$
$$\cup \{b_i A_{i+1} \mid i = 1, \ldots, m-1\}$$
$$\cup \{b_m x \mid x \in \Sigma_T\}.$$

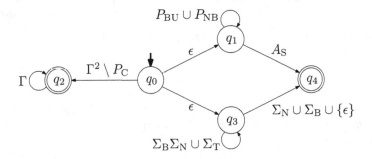

Fig. 1. The GJFA M_G corresponding to a context-free grammar G

For a word $w \in \Gamma^*$ we denote with w_T and $w_{N,B}$ the projections of w to subalphabets Σ_T and $\Sigma_N \cup \Sigma_B$ respectively[1] Let us show that $L(G) = \Sigma_T^*$ if and only if $L(M_G) = \Gamma^*$.

First, suppose that $L(G) = \Sigma_T^*$ and take an arbitrary $w \in \Gamma^*$. Describe a derivation of w_T by G using $v_0, v_1, \ldots, v_d \in (\Sigma_T \cup \Sigma_N)^*$, $d \geq 1$, where

$$v_0 = A_S,$$
$$v_d = w_T,$$
$$v_k = v_{p,k} A_{i_k} v_{s,k},$$
$$v_{k+1} = v_{p,k} u_{i_k} v_{s,k}$$

for each $k \in \{0, \ldots, d-1\}$. For $k \in \{0, \ldots, d\}$, we define inductively a word $w_k \in \Gamma^*$ and a mapping σ_k from each occurrence of $x \in \Sigma_N$ in v_k to an occurrence of the same x in w_k. First, $w_0 = A_S$ and σ_0 is trivial. Next, take $0 \leq k \leq d-1$ and write $w_k = w_{p,k} A_{i_k} w_{s,k}$ such that the A_{i_k} right after $w_{p,k}$ is the σ_k-image of the A_{i_k} right after $v_{p,k}$ in v_k. Then define

$$w_{k+1} = w_{p,k} A_{i_k} b_{i_k} u_{i_k} w_{s,k}$$

and let σ_{k+1} extend σ_k with mapping the occurrences of $x \in \Sigma_N$ within the factor u_{i_k} in v_{k+1} to the corresponding occurrences within the same factor in w_{k+1}. Informally, the words $w_0, \ldots, w_d$ describe the derivation of w_T with keeping all the used nonterminals, i.e., A_{i_k} is rewritten with $A_{i_k} b_{i_k} u_{i_k}$ instead of u_{i_k}. Observe that $(q_1, w_d) \curvearrowright^* (q_1, A_S)$ using the rules labeled with words from P_{BU}. Also observe that, due to Greibach normal form, $w_d \in (\Sigma_T \cup \Sigma_T \Sigma_N \Sigma_B)^*$, i.e., the factors from $\Sigma_N \Sigma_B$ are always separated with letters from Σ_T.

Distinguish the following cases:

- If w does not have a factor from $\Gamma^2 \backslash P_C$, all two-letter factors of w belong to P_C, which implies that w is a factor of a word from $(\Sigma_T t)^*$, where

$$t = A_1 b_1 A_2 b_2 \cdots A_m b_m. \tag{2}$$

  • If w starts with a letter from $\Sigma_T \cup \Sigma_N$ and ends with a letter from $\Sigma_T \cup \Sigma_B$, then $(q_1, w) \curvearrowright^* (q_1, w_d)$ using the rules labeled with words from P_{NB}. Because $(q_1, w_d) \curvearrowright^* (q_1, A_S)$, we conclude that $w \in L(M_G)$.
  • Otherwise, w starts with a letter from Σ_B or ends with a letter from Σ_N. Then

$$w_{N,B} \in \Sigma_B (\Sigma_N \Sigma_B)^* \cup (\Sigma_N \Sigma_B)^* \Sigma_N \cup \Sigma_B (\Sigma_N \Sigma_B)^* \Sigma_N$$

  and we observe that $(q_0, w) \curvearrowright (q_3, w) \curvearrowright^* (q_3, w_{N,B}) \curvearrowright (q_3, u)$ for some $u \in \Sigma_N \cup \Sigma_B \cup \{\varepsilon\}$. As $(q_3, u) \curvearrowright (q_4, \epsilon)$, we get $w \in L(M_G)$.
- If w has a factor $u \in \Gamma^2 \backslash P_C$, write $w = w_p u w_s$ and observe

$$(q_0, w_p u w_s) \curvearrowright (q_2, w_p w_s) \curvearrowright^* (q_2, \epsilon),$$

implying $w \in L(M_G)$.

---

[1] A *projection* to $\Gamma' \subseteq \Gamma$ is given by the homomorphism that maps $x \in \Gamma$ to x if $x \in \Gamma'$ or to ϵ otherwise.

Second, suppose that $L(M_G) = \Gamma^*$ and take an arbitrary $v = x_1 x_2 \cdots x_n \in \Sigma_T^*$ with $x_1, \ldots, x_n \in \Sigma_T$. Let $w = (x_1 t)(x_2 t) \cdots (x_{n-1} t)(x_n t)$, with t defined in (2). We have $w \in L(M_G)$. Observe that:

- The word w does not contain a factor from $\Gamma^2 \backslash P_C$.
- By deleting factors from $\Sigma_B \Sigma_N \cup \Sigma_T$, the word w cannot become a word from $\Sigma_N \cup \Sigma_B \cup \{\epsilon\}$.

Thus, w is accepted by M using a path through the state q_1 ending in the state q_4. In other words, w can be obtained by inserting words from $P_{BU} \cup P_{NB}$ to A_S. During that process, once an occurrence of b_i fails to be preceded by A_i, this situation lasts to the very end, which is a contradiction. It follows that $b_i u_i \in P_{BU}$ can be inserted only to the right of an occurrence of A_i that is not followed by b_i. This corresponds to rewriting A_i with u_i, so we can observe that the whole looping on q_1 (viewed backwards) corresponds to generating $w_T = v$ from A_S using the rules of G. □

Because it is easy to construct a GJFA accepting Σ^*, universality is a special case of both equivalence and inclusion. Thus, the following claim is trivial:

Corollary 6. *Given GJFA M_1 and M_2, it is undecidable both whether $L(M_1) = L(M_2)$ and whether $L(M_1) \subseteq L(M_2)$.*

4   Clearing Restarting Automata with Small Contexts

Recall that the following facts were formulated and proved in [3]:

1. For each $k \geq 3$, the class $\mathcal{L}(k\text{-cl-RA})$ contains a binary language that is not context-free.
2. The class $\mathcal{L}(2\text{-cl-RA})$ contains a language $L \subseteq \Sigma^*$ with $|\Sigma| = 6$ that is not context-free.
3. The class $\mathcal{L}(1\text{-cl-RA})$ contains only context-free languages.

Moreover, for each $k \geq 1$, all the unary languages lying in $\mathcal{L}(k\text{-cl-RA})$ are regular [3]. The present section is devoted to proving the following theorem, which completes the results listed above.

Theorem 7. *The class $\mathcal{L}(2\text{-cl-RA})$ contains a binary language that is not context-free.*

In order to prove Theorem 7, we define two particular rewriting systems:

1. A 1-context rewriting system $R_{uV} = (\{u, V\}, \{u, V\}, I_{uV})$. The set I_{uV} is listed in Table 1.
2. A 2-clearing restarting automaton $R_{01} = (\{0, 1\}, I_{01})$. The set I_{01} is listed in Table 2.

Note that headings of the tables provide identifiers of rules. We write $\rightarrow_{uV}$ for the rewriting relation of R_{uV} and $\dashv_{01}$ for the "generative" relation of R_{01}.

The key feature of the system R_{uV} is:

Table 1. The rules I_{uV}

0	$(\not c, \epsilon \rightarrow uu, \$)$
1	$(\not c, u \rightarrow uuV, \epsilon)$
2	$(\epsilon, Vu \rightarrow uuuV, \epsilon)$
3	$(\epsilon, Vu \rightarrow uuuu, \$)$

Table 2. The rules I_{01}

	a	b	c	d
0	$(\not c, 00, \$)$	-	-	-
1	$(\not c, 10, 00)$	$(\not c, 00, 10)$	-	-
2	$(01, 10, 00)$	$(00, 11, 01)$	$(11, 00, 10)$	$(10, 01, 11)$
3	$(01, 10, 0\$)$	$(00, 11, 0\$)$	-	-

Lemma 8. *Let $w \in L(R_{uV}) \cap \{u\}^*$. Then $|w| = 2 \cdot 3^n$ for some $n \geq 0$.*

The proof is postponed to Sect. 4.1. Next, we define:

1. A length-preserving mapping $\varphi : \{0,1\}^* \rightarrow \{u, V\}^*$ as $\varphi(x_1 \ldots x_n) = \overline{x}_1 \ldots \overline{x}_n$, where

$$\overline{x}_k = \begin{cases} V & \text{if } 1 < k < n \text{ and } x_{k-1} = x_{k+1} \\ u & \text{otherwise} \end{cases}$$

for each $k \in \{1, \ldots, n\}$.

2. A regular language $K \subseteq \{0,1\}^*$:

$$K = \left\{ w \in \{0,1\}^* \mid w \text{ has none of the factors } 000, 010, 101, 111 \right\}.$$

The following is a trivial property of φ and K. Informally, $\varphi(u)$ marks by V the positions where a *defect* occurs in $u \in \{0,1\}^*$. A defect is a position that violates the form $\ldots 00110011 \ldots$, i.e., a position whose neighbours are equal:

Lemma 9. *Let $u \in \{0,1\}^*$. Then $u \in K$ if and only if $\varphi(u) \in \{u\}^*$.*

We index the rules from I_{uV} and I_{01} by the rows of Tables 1 and 2, i.e., by *types* 0 to 3. For a string $w = x_1 x_2 \ldots x_d$, where $x_1, x_2, \ldots, x_d$ are letters, and for integers i, j with $1 \leq i \leq j \leq d$, we denote $w[i, j] = x_i x_{i+1} \ldots x_j$ and $w[i, \ldots] = w[i, d]$.

The next lemma describes how the systems R_{01} and R_{uV} are related. Informally, a rule of the type 2 from I_{01} can be applied only right after a defect in $u \in \{0,1\}^*$. This creates another defect on the right, i.e., a factor $x_1 x_2 y_1 y_2$ of u with defect on x_2 is replaced with $x_1 x_2 z_1 z_2 y_1 y_2$ with defect on y_1. This corresponds to applying the rule $Vu \rightarrow uuuV$ to the defect markers. A rule of the type 1 from I_{01} can introduce a new defect near the beginning of $u \in \{0,1\}^*$, while a rule of type 3 from I_{01} can remove a defect near to the end:

Lemma 10. *Let $u, v \in \{0,1\}^*$. If $u \dashv_{01} v$, then $\varphi(u) \rightarrow_{uV} \varphi(v)$.*

Proof. For $u = v$ the claim is trivial, so we suppose $u \neq v$. Denote $m = |u|$. As u can be rewritten to v using a single rule of R_{01}, we can distinguish which of the rule types is used:

(0) If the rule 0 is used, we have $u = \epsilon$ and $v = 00$. Thus $\varphi(u) = \epsilon$ and $\varphi(v) = $ uu.

(1) If a rule $(\cent, z_1 z_2, y_1 y_2)$ of the type 1 is used, we see that v has some of the prefixes $1000, 0010$ and so $\varphi(v)$ starts with uuV. Trivially, $\varphi(u)$ starts with u. Because $u[1, \ldots] = v[3, \ldots]$, we have $\varphi(u)[2, \ldots] = \varphi(v)[4, \ldots]$ and we conclude that applying the rule $(\cent, \text{u} \rightarrow \text{uuV}, \epsilon)$ rewrites $\varphi(u)$ to $\varphi(v)$.

(2) If a rule $(x_1 x_2, z_1 z_2, y_1 y_2)$ of the type 2 is used, we have

$$u[k, k+3] = x_1 x_2 y_1 y_2,$$
$$v[k, k+5] = x_1 x_2 z_1 z_2 y_1 y_2$$

for some $k \in \{1, \ldots, m-3\}$. As $x_1 x_2 y_1 y_2$ equals some of the factors 0100, $0001, 1110, 1011$, we have

$$\varphi(u)[k+1, k+2] = \text{Vu}.$$

As $x_1 x_2 z_1 z_2 y_1 y_2$ equals some of the factors $011000, 001101, 110010, 100111$, we have

$$\varphi(v)[k+1, k+4] = \text{uuuV}.$$

Because $u[1, k+1] = v[1, k+1]$ and $u[k+2, \ldots] = v[k+4, \ldots]$, we have

$$\varphi(u)[1, k] = \varphi(v)[1, k],$$
$$\varphi(u)[k+3, \ldots] = \varphi(v)[k+5, \ldots].$$

Now it is clear that the rule $(\epsilon, \text{Vu} \rightarrow \text{uuuV}, \epsilon)$ rewrites $\varphi(u)$ to $\varphi(v)$.

(3) If a rule $(x_1 x_2, z_1 z_2, y\$)$ of the type 3 is used, we have

$$u[m-2, m] = x_1 x_2 y,$$
$$v[m-2, m+2] = x_1 x_2 z_1 z_2 y.$$

As $x_1 x_2 y$ equals some of the factors $010, 000$, we have

$$\varphi(u)[m-1, m] = \text{Vu}.$$

As $x_1 x_2 z_1 z_2 y$ equals some of the factors $01100, 00110$, we have

$$\varphi(v)[m-1, m+2] = \text{uuuu}.$$

Because $u[1, m-1] = v[1, m-1]$, we have

$$\varphi(u)[1, m-2] = \varphi(v)[1, m-2],$$

Now it is clear that the rule $(\epsilon, \text{Vu} \rightarrow \text{uuuu}, \$)$ rewrites $\varphi(u)$ to $\varphi(v)$.   □

Corollary 11. *If $u \in L(R_{01})$, then $\epsilon \rightarrow^*_{\text{uV}} \varphi(u)$.*

Proof. Follows from the fact that $\varphi(\epsilon) = \epsilon$ and a trivial inductive use of Lemma 10.   □

Note that $L(R_{01})$ contains, e.g., 00 and 100110. Informally, the claims above imply that $L(R_{01})$ contains only words without defects and that each word from $L(R_{01})$ is obtained from 00 by adding defects to the beginning and pushing them to the end, while the length of the word is tripled for each processed defect. It remains to show that a defect can be always avoided. It turns out to be convenient to describe simultaneous processing of two defects that are close to each other.

The last part of the proof of Theorem 7 relies on the following lemma, whose proof is postponed to Sect. 4.2:

Lemma 12. *For each $\alpha \geq 0$ and $\beta \geq 1$ it holds that*

$$00\,(1100)^{\alpha}\,10\,(0011)^{\beta}\,00 \dashv_{01}^{*} 00\,(1100)^{\alpha+9}\,10\,(0011)^{\beta-1}\,00.$$

Corollary 13. *For each $\gamma \geq 0$ it holds that*

$$0010\,(0011)^{\gamma}\,00 \dashv_{01}^{*} 00\,(1100)^{9\gamma}\,1000.$$

Proof. As the left-hand side equals $00\,(1100)^{0}\,10\,(0011)^{\gamma}\,00$ and the right-hand side equals $00\,(1100)^{9\gamma}\,10\,(0011)^{0}\,00$, the claim follows from Lemma 12 applied γ times.   □

Corollary 14. *The language $L(R_{01}) \cap K$ is infinite.*

Proof. We show that for each $k \geq 0$,

$$00\,(1100)^{\frac{2 \cdot 9^{k}-2}{4}} \in L(R_{01}).$$

In the case of $k = 0$ we just check that $00 \in L(R_{01})$. Next, we suppose that the claim holds for a fixed $k \geq 0$ and show that

$$00\,(1100)^{\frac{2 \cdot 9^{k}-2}{4}} \dashv_{01}^{*} 00\,(1100)^{\frac{2 \cdot 9^{k+1}-2}{4}}.$$

Using the rules 1a and 1b we get

$$00\,(1100)^{\frac{2 \cdot 9^{k}-2}{4}} \dashv_{01} 1000\,(1100)^{\frac{2 \cdot 9^{k}-2}{4}} \dashv_{01} 001000\,(1100)^{\frac{2 \cdot 9^{k}-2}{4}},$$

while Corollary 13 continues with

$$0010\,(0011)^{\frac{2 \cdot 9^{k}-2}{4}}\,00 \dashv_{01}^{*} 00\,(1100)^{\frac{2 \cdot 9^{k+1}-18}{4}}\,1000.$$

Finally, denoting $p = 00\,(1100)^{\frac{2 \cdot 9^{k+1}-18}{4}}$, using rules 3b, 2a, 2b, 2d, 2c, and 3a respectively, we get

$$p1000 \dashv_{01} p100\underline{110} \dashv_{01} p11\underline{0}00110 \dashv_{01} p\,(1100)\,\underline{110}110 \dashv_{01} p\,(1100)\,1100\underline{1}110 \dashv_{01}$$

$$\dashv_{01} p\,(1100)\,(1100)\,110\underline{0}10 \dashv_{01} p\,(1100)\,(1100)\,(1100)\,1\underline{1}00 = 00\,(1100)^{\frac{2 \cdot 9^{k+1}-2}{4}}.$$

                                                                          □

We conclude the proof of Theorem 7 by pointing out that Lemmas 8, 9, and 10 say that for each $w \in \{0,1\}^*$ we have

$$w \in L(R_{01}) \cap K \Rightarrow \varphi(w) \in L(R_{uV}) \cap \{u\}^* \Rightarrow (\exists n \geq 0) |w| = 2 \cdot 3^n.$$

This, together with the pumping lemma for context-free languages and the infiniteness of $L(R_{01}) \cap K$, implies that $L(R_{01}) \cap K$ is not a context-free language. As the class of context-free languages is closed under intersections with regular languages, $L(R_{01})$ is not context-free either.

4.1   Proof of Lemma 8

We should show that $w \in L(R_{uV}) \cap \{u\}^*$ implies $|w| = 2 \cdot 3^n$ for some $n \geq 0$. Let $\Phi : \{u, V\}^* \to \mathbb{N}$ be defined inductively as follows:

$$\Phi(\epsilon) = 0,$$
$$\Phi(u^k w) = k + \Phi(w),$$
$$\Phi(Vw) = 1 + 3 \cdot \Phi(w)$$

for each $k \geq 1$ and $w \in \{u, V\}^*$. Observe that we have assigned a unique value of Φ to each word from $\{u, V\}^*$. Next, we describe effects of the rules of R_{uV} to the value of Φ.

(0) The rule 0 can only rewrite $w_1 = \epsilon$ to $w_2 = uu$. We have $\Phi(w_1) = 0$ and $\Phi(w_2) = 2$.
(1) The rule 1 rewrites $w_1 = uw$ to $w_2 = uuVw$ for some $w \in \{u, V\}^*$. We have $\Phi(w_1) = 1 + \Phi(w)$ and $\Phi(w_2) = 3 + 3 \cdot \Phi(w)$. Thus, $\Phi(w_2) = 3 \cdot \Phi(w_1)$.
(2) The rule 2 rewrites $w_1 = \overline{w}Vuw$ to $w_2 = \overline{w}uuuVw$ for some $w, \overline{w} \in \{u, V\}^*$. We have
$$\Phi(Vuw) = \Phi(uuuVw) = 4 + 3 \cdot \Phi(w).$$
It follows that $\Phi(w_1) = \Phi(w_2)$.
(3) The rule 3 rewrites $w_1 = \overline{w}Vu$ to $w_2 = \overline{w}uuuu$ for some $\overline{w} \in \{u, V\}^*$. We have $\Phi(Vu) = \Phi(uuuu) = 4$ and thus $\Phi(w_1) = \Phi(w_2)$.

Together, each $w \in L(R_{uV})$ has $\Phi(w) = 2 \cdot 3^n$ for some $n \geq 0$. As $\Phi(w) = |w|$ for each $w \in \{u\}^*$, the proof is complete.   □

4.2   Proof of Lemma 12

We should prove that

$$00 \, (1100)^{\alpha} \, 10 \, (0011)^{\beta} \, 00 \dashv_{01}^* 00 \, (1100)^{\alpha+9} \, 10 \, (0011)^{\beta-1} \, 00$$

for $\alpha \geq 0, \beta \geq 1$. Let $p = 00 \, (1100)^{\alpha}$, $q = (0011)^{\beta-1} \, 00$, and derive the claim as follows:

$$
\begin{array}{lll}
p10 \, (0011) \, q & \dashv_b \ p100\underline{11}011q & \dashv_a \\
p1\underline{1000}011011q & \dashv_b \ p \, (1100) \, \underline{11}011011q & \dashv_d \\
p \, (1100) \, 1100\underline{11}1011q & \dashv_d \ p \, (1100)^2 \, 11100\underline{111}q & \dashv_c \\
p \, (1100)^2 \, 11\underline{00}100111q & \dashv_a \ p \, (1100)^3 \, 1\underline{1000}0111q & \dashv_b \\
p \, (1100)^4 \, \underline{11}0111q & \dashv_c \ p \, (1100)^4 \, 11011\underline{00}1q & \dashv_d
\end{array}
$$

$$p\,(1100)^4\,11\underline{00}111001q \;\;\dashv_c \;\; p\,(1100)^5\,11\underline{00}1001q \;\;\dashv_a$$
$$p\,(1100)^6\,1\underline{10}001q \;\;\dashv_a \;\; p\,(1100)^7\,01\underline{10}q \;\;\;\;\;\;\dashv_b$$
$$p\,(1100)^7\,\underline{11}0110q \;\;\dashv_d \;\; p\,(1100)^7\,11\underline{00}1110q \;\;\dashv_c$$
$$p\,(1100)^8\,11\underline{00}10q,$$

where uses of particular rules of the type 2 are indicated by typing $\dashv_a, \dashv_b, \dashv_c, \dashv_d$ instead of $\dashv_{01}$. $\qquad\qquad\qquad\qquad\qquad\qquad\qquad\qquad\qquad\qquad\qquad$ □

5   Conclusions and Remarks

We made a progress in studying basic properties of two recently introduced formalisms. Even if these particular models do not find application in practice, our results may be of key importance for designing suitable modifications.

The maximum length of labels is a key property of a GJFA. It remains open whether our undecidability results hold if restricted to GJFA with labels of a fixed maximum length. In *jumping finite automata*, i.e., GJFA with labels of length one, the problems become decidable (see [5] for a thorough survey).

Note that there is a group of older models that can be, in fact, put to a common framework with GJFA and cl-RA, immediately sharing some properties following from our new results:

- *Insertion systems* [16] were introduced in the scope of DNA computing. They generate sequences by inserting factors according to contexts of restricted lengths. Their generalization to *graph-controlled* [1] insertion systems together with contexts of zero length corresponds to the expressive power of GJFA. Using the notation of [1], we have $\mathrm{LStP}_*\left(\mathrm{ins}_*^{0,0}\right) = \mathbf{GJFA}$. Another (historical) work introduces *regular control semi-contextual grammars without appearance checking* [11]. Again, the variant with forbidden contexts (with a language class denoted by $\mathcal{C}_0$) is equivalent to GJFA. Our results imply that universality, inclusion, and equivalence are undecidable for these models as well.
- Up to explicit endmarking, insertion systems and the basic variant of semi-contextual grammars [15], both with contexts bounded by some $k \geq 1$, are equivalent to k-cl-RA. More precisely, each language from the class denoted by INS_*^k or $\mathcal{J}_k$ is accepted by a k-cl-RA, while for each k-cl-RA M, the language $\mathfrak{c}L(M)\$ $ lies in $\mathrm{INS}_*^k = \mathcal{J}_k$. Thus, we can conclude that the class $\mathrm{INS}_*^2 = \mathcal{J}_2$ contains non-context-free binary languages.

The remarks above are hard to present in more depth because the original definitions of insertions systems and semi-contextual grammars use non-compatible notational paradigms. Once these definitions are understood, the claims are very easy to check (see [17]).

References

1. Alhazov, A., Krassovitskiy, A., Rogozhin, Y., Verlan, S.: Small size insertion and deletion systems. In: Martin-Vide, C. (ed.) Scientific Applications of Language Methods, pp. 459–524. Imperial College Press (2010)

2. Černo, P.: Clearing restarting automata and grammatical inference. In: Heinz, J., Colin de la Higuera, T.O. (eds.) Proceedings of the Eleventh International Conference on Grammatical Inference. JMLR Workshop and Conference Proceedings, vol. 21, pp. 54–68 (2012)
3. Černo, P., Mráz, F.: Clearing restarting automata. Fund. Inf. **104**(1), 17–54 (2010)
4. Ehrenfeucht, A., Haussler, D., Rozenberg, G.: On regularity of context-free languages. Theoret. Comput. Sci. **27**(3), 311–332 (1983)
5. Fernau, H., Paramasivan, M., Schmid, M., Vorel, V.: Characterization andcomplexity results on jumping finite automata. Accepted to Theoretical Computer Science (2015). http://arxiv.org/abs/1512.00482
6. Fernau, H., Paramasivan, M., Schmid, M.L.: Jumping finite automata: characterizations and complexity. In: Drewes, F. (ed.) CIAA 2015. LNCS, vol. 9223, pp. 89–101. Springer, Heidelberg (2015)
7. Haussler, D.: Insertion languages. Inf. Sci. **31**(1), 77–89 (1983)
8. Hopcroft, J.E., Motwani, R., Ullman, J.D.: Introduction to Automata Theory, Languages, and Computation, 2nd edn. Addison-Wesley (2003)
9. Ito, M., Kari, L., Thierrin, G.: Insertion and deletion closure of languages. Theoret. Comput. Sci. **183**(1), 3–19 (1997)
10. Jančar, P., Mráz, F., Plátek, M., Vogel, J.: Restarting automata. In: Reichel, H. (ed.) FCT 1995. LNCS, vol. 965, pp. 283–292. Springer, Heidelberg (1995)
11. Marcus, M., Păun, G.: Regulated Galiukschov semicontextual grammars. Kybernetika **26**(4), 316–326 (1990)
12. Meduna, A., Zemek, P.: Jumping finite automata. Int. J. Found. Comput. Sci. **23**(7), 1555–1578 (2012)
13. Meduna, A., Zemek, P.: Chapter 17 Jumping Finite Automata. In: Regulated Grammars and Automata, pp. 567–585. Springer, New York (2014)
14. Mráz, F., Plátek, M., Vogel, J.: Restarting automata with rewriting. In: Král, J., Bartosek, M., Jeffery, K. (eds.) SOFSEM 1996. LNCS, vol. 1175, pp. 401–408. Springer, Heidelberg (1996)
15. Păun, G.: Two theorems about Galiukschov semicontextual languages. Kybernetika **21**(5), 360–365 (1985)
16. Păun, G., Rozenberg, G., Salomaa, A.: Insertion-deletion systems. In: DNA Computing: New Computing Paradigms. Texts in Theoretical Computer Science, pp. 187–215. Springer, Heidelberg (1998)
17. Vorel, V.: On basic properties of jumping finite automata. Int. J. Found. Comput. Sci., conditionally accepted in 2015. http://arxiv.org/abs/1511.08396

Author Index

Printed in the United States
By Bookmasters

Printed in the United States
By Bookmasters